R0021952822

CHICAGO PUBLIC LIBRARY
HAROLD WASHINGTON LIBRARY CENTER

R0021952822

D1794173

Submicroscopic Cytochemistry

VOLUME II

Membranes, Mitochondria, and Connective Tissues

Contributors

ISIDORE GERSH
ZELMA MOLNAR
GIOVANNI L. ROSSI

Submicroscopic Cytochemistry

Volume II

MEMBRANES, MITOCHONDRIA, AND CONNECTIVE TISSUES

Edited by

Isidore Gersh

*Laboratories of Anatomy
Department of Animal Biology
School of Veterinary Medicine
University of Pennsylvania
Philadelphia, Pennsylvania*

ACADEMIC PRESS   New York and London   1973
A Subsidiary of Harcourt Brace Jovanovich, Publishers

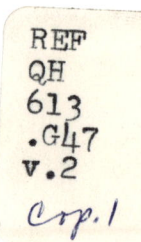

COPYRIGHT © 1973, BY ACADEMIC PRESS, INC.
ALL RIGHTS RESERVED.
NO PART OF THIS PUBLICATION MAY BE REPRODUCED OR
TRANSMITTED IN ANY FORM OR BY ANY MEANS, ELECTRONIC
OR MECHANICAL, INCLUDING PHOTOCOPY, RECORDING, OR ANY
INFORMATION STORAGE AND RETRIEVAL SYSTEM, WITHOUT
PERMISSION IN WRITING FROM THE PUBLISHER.

ACADEMIC PRESS, INC.
111 Fifth Avenue, New York, New York 10003

United Kingdom Edition published by
ACADEMIC PRESS, INC. (LONDON) LTD.
24/28 Oval Road, London NW1

Library of Congress Cataloging in Publication Data

Gersh, Isidore, DATE
    Submicroscopic cytochemistry.

    Includes bibliographies.
    CONTENTS: v. 1.   Proteins and nucleic acids.–v. 2.
Membranes, mitochondria, and connective tissues.
    1. Cytochemistry.   2. Ultrastructure.   3. Connec-
tive tissues.   I. Title. [DNLM: 1. Histocyto-
chemistry.   2. Molecular biology. QS504 G381s]
QH613.G47   574.8'76   73-904
ISBN 0–12–281402–9 (v. 2)

PRINTED IN THE UNITED STATES OF AMERICA

To Dr. Louis B. Flexner
for his inspiration and support over a period of forty years

Contents

List of Contributors	xi
Preface	xiii
Acknowledgments	xv
Contents of Volume I	xvii

1. Cytochemical Studies on Lipid and Protein Components of Membranes in Pancreatic Acinar and Hepatic Cells of the Mouse

Isidore Gersh

Methods	4
Results	16
Discussion	23
Summary	26
References	27

2. Densitometric Studies of Lipid Membranes in Pancreatic Acinar Cells of the Mouse

Isidore Gersh

Methods	34
Results and Discussion	34
Summary	41

3. Cytochemical Study of Lipid, Protein, and Nucleic Acids of Mitochondria in Exocrine Pancreatic and Hepatic Cells of the Young Adult Mouse

Isidore Gersh

Methods	43
Observations	48
Discussion	55

Summary	61
References	61

4. Possible Precursor Granules in Fibroblasts
Giovanni L. Rossi, Isidore Gersh, and Zelma Molnar

Materials and Methods	63
Observations	64
Discussion	69
Summary	84
References	84

5. Nucleic Acids in Chondrocytes of Epiphyseal Plate of the Rat Tibia
Isidore Gersh

Part I—Submicroscopic Distribution of Nucleic Acids	88
Part II—Microscopic Distribution of Nucleic Acids	105
Discussion	117
Summary	124
References	124

6. Possible Precursor Granules in Chondrocytes and Osteoblasts
Zelma Molnar, Isidore Gersh, and Giovanni L. Rossi

Part I—Possible Precursor Granules in Chondrocytes	127
Part II—Bone Cells	137
Discussion	141
Summary	147
References	147

7. Morphochemical Study of the Matrix of Epiphyseal Plate and Joint Cartilage and the Origin of Protein–Polysaccharide Complex
Isidore Gersh

Methods	150
Observations	154
Discussion	166
Summary	174
References	174

8. The Movement of Ferrocyanide in Cartilage Matrix
Isidore Gersh

Methods	178
Observations	182
Discussion	183
Summary	185
References	185

9. Relation of the Walls of Large Matrix Compartments of Epiphyseal Cartilage to the Formation of Calcium Crystals

Isidore Gersh

Method	189
Observations	191
Discussion	195
Summary	203
References	203

10. Vascularity and Protein—Polysaccharide Complex in Tendons of Young Rats

Isidore Gersh

Methods	206
Observations and Discussion	210
Summary	211
References	211

11. Summary, Synthesis, and Speculations

Isidore Gersh

State of the Genetic Material	214
Changes in the State of the Genetic Material	214
Association of RNA with DNA	220
Arrangement of Protein in the Nucleus as Pseudovacuoles	221
Protein and the Organization of DNA	222
Cytoplasm	222
Connective Tissues	229
Coda	233
References	235

Author Index	237
Subject Index	242

List of Contributors

Numbers in parentheses indicate the pages on which the authors' contributions begin.

ISIDORE GERSH (1,31,42,63,86,127,149,177,187,206,213), Laboratories of Anatomy, Department of Animal Biology, School of Veterinary Medicine, University of Pennsylvania, Philadelphia, Pennsylvania

ZELMA MOLNAR (63,127), Departments of Pathology, University of Chicago and Loyola University, Chicago, Illinois, and Veterans Administration Hospital, Hines, Illinois

GIOVANNI L. ROSSI (63,127), Institut für Tierpathologie, Universität Bern, Bern, Switzerland

Preface

In this age of specialization and fragmentation the opportunity for an individual to carry out a comprehensive laboratory study is rare indeed. Through a series of accidental happenings, I found myself in a position which impelled me to continue a long-term study of a general type. It is difficult to express briefly the fascination and absorbing interest in developing subtle interrelationships of methods and ideas and in molding complex materials into a coherent view of some aspects of cellular activity. The first eight years of this study were devoted largely to developing and testing cytochemical methods to the point of satisfaction, and the next five years chiefly to applying them to biological studies. The analytic processes continued up to and even after the time of writing of the last chapter.

Three major topics are considered in this two-volume work: (1) the pattern of distribution of nucleic acids at the molecular level in various cells and states of activity and in relation to protein synthesis; (2) the molecular and macromolecular organization of cellular membranes; and (3) the origin and distribution of the major macromolecular aggregates of connective tissue. The first major topic comprises Volume I; the remaining ones, Volume II. All three are integrated in the last chapter of this volume.

There is a high premium in this work on morphology, that is, on the distribution of cell components and their organization in cells at the macromolecular level. Such studies required that new methods be developed to preserve these molecules and molecular aggregates very nearly in the position they occupied in the living state and to identify them more or less selectively. The crucial tests were chiefly for proteins, nucleic acids, lipids, and acid mucopolysaccharides. A glance at the Contents will indicate the wide range of biological problems which could be approached with the new methods. The last chapter is the most general one, and serves to bind together all the others. Here morphology, through cytochemistry, is integrated with parts of genetics, biochemistry, cytophysiology, developmental biology, and pathology.

The main emphasis in this work is on the presentation of laboratory findings and theoretical aspects involved or derived from them. These are related, whenever possible, to the main body of thought on the various topics discussed by specific or general references. The latter are so varied, competent, and numerous that it was felt they would suffice in most instances to relate the original work to contributions from the literature available to most workers. Since there was no need for encyclopedic reviews of the various fields, the references cited are minimal, except in certain specific instances where the reviews were incomplete or otherwise unsatisfactory. An effort was made to update the literature citations up to the time the manuscript went to the publisher.

A few words should be written about prints. Nearly all negatives were printed directly at a magnification of two times. Where the contrast was marked, as with apatite crystals (Chapter 9, Volume II), some prints were magnified five times. Only a few negatives were printed with some dodging or burning. A small number of negatives were double printed to give enhanced, selective contrast to some specific component. These were printed from the original negative, usually on a soft paper. This was then photographed on a second EKTAPAN (Kodak) film, which was printed again usually on a soft paper. The final magnification was two to four times the original electron micrograph. Such prints are labeled "G2 print" in the figure legend. In taking some of the electron micrographs for such prints, an appreciable astigmatism was introduced in order to improve resolution and contrast. This method was suggested by F. S. Sjöstrand ("Electron Microscopy of Cells and Tissues," Volume I, p. 127, Academic Press, New York).

Isidore Gersh

Acknowledgments

The original research reported in this book was supported in part by grants from the Commonwealth Fund of New York, the U.S. Public Health Service, and the American Cancer Society. Two small intramural grants helped to tide me over a difficult period. The financial and moral support of Drs. Lester Evans, Roderick Heffron, and Charles O. Warren, executive officers of the Commonwealth Fund, encouraged the initiation of the long search for cytochemical methods adequate for solution of some biological problems. A separate grant by the Fund made possible the preparation of the manuscript.

I am indebted to the following for their essential and loyal technical work: Faustina Manelis and Elizabeth Vilkas, of Chicago; Vivien Catlin, Eugénie Ford, Patsy Jo Terrell, and Eliana Muñoz, formerly of Philadelphia; and George J. Grigonis, Jr., also of Philadelphia. It was a great comfort when Diane Weinstock took over the typing. Diagrams were developed by Mary Jo Larsen, who was of great aid, in addition, in connection with other illustrations. I wish to thank Janice Heald for facilitating work in the library and for checking the references.

I am pleased to recall the perceptive advice of Dr. Herbert S. Anker, Department of Biochemistry, University of Chicago, and of numerous other chemists. For advice on matters pertaining to Chapter 2, Volume II, I am indebted to Dr. Donald A. Abt, of the Department of Medicine, and to Dr. James A. Dvorak, when he was in the Department of Pathobiology in the School of Veterinary Medicine, University of Pennsylvania.

In addition to working jointly on the material described in five chapters, my wife has helped me significantly to interpret observations in other chapters from a genetic point of view. She also helped me clarify the analyses of the various problems studied and to express the thoughts involved. I enjoyed the collaboration of several colleagues, past and present, who are coauthors of several chapters: Dr. Peter L. Amenta, Department of Anatomy, Hahnemann Medical College, Philadelphia; Dr. Peter J. Hand, Department of Animal Biology, University of Pennsylvania, Philadelphia; Dr. Zelma Molnar, Department of Pathology, University of Chicago, Chicago; and Dr. Giovanni L. Rossi, Institut für Tierpathologie, Universität Bern, Bern.

This work was started when I was Professor in the Department of Anatomy at the University of Chicago, and was continued with very little interruption at the University of Pennsylvania.

Contents of Volume I

Proteins and Nucleic Acids

Submicroscopic Cytochemistry
  Isidore Gersh

Distribution of Protein in Hepatic and Exocrine Pancreatic Cells of the Mouse
  Isidore Gersh

Methods for the Fixation and Staining of DNA and RNAs Applied to Exocrine Pancreatic and Hepatic Cells of the Mouse
  Isidore Gersh and G. L. Rossi

Cytochemical Study of Nucleic Acids in Salivary Gland Chromosomes of *Drosophila* Larvae
  Eileen S. Gersh and Isidore Gersh

Distribution of Protein in Nuclei of Salivary Glands of *Drosophila* Larvae
  Isidore Gersh and Eileen S. Gersh

DNA Patterns of Cells of Imaginal Disks of *Drosophila* Larvae
  Eileen S. Gersh and Isidore Gersh

Submicroscopic Distribution of Nucleic Acids in the Barr Body of Chondrocytes in the Proximal Tibial Epiphysis of Female Rats
  Isidore Gersh

Preliminary Study of the Specificity of the DNA Molecular Patterns of Differentiated Cells of the Young Adult Mouse
  Isidore Gersh

Changing DNA Molecular Patterns during Cell Maturation in the Mouse
  Isidore Gersh

Alteration of DNA Molecular Patterns with Activation of Certain Cells in the Mouse
  Isidore Gersh

DNA Molecular Pattern during Metaplasia of Stratified Squamous Epithelium of Mouse Vagina to Mucin-Producing Cells
  Isidore Gersh

Changes in Nuclear DNA Molecular Pattern of Spinal Ganglion Cells during Regeneration of Spinal Nerves of the Cat
  Isidore Gersh and Peter J. Hand

Organization of DNA during the Mitotic Cycle in Eggs of *Drosophila melanogaster*
  Eileen S. Gersh and Isidore Gersh

Persistence of Individuality of Chromosomes during Interphase, and the Role of the Nuclear Membrane
  Peter S. Amenta, Isidore Gersh, and Eileen S. Gersh

Author Index—Subject Index

1

Cytochemical Studies on Lipid and Protein Components of Membranes in Pancreatic Acinar and Hepatic Cells of the Mouse

Isidore Gersh

Research on cell membranes has been very lively. Some of the activity has revolved on differences between various kinds of membranes in their position in cells, in their geometry, thickness, chemical composition, enzymic activity, immunological and electrical properties, and in their relations to studies on transport and permeability. These aspects are reviewed in various books and symposia *(9, 13, 14, 66, 67)*. One thread which runs through a seemingly chaotic jumble of data is the unit membrane hypothesis *(58)*. According to this hypothesis, virtually all cellular membranes of all cells comprise a lipid bimolecular leaflet with protein or polysaccharide on the external surfaces of the lipid layer in association with the hydrophilic ends of the oriented lipid molecules. Proof for the existence of such organized lipid layers in myelin and retinal rods, and in artifical layers seems quite overwhelming, but is, in general, lacking for almost all other cells *(30, 31)*. Recently, the proposed unique structure of the *in vitro* membranes was challenged by studies which showed that lipids may be organized in several regular manners, which may be in equilibrium with the bimolecular leaflet *(20, 40, 41)*. The possibility arose that unit membranes (except for myelin sheath and retinal rods) may be an artifactual rearrangement of lipid molecules, which takes place during fixation of cells, to its most stable form, the bimolecular leaflet *(40)*. An additional uncertainty is that lipid molecules may be oriented at least in some membranes, artificial and cellular, in the form of a lipid globular pattern with associated intercalated protein *(19, 21, 24, 38, 39, 66, 67, 72, 76, 77)*. Numerous modifications of the unit membrane hypothesis have been suggested and criticized in reviews of this topic *(16, 32, 42, 62, 63, 72, 75, 76, 79)*. Some studies of molecular distortions, which take place during the fixation of membranes, have also been reported *(21, 44, 50, 51)*. Danielli *(11)* showed that lipid layers must exist as bimolecular leaflets, but the thermodynamic analysis may not help in ascertaining whether such leaflets occur in living cells. As far as cellular membranes are concerned, a basic cause for the ambiguities enumerated above is that none of the morphological methods used in electron microscopy is sufficiently specific chemically to be regarded as a positive identification of phospholipid *(30, 31)*.

Numerous recent reports emphasize structural proteins of membranes. These are regarded by some as noncatalytic structural proteins which are self-assembled to form membranes in conjunction with phospholipids. Others regard the structural proteins of membranes as capable of enzymic activity. In both cases, the membranes are regarded as largely hydrophobic in nature, because of a preponderance in the inner parts of polypetide chains of nonpolar amino acids, such as glycine, alanine, leucine, valine, and proline. Phospholipids are considered important in stabilizing the folded configuration of many proteins through weak interactions involving chiefly short-range forces, especially electrostatic and hydrophobic bonds *(59)*. Some of the forces involved have been discussed and numerous models incorporating these newer data have been published *(3, 4, 8, 26, 33, 34, 45–47, 73, 74, 78)*.

Certainly, recent work on isolated membranes includes analyses of both lipid and protein components, with special emphasis on enzymes *(4, 26, 33, 45–47, 75)*.

Isolated membrane proteins were found to be a complex array of varying molecular weight. They were rapidly labeled, through with varying rates of turnover, suggesting that the membrane proteins were heterogeneous *(27, 28)*. In general, such membranes are composed of protein molecules, whose interior is largely hydrophobic and whose surface is charged and hydrophilic. The surface of the protein molecules is such that, in the presence of phospholipids, they aggregate as mono- or bilayered sheets. The phospholipid molecules are arranged in this sheet with their charged head on or near the outer surface of the protein sheet, and their aliphatic chains extending internally between the hydrophobic polypeptide coils of the protein molecules. Some of the lipids may also form small bilaminar mosaic patches between the predominantly protein molecules. This is the general view of some biochemists, who have separated various kinds of mitochondrial particles whose enzymic activity has been studied before and after reconstitution *(2, 10, 20, 22, 23, 54–56, 60)*. A generally cautious note on the technical limitations in the study of membrane proteins and on the possible physiological roles of such membranes was expressed in a very recent review *(25)*. While these fractionations have been controlled by electron microscopy, the difficulty has been to ascertain whether the particulates preexist as such in living cells, and whether the reconstituted "membranes" are, indeed, similar to the original membranes in living cells.

On the other side, is the work of Sjöstrand and Barajas *(70)*, who have identified globular molecules in certain membranes, and have interpreted them as their basic elements. They appear largely as nonstainable (hydrophobic) regions enclosed by extremely fine stained (hydrophilic) walls, giving the impression that each one is a globule. Their relations to each other and to lipids are interpreted more or less as in the preceding paragraph. The hope is projected that "complexes of molecules could be recognized and possibly identified due to the characteristic distribution of stain sites within the complex" *(68)* and that protein molecules could be identified in thin sections provided that (1) the distribution in space of sites that bind the stain were known, and (2) the conformation of protein molecules could be maintained sufficiently close to their native conformation to allow recognizing the characteristic of stain sites *(67)*. Again, "It is quite conceivable that in the future at least certain types of protein molecules with, for instance, a characeristic arrangement of identical subunits will be identified in tissue sections and their location in the cellular structures determined in this way" *(67)*. These are, in my opinion, extremely remote possibilities which will be difficult or impossible to realize because of the molecular distortions which are inevitable with the addtion or removal of water and with cross-linking, whether too much or too little. Moreover, the enzymic properties of these globular molecules still remain to be demonstrated.

Thus, it seems that the biochemical and morphological emphasis on the important roles of proteins in membranes are convergent. However, it should be kept in mind that, on the one side, the reality of the biochemically characterized particulates of membranes in living cells is uncertain, and, on the other side, the morphological particles (globules) of membranes in cells are not biochemically characterized. Another discussion of membrane proteins is presented in this volume (Chapter 3) which deals with membranes in mitochondria.

The purposes of the work reported here are: (1) to test the question whether lipid membranes of cells as commonly conceived are real or artifactual; (2) to identify the lipid nature of such membranes as might exist by a chemical method which does not involve the use of osmium tetroxide or potassium permanganate; (3) to test the hypothesis of the unit membrane; (4) to ascertain whether there might be differences in the lipid components in different parts of the membrane as well as between different membranes; and (5) to compare the distribution of lipid and protein components of membranes.

The most reactive components of the fatty acid moiety of phospholipids are the carboxyl ester linkage and the unsaturated carbon bonds of the aliphatic chain *(12, 35, 37, 53, 57, 80)*. There are many reactions for the former, but the most promising seemed to be conversion to an amide by use of an amine of high molecular weight to increase contrast in electron micrographs. There are even more reactions for marking the unsaturated double bonds of the aliphatic chain, but in the interests of simplicity it was thought wise to use the same amine used to form amides. At the same time, efforts were made to define the conditions of the test so as to prevent or reduce molecular rearrangements or displacements of lipid molecules. I cannot claim complete success in overcoming these technical requirements, either in attaining an absolute specificity or in preventing all molecular displacements, but the results are sufficiently satisfactory to merit presentation.

A discussion of the general aspects of staining freeze-dried specimens for protein has been presented in Vol. I, Chapter 2.

The results will be presented in three parts. In the first (this chapter), the morphological distribution of lipid and protein components of membranes of hepatic and exocrine pancreatic cells will be described. In the second (Chapter 2), certain of these morphological patterns will be analyzed quantitatively based on densitometric tracings, in an effort to construct three-dimensional (but not molecular) models of lipid components of the rough endoplasmic reticulum (RER), the nuclear membrane, and Golgi structures. In the third part (Chapter 3), observations on lipids and proteins of mitochondria will be presented.

Methods

LIPIDS

The methods to be described were developed to overcome certain difficulties encountered in lipid cytochemistry. These include extraction of lipids (with probable displacement of the remaining lipids), the flowing and coalescence of lipids because of the low melting point of many lipids and their intersolubility, and the tendency of certain lipids to separate from their associated protein when water is removed as during freezing and drying, and thus to be displaced significantly *(14, 15, 35–37)*. There is also a strong tendency for lipids to seek the most stable form of molecular orientation *(40)*. Finally, the chemical reaction used to characterize lipids in the electron microscope should be clearly distinquishable as increased contrast over a low or near zero background.

The methods to be described overcome these difficulties to a very large degree. The specimens were frozen ultrarapidly without ice crystal formation. [The general acceptance of freeze-etching and of freeze-fracturing carries with it the implication that specimens can be frozen sufficiently rapidly to preclude ice crystal formation *(7, 29, 49)*.] The artifacts that were regarded as having been caused by ice crystals *(5, 61, 65, 69, 71)* should be attributed instead to the entrapment of noncondensable and insoluble gases during the infiltration of the specimen with fluids, after the specimens have been dried. In nearly all specimens, which were properly infiltrated with the plastic monomer mixture, submicroscopic holes (now attributed to gas entrapment) were altogether absent. Bullivant *(5)* also attributed the holes to improper infiltration with the embedding materials. Evidence is presented in Chapters 1 and 2 (Vol. I) that shows that no ice crystals are formed, during freezing and drying, of a size sufficiently large to be visible with the electron microscope. When suitably postfixed and stained, the spaces formerly claimed to be empty (i.e., occupied only by ice crystals before drying) were shown to be filled with protein.

During the time that water was subliming from the specimen at some temperature lower than $-30°C$, the reactive groups of the lipids (as they were exposed) reacted *in vacuo* with vapors, which raised markedly the melting point of the affected lipids and decreased their solubility. There is no evidence that ice crystals formed at this temperature, since no holes could be observed in electron micrographs of sections. The reactive groups were the carboxyl ester linkage and the unsaturated carbon bonds of the aliphatic chain, as well as any preexistent active aldehyde or ketone group. The specimen was then infiltrated directly with a water-soluble plastic, and the plastic was polymerized at a low temperature.

The chief test reagent employed for the identification of lipids is 2,4-diiodophenylhydrazine (IIPH) *(52)*:

$$\underset{I}{\underset{\displaystyle\bigcirc}{}}\underset{I}{\overset{NH_2}{\underset{|}{NH}}}$$

Equations (1) and (2) illustrate the general reaction with carboxyl ester linkages to form amides of markedly higher melting points and reduced solubility in water.

$$R'COOR'' + H_2NR \rightarrow R'CONHR + HOR'' \quad (1)$$

Fatty acid ester Primary amine Fatty acid amide

$$R'COOR'' + H_2NNHR \rightarrow R'CONHNHR + HOR'' \quad (2)$$

Fatty acid ester Hydrazine Fatty acid hydrazide

The specimens were then treated with ozone at a low temperature to convert unsaturated aliphatic carbon bonds to ozonides [Eq. (3)] which, on further treatment with hydrazine, formed hydrazones or osazones *(1, 12, 53, 57, 80)* [Eq. (4)].

$$R'''-\overset{H}{\underset{|}{C}}=\overset{H}{\underset{|}{C}}-R'''' + {}^+O-O-O \longrightarrow \underset{O}{\overset{R'''-\overset{H}{\underset{|}{C}}-\overset{H}{\underset{|}{C}}-R''''}{\underset{\diagdown\;\diagup}{O\qquad O}}} \quad (3)$$

Aliphatic unsaturated carbon compound   Ozone   Aliphatic ozonide

$$\underset{O}{\overset{R'''-\overset{H}{\underset{|}{C}}-\overset{H}{\underset{|}{C}}-R''''}{\underset{\diagdown\;\diagup}{O\qquad O}}} + H_2NNHR \longrightarrow R'''-\overset{H}{\underset{\underset{R}{\underset{|}{HN}}}{\underset{|}{\overset{\|}{N}}}}C + \overset{H}{\underset{\underset{R}{\underset{|}{NH}}}{\underset{|}{\overset{\|}{N}}}}C-R'''' \quad \text{or} \quad R'''-\overset{}{\underset{\underset{R}{\underset{|}{HN}}}{\underset{|}{\overset{\|}{N}}}}C-\overset{}{\underset{\underset{R}{\underset{|}{NH}}}{\underset{|}{\overset{\|}{N}}}}C-R'''' \quad (4)$$

Ozonide   Hydrazine   Aliphatic hydrazones   Aliphatic osazone

The end result is that wherever reactions have taken place which lead to the formation of hydrazides, hydrazone or osazone derivatives, the mass of the original lipid in the specimen has been increased. The site where this occurs is recorded as one of increased contrast if the site is resolvable and if the increase in mass is sufficiently great. The numerous controls which assure a degree of specificity of the reaction for lipids will be enumerated after the methods are described in detail. A preliminary note has been published *(18)*.

The reactions were all performed *in vacuo* under controlled temperature conditions in a glass system. Thin slices of mouse pancreas or liver were frozen ultrarapidly by the method described in Chapter 2, (Vol. I) to avoid the formation of ice crystals. The specimen tube adjacent to stopcock (B) (Fig. 1.1) was cooled to the temperature of liquid nitrogen, and the frozen specimen attached to aluminum foil was dropped in its glass vial into it. The whole assembly between (B), (C), and (D) was evacuated to about 0.1 mm mercury. Meanwhile, the water vapor trap adjacent to stopcock (C) was immersed in an alcohol–dry ice bath at about −80°C. In this way, the glass line was evacuated without sublimation of water

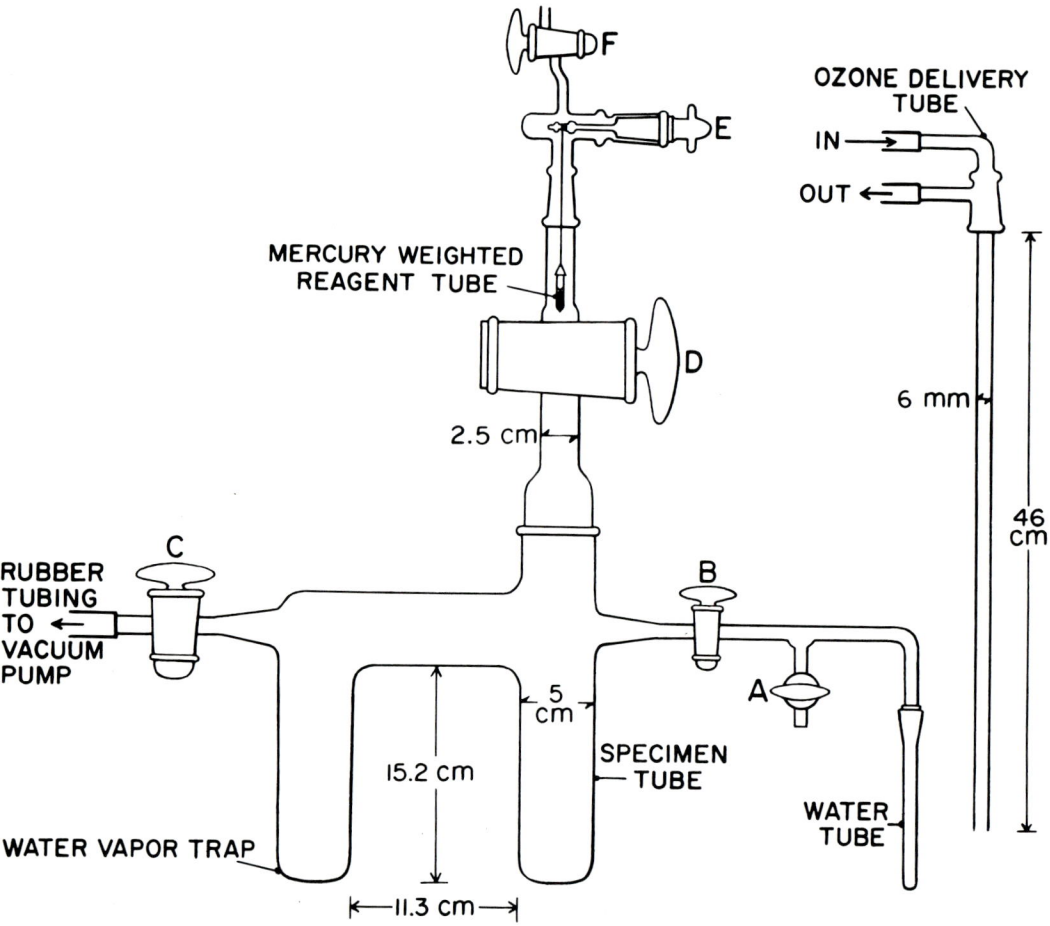

Fig. 1.1. Diagram of apparatus used in the preparation of freeze-dried specimens for the study of their lipids with the electron microscope. See text for details.

from the specimen because of its low temperature, following which stopcock (C) was closed.

In the meantime, the superstructure had been disassembled in order to load the upper part of the mercury-weighted reagent tube with diiodophenylhydrazine (IIPH). The superstructure was reassembled, and evacuated through (F) after which the stopcock was closed. The two vacuum chambers were joined by turning the large-bore stopcock (D), and the weighted reagent tube was lowered through (D) by turning the winchlike joint at (E) until it reached the bottom of the specimen chamber. It was found that when water vapor, most of which condensed on the walls of the specimen chamber, was introduced into the vacuum, the results were more uniform and consistent, and shrinkage (e.g., of mitochondria) was greatly reduced. It is possible that water vapor acts as a catalyst in the subsequent reactions. Water vapor was introduced into the vacuum in the following way: The water tube was immersed in liquid nitrogen and the atmosphere between the ice and stopcock (B) was evacuated through the tube containing stopcock (A). At this point, stopcock (A) was closed, the ice was permitted to warm, and some of the vapor was allowed to enter the main vacuum line through (B), which was closed again.

The liquid nitrogen surrounding the specimen tube was replaced by an alcohol bath at $-30°C$. During the next 10 hours, in the closed vacuum system, vapors of IIPH reacted with lipid components at $-30°C$ and during an additional 5 hours at $0°C$. The reagent tube was then lifted through (D) into the superstructure by turning the winch-cock (E). Stopcock (D) was closed, the specimen tube was evacuated for about 15 hours at $0°C$, after which the temperature of the water trap was raised to about $24°C$. This step was designed to remove solid-phase reagent which may have been deposited on the surface of the vacuum line or in the specimen through nucleation and crystal growth, as well as water condensed on the walls of the water vapor trap.

After this, the vacuum was broken by dry air admitted through a 35 cm column of granulated silica gel through (A). The specimen tube was then chilled to $-80°C$ and ozone was passed over the specimen for 20 minutes at about $-80°C$ by means of a glass ozone delivery tube (Fig. 1.1), which replaced the superstructure. The ozone was generated from a mixture of 95% helium and 5% oxygen passing through a Welsbach Model T-23 Laboratory Ozonator (purchased from The Welsbach Corporation, 1500 Walnut Street, Philadelphia, Pennsylvania 19102). The concentration of the ozone as determined by the commonly used iodometric method was 0.0143 mg/liter, calculated at $25°C$ and 760 mm mercury pressure, and the pressure of the ozone stream was at 5.5 psi. The stream passed through a water vapor trap at $-80°C$ to remove excess water before entering the specimen chamber, and on exit from it was led through a large volume of a concentrated solution of potassium iodide. Under these conditions of temperature, concentration, and time, ozone acted specifically on double bonds in aliphatic chains without notably disturbing the fine structure of the specimen.

The specimen tube was returned to $0°C$, the superstructure was replaced, and the whole system was reevacuated. The reagent was reintroduced into the speci-

TABLE 1.1

Reagents Tested and Used as Controls

Acridine	Ethylamine
Aminoethylbenzene	Hexabrombenzene
3-Aminoquinoline	p-Iodoaniline
1-Amino-1,2-diphenylethane	Iodonium bromide
Aniline	p-Nitrophenylhydrazine
1,4-Bis(aminomethyl)cyclohexane	Bis(2-nitro-4-hydrazinophenyl)methane
2,6-Diiodo-4-nitroaniline	Nonylamine
N,N-Dimethyl-p-phenylenediamine	Octadecylamine
2,4-Dinitrophenylhydrazine	Octylamine
Diphenylamine	Phenylethylamine
Diphenylthiocarbazone	Phenylhydrazine
Dodecylamine	Tyramine

men chamber by lowering the mercury weighted tube as before. Water vapor was also introduced into the system, and stopcock (C) was closed. The IIPH reacted with the specimen for 5 hours at 0°C and a final 2.5 hours at room temperature. This time the reagent reacted with aldehyde and/or ketone groups generated at the site of ozonized aliphatic double bonds. As before, the reagent tube was removed and the excess reagent in vapor or solid form was evacuated during 15 hours through (C).

Finally, the water-soluble plastic Durcupan (components ABCD freshly prepared and well stirred) was introduced into the specimen chamber through stopcocks (A) and (B). This took place slowly, with the vacuum pump connected. When the specimen vial was covered, the vacuum was broken. The vacuum was made and broken two times, and then the superstructure was removed. The specimen vial was poured out of the specimen tube into a Petri dish. The specimen was removed from the vial and the aluminum foil and cut into smaller pieces under a dissecting microscope. These were transferred to a covered preparation dish (37 × 19 × 25 mm) containing a fresh mixture of the water-soluble components of Durcupan. The dish was wrapped in Parafilm and placed on a platform, which tilts 4 cm at a distance of 15 cm from its hinge, at a rate of four times per minute. Infiltration of the specimens continued at 16°C for about 36 hours. At this time, the specimens were placed in freshly prepared plastic mixture in BEEM capsules. These remained at room temperature for 1 day, at 37°C for 2 days, and at 50°C for a final day. The water-soluble monomer mixture of Durcupan was found, in general, to be superior to the water-insoluble mixture, perhaps because of the lower viscosity and, consequently, more thorough penetration of the specimen during infiltration. It should be noted also that water-soluble Durcupan has been found to extract very little phospholipid *(44)*.

The polymerized blocks were trimmed and sections were cut with glass knives, floated on water, mounted on grids, and examined with an Hitachi HU-11A electron microscope operated at an accelerating voltage of 50 kv. Electron micrographs were enlarged × 2 in printing.

Methods

An abbreviated procedure eliminated the ozonation step. Specimens were dried in the presence of IIPH and water vapors for 10 hours at −30°C and 5 additional hours at 0°C. Excess vapors and solids were evacuated for 3 hours at room temperature. Infiltration with water-soluble Durcupan and polymerization took place as described above.

The reagent (IIPH) was selected from among a large number of possibilities and controls (Table 1.1). In this list are compounds (1) which have a high vapor

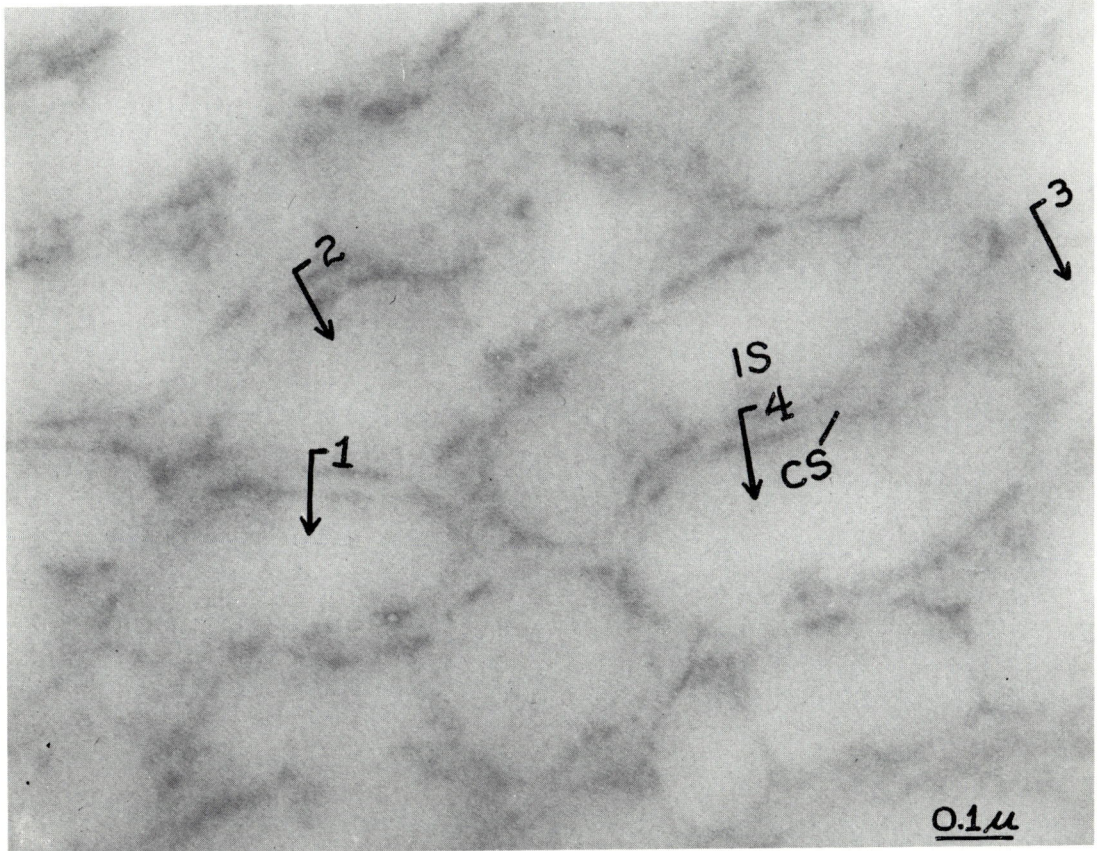

Fig. 1.2. Electron micrograph illustrating lipid component of rough endoplasmic reticulum (RER) in pancreatic acinar cells, stained for lipid by the basic method described on p. 8. Note discontinuities in lipid component of RER membranes, which are paired and enclose the cisternal space (CS), and differences in apparent thickness and density from point to point. They are sometimes joined by positively stained lines or "membranes" which traverse the intercisternal (IS) space rich in ribosomes (which are unstained in this preparation). There is an overall unresolvable background density. The lines mark numbered sites where tracings were made with a microdensitometer. × 100,000.

pressure but are unreactive, (2) which have a vapor pressure which was too low to be detected, (3) which are primary, secondary, and tertiary amines, (4) which are primary amines of low molecular weight and thus could not increase the mass sufficiently after the reaction, (5) which are fluid or unstable, and (6) which may combine with acids or heavy metals. Of these compounds, only primary amines caused increased density and contrast on electron micrographs, and of these diiodophenylhydrazine seemed most satisfactory.

Of the numerous additional control experiments which were made over an 8-year period, the following are cited:

1. Soxhlet extraction with chloroform: methanol (1:1) prior to the test. Dinitrophenylhydrazine was used instead of diiodophenylhydrazine, because of

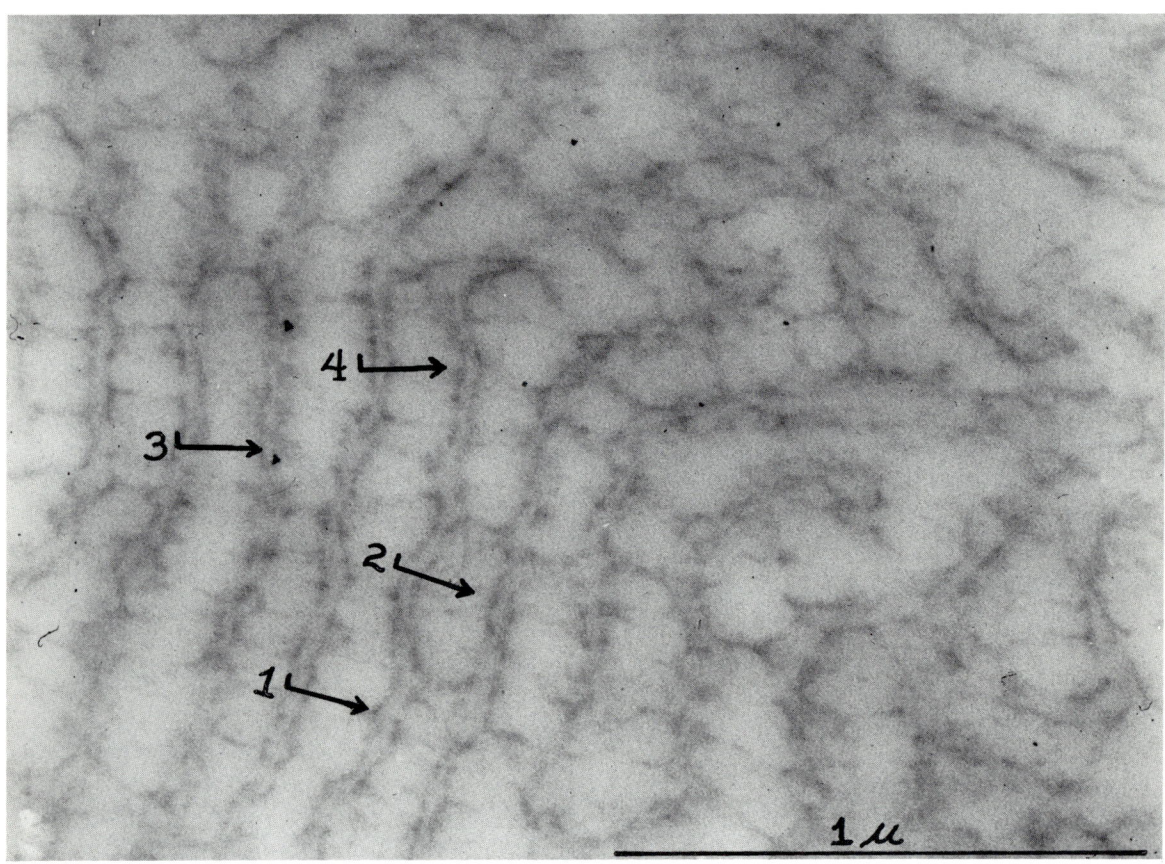

Fig. 1.3. Electron micrographs at a lower magnification, of a similar preparation. × 70,000.

Methods

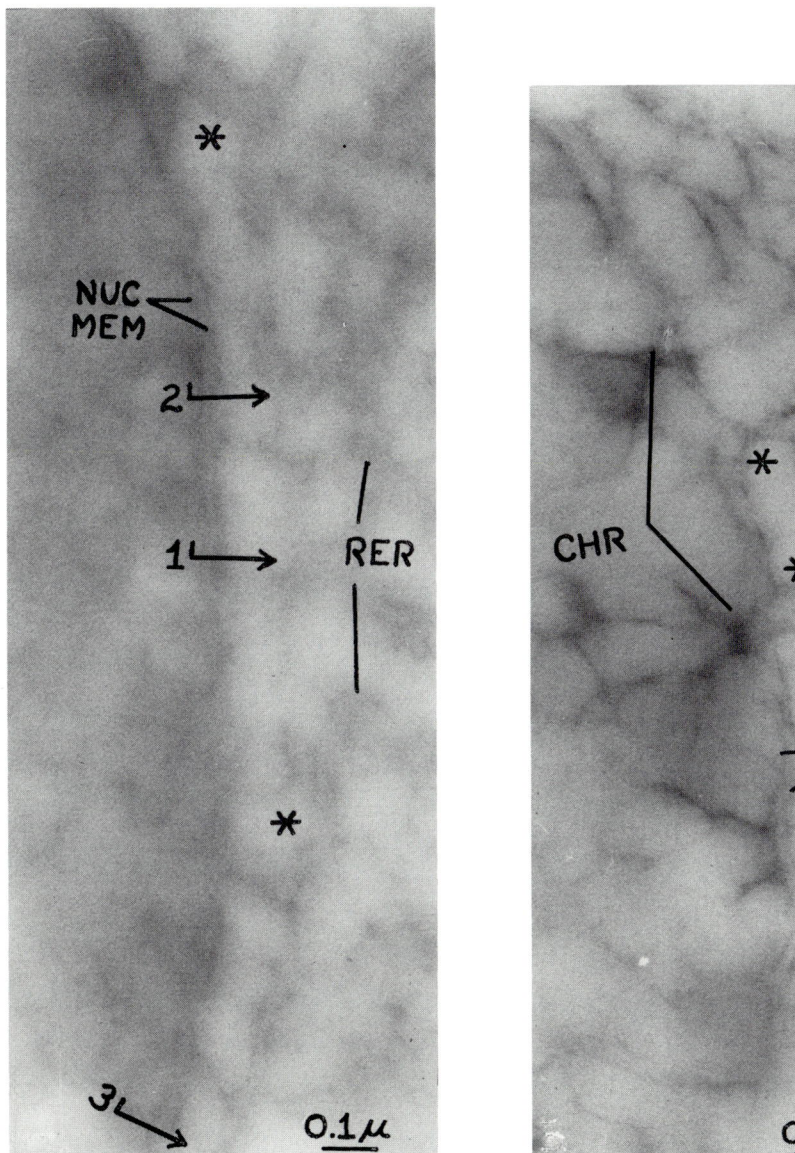

FIGS. 1.4 (LEFT) and 1.5 (RIGHT). Electron micrographs illustrating lipid component of nuclear membrane (NUC MEM) and relations to RER stained by the basic method. The lipid component of the outer nuclear membranes seem to be attached to or joined with the lipid component of the RER at points marked with an asterisk. CHR, Chromatin. × 70,000.

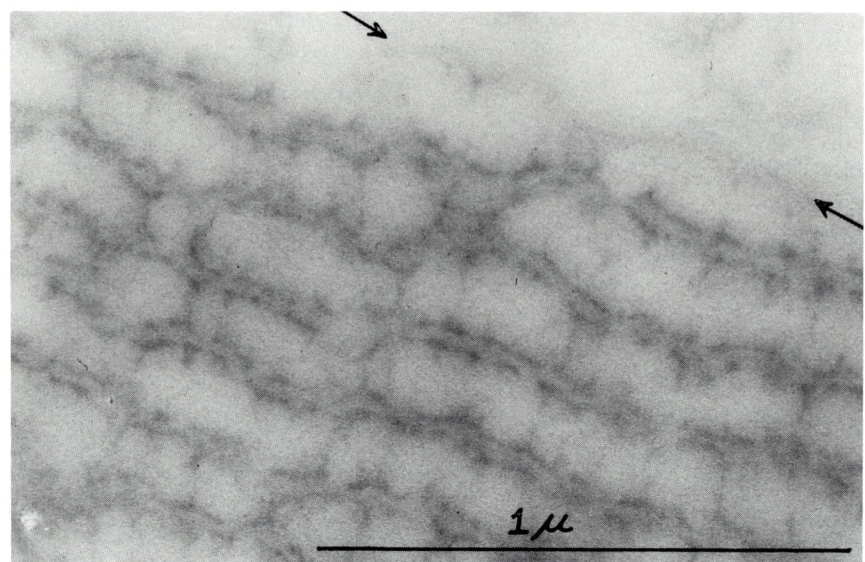

Fig. 1.6.

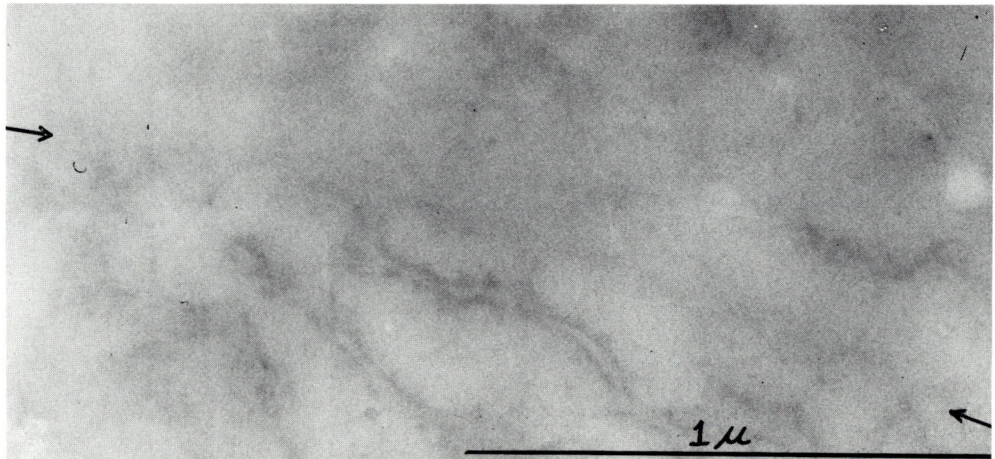

Fig. 1.7.

Figs. 1.6–1.8. Lipid component of cell membranes and of adjacent RER after staining with the basic method. The lipid component of the cell membranes, which stretches between the arrows, is weak compared with that of the RER. At some points, the lipid component of the RER appears to become continuous with that of the cell membrane. The lines mark numbered sites where densitometric tracings were made. × 70,000.

Methods 13

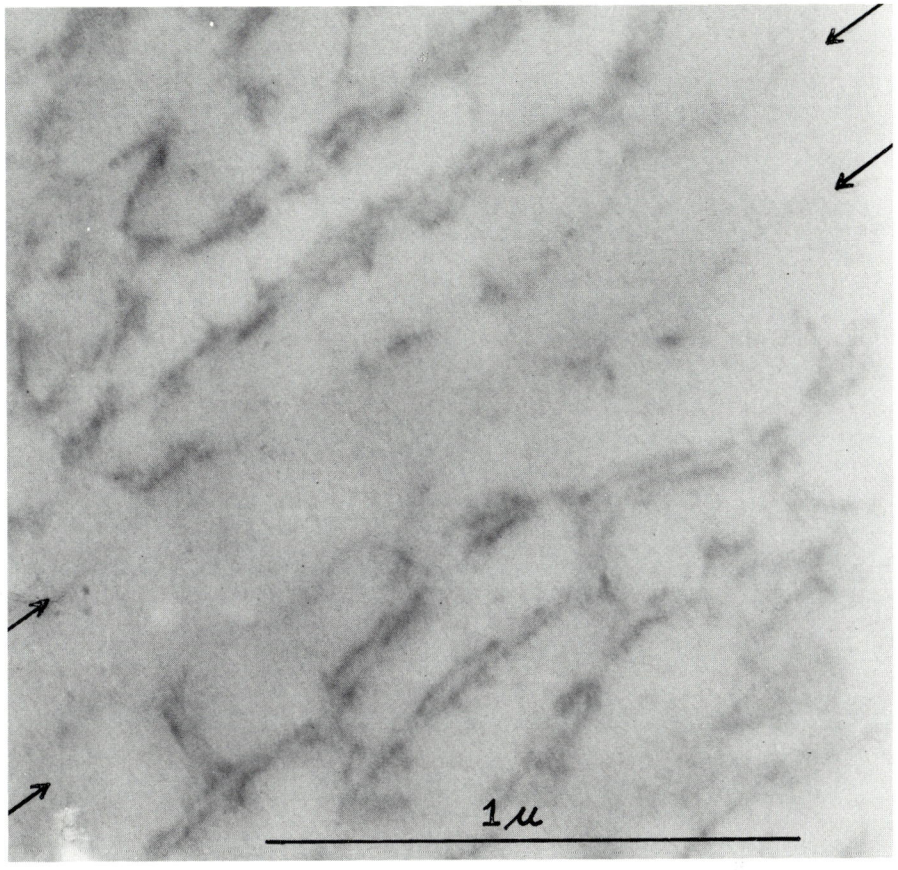

Fig. 1.8.

the high extinction of the yellow fatty acid derivatives in visible light. Specimens appeared barely colored grossly, and this was not detectable at all in sections observed with the light microscope, as compared with the brilliant color observed grossly and microscopically in the unextracted specimen. It should be pointed out that the yellow color could not be the result of its solubility in lipids under the conditions of the test, since the lipids were solid at $-30°C$, when exposed to the reagent (with excess reagent adherent to glass or tissue in the vacuum line removed prior to examining the specimen in the light microscope) or to infiltration with the reagent. In the electron microscope, no regions of increased contrast were observed in the extracted specimen, as compared with the clear lines of the RER in unextracted specimens.

2. There was a safety factor of hours between the time and temperature selected for ozonization of the aliphatic bonds and of aromatic double bonds. After the longer exposures to ozone, there was enhanced nonspecific contrast.

14　　　　　　　　　　　　　　　　　　　　1. Lipid and Protein in Membranes

3. The weak reaction of cytoplasmic ribosomes which are rich in RNA, and of DNA in the nucleus show that the combination with nucleic acids, if it occurs, is readily broken during the subsequent steps, as neither shows enhanced contrast.

4. The extremely weak cytoplasmic reaction after prior extraction of the specimen with chloroform:methanol (1:1) indicates that if carboxyl groups of

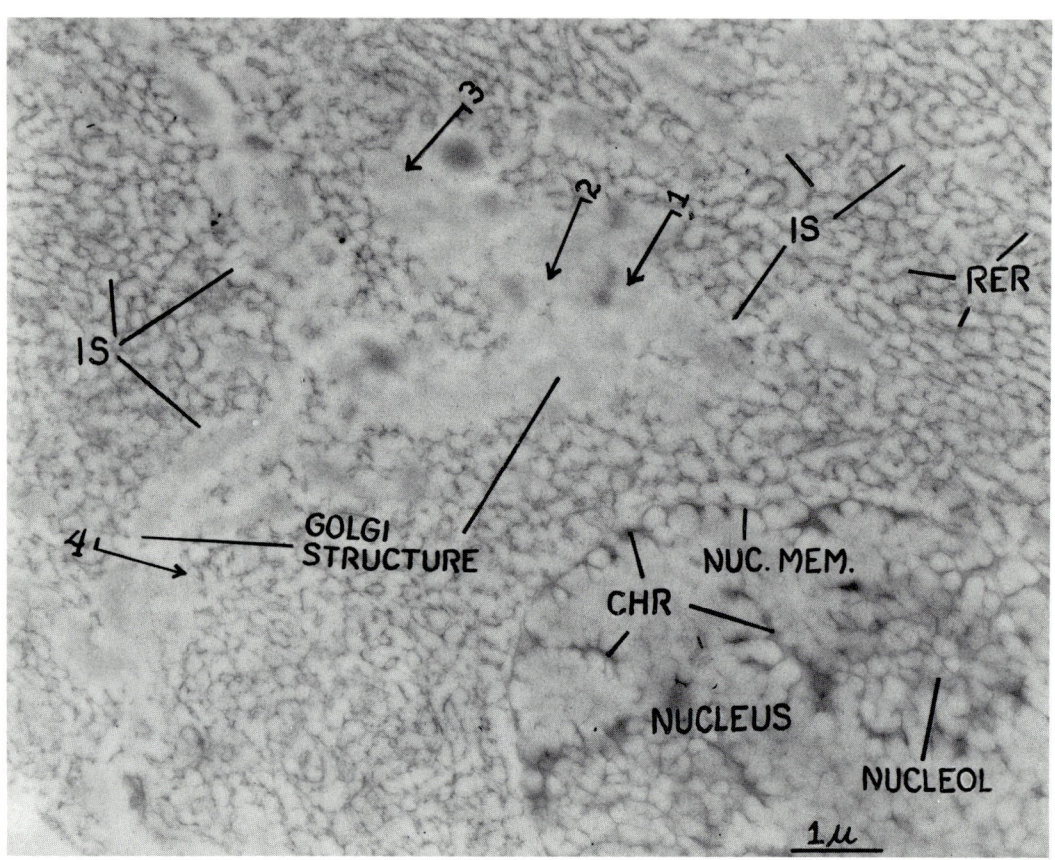

Fig. 1.9.

Figs. 1.9 and 1.10. Low power electron micrographs to show general relations of lipid components of Golgi complex to those of the RER, nuclear membranes (NUC MEM), and nucleus stained by the simplified method. The Golgi complex consists of mostly reticular structures of low but variable density, with some spherical dense bodies, at least some of which are enclosed in a membrane. The surface of the greater part of the former is in contact with the unstained intercisternal space (IS) of the RER. Strands of the Golgi complex interdigitate intimately with the RER. In Fig. 1.10, some of the RER interdigitations into the Golgi complex appear separated from the main body of the RER, but this is probably related to the plane of section. Intranuclear lipid is notable in regions rich in nuclear protein, especially the chromatin masses (CHR) and nucleoli (NUCLEOL), but this distribution does not correspond with that of the nucleic acids. At higher magnification (as in Fig. 1.5), it is clear that the lipid distribution pattern in the nucleus cannot be confused with that of the RER. The lines mark numbered sites where tracings were made with a microdensitometer. × 12,000.

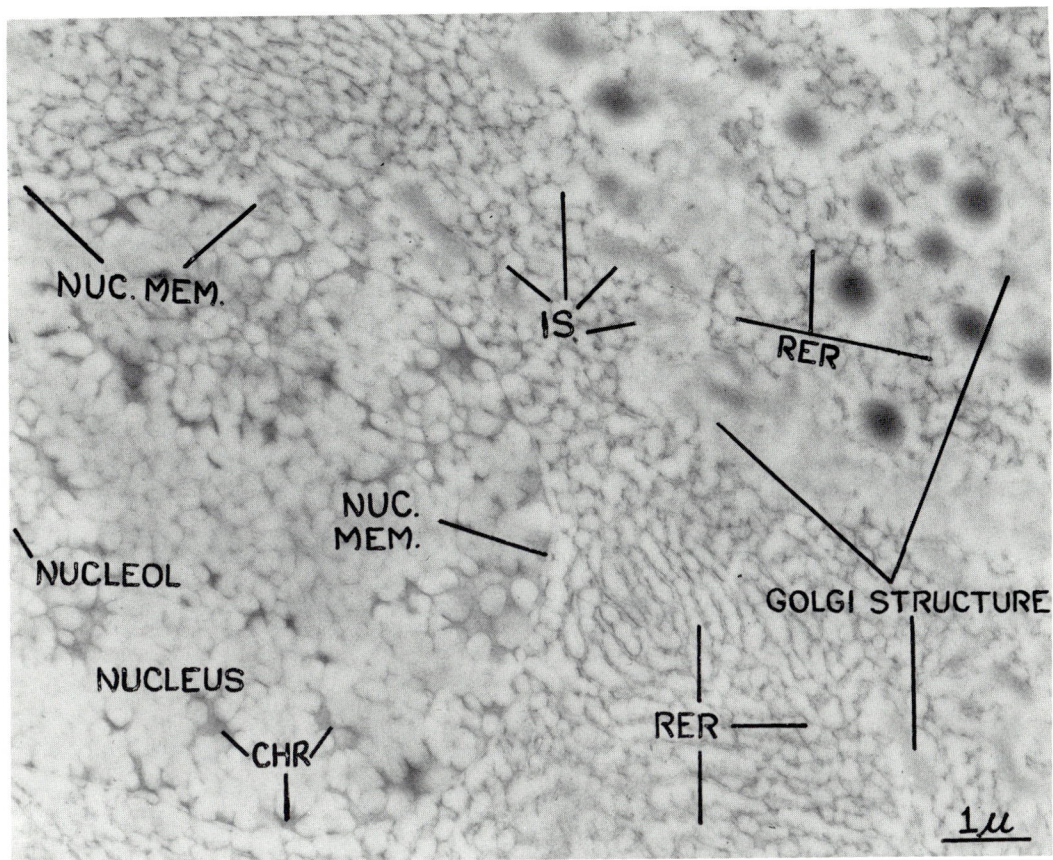

Fig. 1.10.

terminal amino acids react at all with the hydrazines, the reaction is not recordable on the electron micrograph, or does not persist through subsequent steps.

5. In Chapter 2 (Vol. I), it was shown that various organic and organometallic reagents in vapor form combine with proteins alone or with both proteins and nucleic acids of freeze-dried cells. The evidence clearly shows that the regions of increased mass caused by these reagents corresponds only to some extent with regions containing reactive lipids noted in this report.

Two further observations support the evidence that lipids have been demonstrated with great precision. The first has already been reported (18) and concerns the use of a cross-linking hydrazine derivative synthesized by Dr. T. C. Myers, of the University of Illinois [bis(2-nitro-4-hydrazinephenyl)methane]. It was postulated that this compound forms a cross-linked polymer of indefinite extent. Such a large polymer would surely be minimally displaced during infiltration, etc. It is significant under these circumstances that the morphological distribution of lipid which appeared in the RER was indistingushable from that which appeared after the use of diiodophenylhydrazine.

A second observation speaks also for the precision with which lipids are demonstrated: a regular periodicity of about 226 Å has been demonstrated in the retinal rods of guinea pig retina by the technique described. This compares well with the 240 Å periodicity derived by Sjöstrand *(64)* from electron micrographs of chemically fixed retina. This period can thus be regarded as confirming that ascertained by X-ray diffraction. In the same way that this correspondence validated the immersion methods of fixation and staining, the same correspondence must be regarded as validating the method of freeze-drying and of postfixation and staining with vapors of diiodophenylhydrazine as described in this chapter.

PROTEINS

To ascertain the relations of lipid material stained with IIPH to proteins, ultrathin sections of frozen-dried specimens treated as described above were stained with uranyl acetate on grids before study with the electron microscope. Most of these sections were transparent when viewed while they were floating on the sectioning boat. They were photographed with a Hitachi-11A electron microscope, operating at an accelerating voltage of 50 kv, with pointed filaments. All plates were taken at a magnification of 100,000, and developed with fine grain developer. Prints were made at a magnification of × 2 or × 3.

Finally, some specimens (not treated with IIPH) were treated with one of a variety of organic and organometallic reagents in vapor form to postfix and stain proteins. Only some of these were favorable for the study of "globular" protein in membranes, and these are listed in Chapter 2 (Vol. I), where technical details may be found. Sections of such specimens were viewed as described in the preceding paragraph.

Results

LIPID COMPONENTS OF PANCREATIC ACINAR CELLS

The RER is distributed as paired sheets or plates throughout most of the cytoplasm. Each sheet or layer appears on micrographs as a discrete, thin, discontinuous line of single thickness but of variable density (Fig. 1.2). The discontinuities may, but usually do not match up with similar gaps of the line opposite. The lines are arranged in pairs and constitute the walls of the cisternae. The intracisternal space is much paler than the walls, and appears structureless. The space between the pairs of lines (corresponding to the intercisternal space) is larger than the intracisternal space. Frequently, dense lines extend across the intercisternal space either singly or as pairs, connecting the intracisternal space of adjacent cisternae (Fig. 1.3).

The lipid of the nuclear membrane is also in two layers, each of which appears in micrographs as a single line (Figs. 1.4 and 1.5). The outer layer is attached by strands and pairs of lines with adjacent RER, and one might mistakenly assert that there seems to be no marked difference in thickness or density between the adjacent lipid lines. The inner layer appears as a similarly dense line which ap-

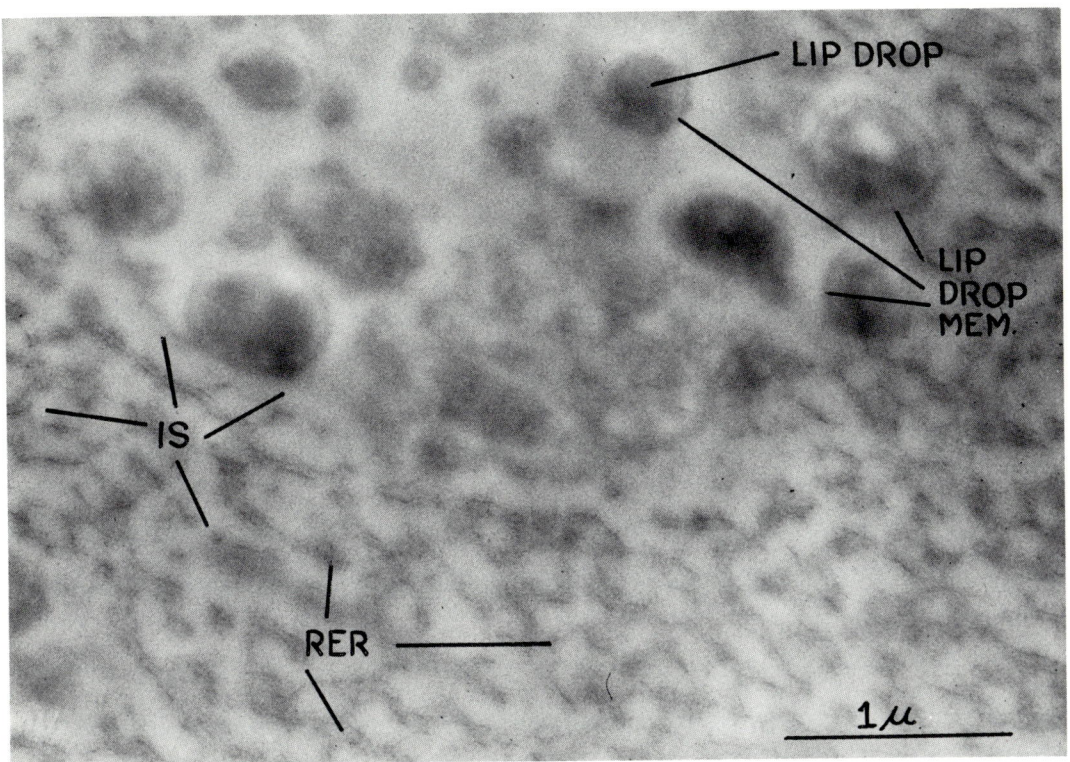

Fig. 1.11. Electron micrograph of lipid components of pancreatic acinar cells prepared by the simplified method. Strands and cords of Golgi material form reticulated nets of low density which are attached by thin strands to the RER. However, most of the surface is in relation to the pale, intercisternal spaces (IS) of the RER. The spherical structures, which are on the whole denser than the remainder of the Golgi complex are enclosed in a lipid-rich membrane. LIP DROP, lipid droplet; LIP DROP MEM, lip droplet membrane. × 30,000.

pears to fuse with intranuclear dense lines and zones which correspond with parts of the nucleus rich in nucleoprotein (CHR and NUCLEOL in Figs. 1.9 and 1.10). In some small regions, the lines clearly fuse, and in other small regions, they seem to be broken at the same level, without appearing to be joined, perhaps because of bulges in the nuclear membrane.

The lipid component of the cell membrane is visible in electron micrographs only occasionally as a very slight increase in density over the background (Figs. 1.6–1.8). Like each layer of the RER and the nuclear membrane, the increased density appears in electron micrographs as a single line. Continuous with the cell membrane are paired RER lines with their (relatively) markedly greater density. Sometimes this density was observed to extend on the surface of the cell (Fig. 1.8).

The lipid component of the Golgi apparatus is usually less dense than that of the RER and is extremely variable (Figs. 1.9–1.12). In some parts, the lipid is distributed as branched, reticulated rods or sheets which may appear solid or may have a paler lumen, which may run from cord to cord, and may be continuous with the intracisternal space of the RER. The decrease in density from RER to the Golgi channel is abrupt (Figs. 1.11 and 1.12). The solid cords may bulge in some parts so as to appear in sections as "solid" short rods and/or spheres. Some of the latter appear to be surrounded by lipid membranes.

Of great interest, is the fact that the cords, regardless of their structure, are enclosed in a generally poorly outlined lipid deposit which is directly applied to and continuous with the intercisternal spaces of the RER with their characteristic ribosomes.

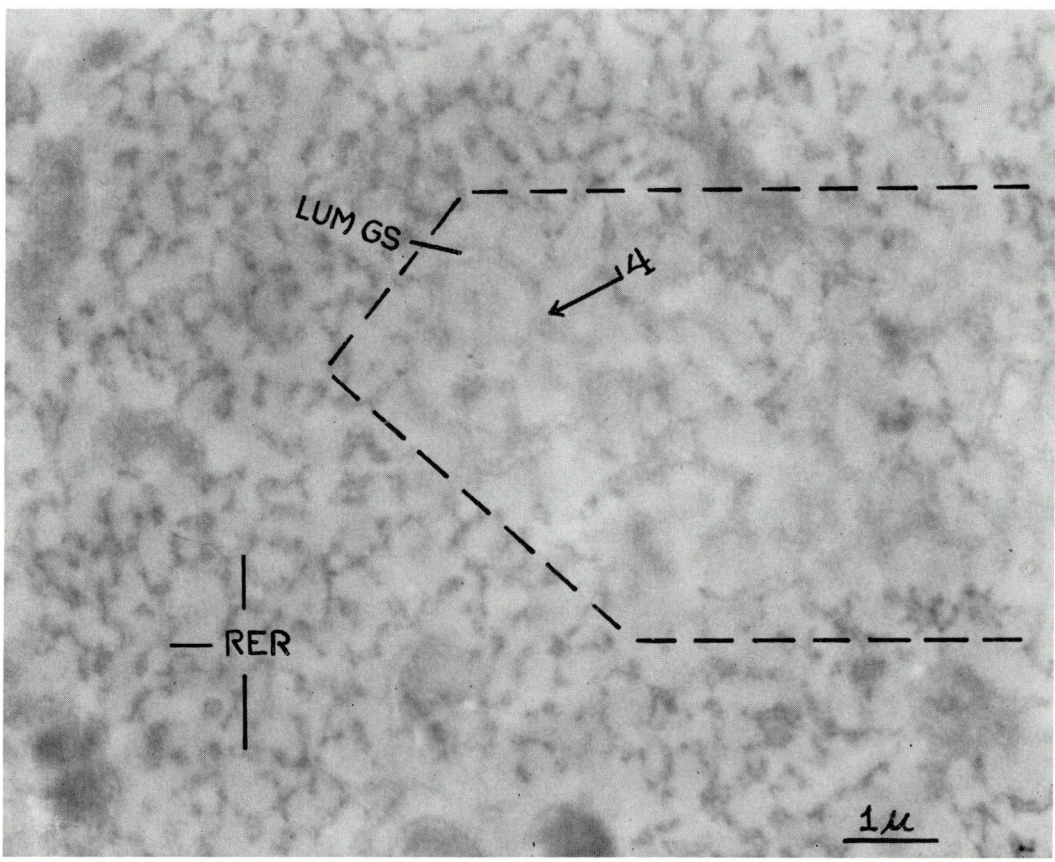

FIG. 1.12. Electron micrograph of lipid components of pancreatic acinar cells prepared by the simplified method. RER refers to lipid component of RER. Near the dotted line, the cords and tubes of the Golgi structures (LUM GS) are continuous with the RER. This marks the zone also where the Golgi structures impinge on the intercisternal spaces of the RER. The arrow marks the site where tracings were made with the microdensitometer. × 12,000.

Results

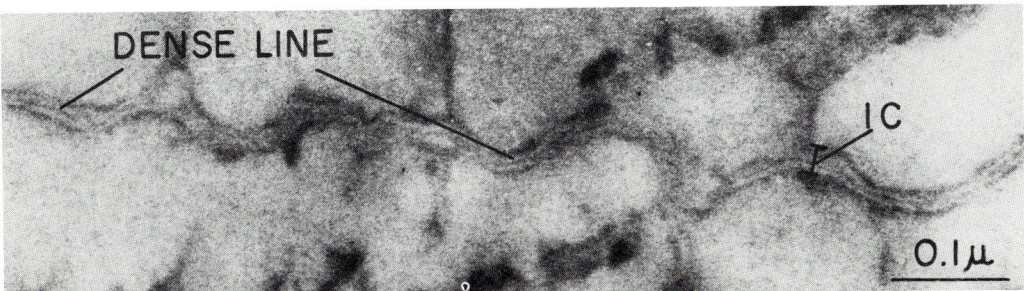

Fig. 1.13. Protein stain of pancreas, postfixed with vapors of IIPH and section stained with uranyl acetate. The intracisternal space (IC) consists of two clear layers of globular molecules separated by an apparent dense line. × 150,000. G2 print. See Preface for definition of term.

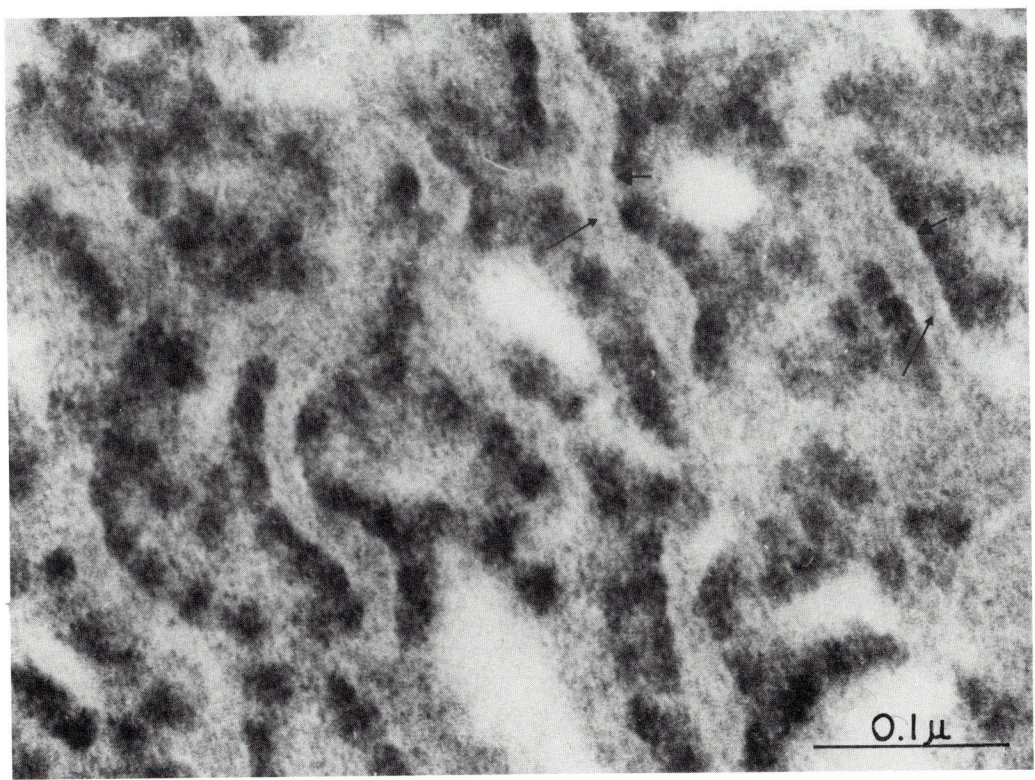

Fig. 1.14. Liver; same postfixation and staining procedure as in Fig. 1.13. The clear intracisternal space is lined by ribosomes. At points marked by arrows, the clear dense central line can be resolved into wavy lines. These are referred to in the text as the hydrophilic surface of apposed globular molecules. The protein (globular) molecules are clear at some points marked by arrowheads. × 258,000. G2 print.

20   1. Lipid and Protein in Membranes

Lipid Components of Hepatic Cells

Study of the lipid distribution in hepatic cells did not seem as profitable as in pancreatic acinar cells. The observations in liver cells were of the same nature as in acinar cells, as far as they went. Lipids in the walls of the RER, in the nuclear membrane, and in the nucleus were observed, but they were not as dense as in pancreas. The lipid component could not be identified in Golgi apparatus or in the cell surface. The background density was readily appreciable.

Protein Components of RER, Nuclear Membrane, and Cell Membrane

The protein components of all three structures resemble each other morphologically.

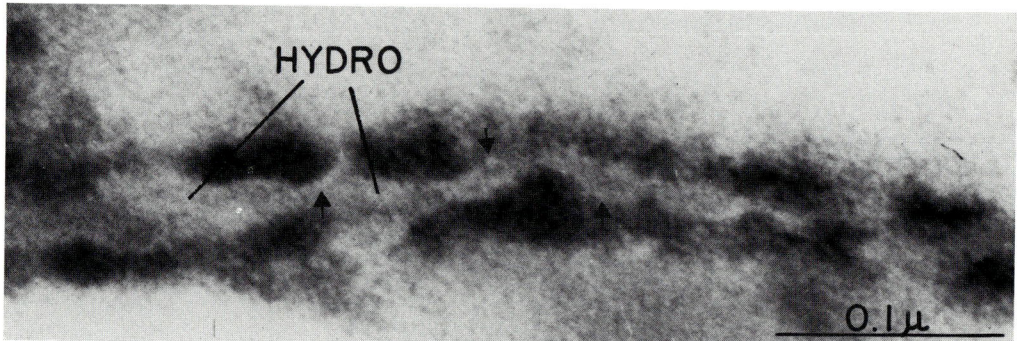

Fig. 1.15. ▲   ▼ Fig. 1.16.

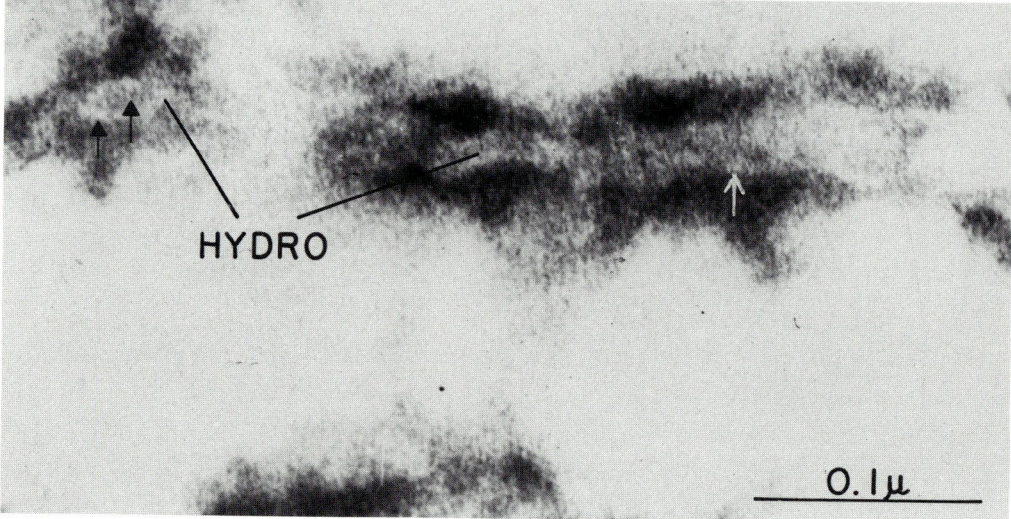

Figs. 1.15 and 1.16. Pancreas prepared as in Fig. 1.13. The clear intracisternal space is divided in two by the hydrophilic surfaces of the protein (globular) molecules (marked by arrow heads) as in Fig. 1.14. The dark central line is resolved as the hydrophilic apposed surfaces (HYDRO) of globular molecules. × 300,000. G2 print.

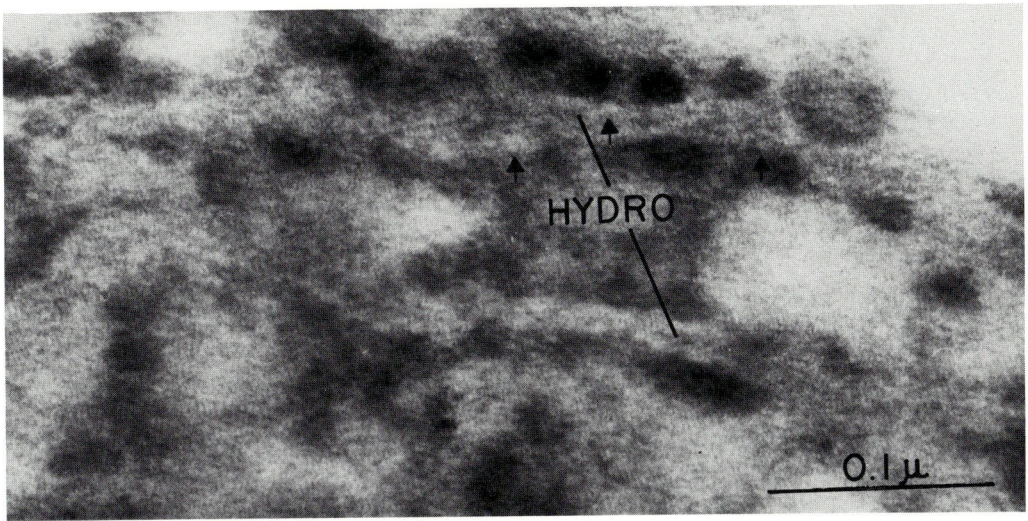

Fig. 1.17. Pancreas postfixed with vapors of HEFDOD and section stained with uranyl acetate. The intracisternal space appears pale because of the lipophilic, globular molecules (marked by arrowheads), which are largely unstained except for the hydrophilic apposed surface (HYDRO). × 300,000. G2 print.

In contrast with the homogeneous distribution of lipid in the intracisternal spaces, which are weakly stained with vapors of IIPH, is the very remarkable degree of internal structure of similar sections stained additionally on the grid with uranyl acetate. At high magnification (Fig. 1.13) the intracisternal space appears to be divided longitudinally by a thin, dense line. At a still higher magnification, the line is observed to be irregular in thickness. Moreover, the line is resolved as

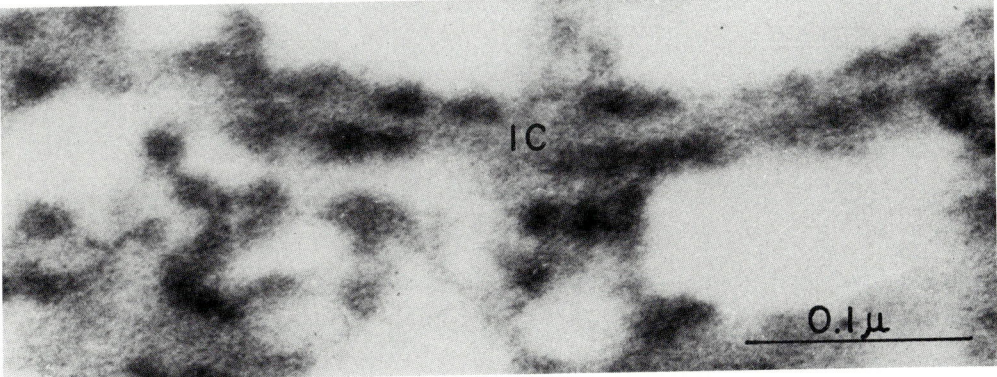

Fig. 1.18. Pancreas, stained with alcoholic $PtBr_4$ and section stained with uranyl acetate. The globular protein molecules of the intracisternal space (IC) are stainable after this particular kind of denaturation. × 300,000. G2 print.

the more densely stained parts of pale globules whose spheroidal outlines extend to the cisternal walls (Figs. 1.13–1.16). Interpreted in three-dimensional terms, the intracisternal space consists of closely packed spheroidal structures about 50 Å in diameter, each with a pale center and stained outlines, which together form a continuous sheet. The stain is more intense where the spheroidal structures abut on the equivalent structure of the opposite half, and where they abut on the lipid wall. The latter can be made out only in the bald parts of the cisternal wall, where the lipid layer is discontinuous. It can also be observed after other protein stains. (See HEFDOD and Zr hfac below.)

The lipid distribution in the walls of the intracisternal space is in marked contrast with the distribution of protein and RNA. The lipids occur as discontinuous slabs of quite variable thickness, but which are, on the average, 77 Å thick (Chapter 2, this volume). On the other hand, the protein is all pervasive, and is spread over lipid-rich as over lipid-poor parts. There are variations in density of the protein, if not discontinuities, and these are related to ribosomes (whose proteins are stained, as well as the newly synthesized protein molecules). These were described in Chapter 2 (Vol. I). The nucleic acids are, of course, discontinuous, as ribosomal, tRNA, or mRNA particles. On the biochemical side, one must visualize also the presence in the wall of the RER of the whole enzymic and nucleo-

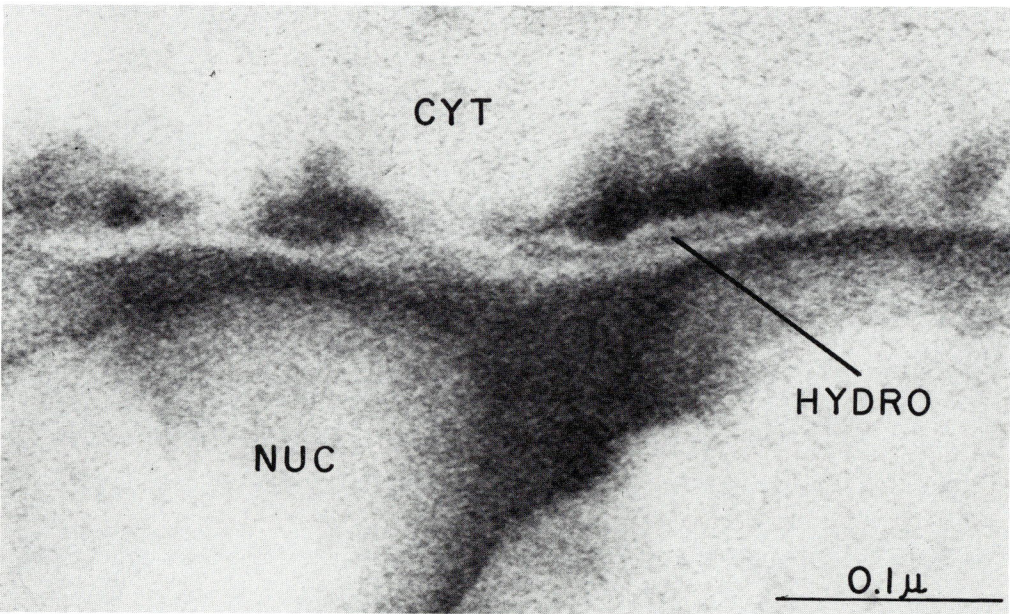

Fig. 1.19. Pancreas, postfixed and stained as in Figs. 1.13–1.16. The globular structure of the nuclear membrane resembles that of the intracisternal space of the RER, with a central dense line caused by the apposed hydrophilic (HYDRO) surface of the lipophilic (globular) molecules of each surface of the nuclear membrane. NUC, nucleus; CYT-cytoplasm. × 300,000. G2 print.

Discussion

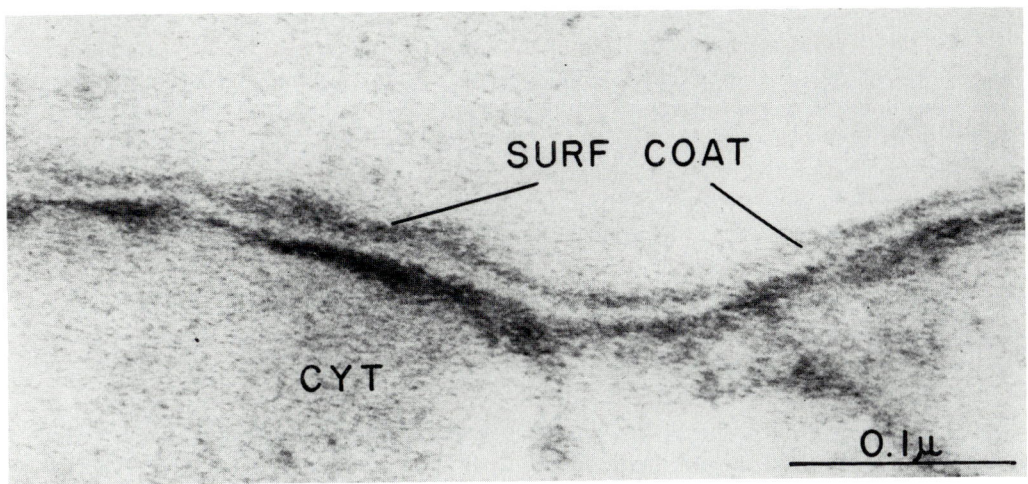

Fig. 1.20. Pancreas, postfixed and stained as in Fig. 1.19. The globular molecules form a single layer between the remainder of the cytoplasm (CYT) and the surface coat (SURF COAT). × 300,000. G2 print.

tide apparatus for protein synthesis, as well as for other essential metabolic enzymic activities.

After postfixation with vapors of HEFDOD (Fig. 1.17), followed by section staining with uranyl acetate, the protein distribution is the same as that described above. In contrast with this distribution of the stainable protein in the IC space, is that observed after other methods of postfixation and staining. After Zr hfac, the peripheral parts of the protein globules may be stained more deeply than the central parts. After postfixation with alcohol and staining with alcoholic platinum tetrabromide, the staining is maximal, since all parts of the protein globules are stained equally (Fig. 1.18). In this condition, the protein in the intracisternal space appears homogeneous and dense. The morphology of the protien globules (molecules) of the nuclear membrane closely resembles that of the intracisternal space (Fig. 1.19).

The protein components of the cell membrane are occasionally recorded in sections as a "single" layer of "globular" or hydrophobic molecules between the superficial layer of the cytoplasm and a surface coat, both of which are readily stainable (Fig. 1.20). Morphologically, the protein component of the cell membrane corresponds to only one-half of either the intracisternal wall or nuclear membrane.

Discussion

Reasons were given (pp. 5–15), during the consideration of the rationale for the various steps in the method for visualizing cellular lipid, to support the beliefs that the method is probably more or less specific and that the distribution of lipid observed is probably real. It should be recalled that the interpretation of the electron micrographs is somewhat equivocal because we do not know to what degree the reaction is incomplete, exactly which reactive groups have been active, or the degree to which steric hindrance or other factors interfere with the completeness of the reaction. That the reaction is incomplete is almost certain because of the low temperature at which the reaction takes place. This temperature was selected deliberately in order to prevent displacement of reacting lipids such as would certainly take place in freeze-dried specimens treated with the reagent at room temperature or higher. A complete reaction would be disadvantageous, since it would have made the specimen impenetrable to infiltration by the plastic monomer mixture. That the reaction, incomplete though it is, does take place is clear from a consideration of the numerous controls described earlier. In addition, it is frequently difficult to interpret two-dimensional density images of a photographic plate in terms of the three dimensions of the living cell. The last difficulty has been overcome, though incompletely, by densitometric studies of the electron micrographic plates. These studies are presented in the next chapter.

Phospholipids in cells appear to be largely associated with proteins (24). The properties of the lipid components cannot be separated from those of lipoproteins or other proteins and their aggregates. The chief effects of the lipid portion would be through their charges in orienting adjacent proteins and other charged substances and through the relations of their hydrophobic chains to proteins containing groups of hydrophobic amino acids. A similar suggestion has been made by Finean (15). The molecular orientation effects of the lipid portion could possibly extend for an appreciable distance through their effects on lipid or lipoproteins, vaguely represented by the diffuse background reaction.

The lipid portion differs according to the membrane it is part of. The lipid portion of the RER is denser (reacts more completely or is present in high concentration) than that in the nuclear membrane, which, in turn, appears denser than that of the cytoplasmic membrane, where it is only occasionally dense enough to be recorded on a photographic plate. The orientation of the lipid molecules on the surface of the cell would depend on the molecular components and orientation of peripheral molecules of the cytoplasm, as well as on similar properties of the ground substance or intercellular substance outside the cell. It seems clear that the membrane lipids of the RER are either interrupted in patches, or that different (less reactive or nonreactive) lipids alternate with patches of very reactive lipids. The lipid component of nuclear membranes is generally more or less homogeneous structurally.

It should be emphasized that the unit membrane hypothesis is not substantiated. All membranes appear as single lines in electron micrographs, and not as bilaminar structures. The nature of the reaction is such that the abbreviated procedure described on p. 8 should favor the appearance of bilaminar structures if they occurred in membranes because of the concentration of double bonds at the choline end of the lipid molecules. The fact that all the RER and nuclear membranes appeared as single lines in electron micrographs, regardless of the method

Discussion

employed, argues strongly against the unit membrane hypothesis. Moreover, the lipid layers are discontinuous and of variable thickness, whereas the unit membrane hypothesis calls for continuous layers of uniform thickness. Finally, the most prominent feature of the membranes is the occurrence of closely packed protein molecules, leaving little space internally for lipid molecules. This view accounts for the pale staining of the intracisternal lipids, though other interpretations are possible. The evidence suggests that the unit membrane arises from a redistribution of substances during fixation.

The structure of the RER and nuclear membranes as visualized in this chapter resembles closely that which appears in electron micrographs by Bullivant (6), Malhotra (43), and especially by Sjöstrand and his students (67–69, 71). The protein components of the RER and nuclear membranes do, indeed, seem to be distributed as Sjöstrand had suggested. I believe the wide range of methods of postfixation and staining lend strength to this interpretation of the structure. However, no evidence could be adduced from the observations as to the nature of the intracisternal protein, i.e., as to whether the protein is enzymic or structural or shares both properties. Certainly the evidence in the literature suggests that some enzymic activity is present on the outer layers of the cisternal wall, associated with the ribosomes.

Study of the lipid component of the cell membrane is incomplete despite a deliberate attempt to concentrate on this structure. Very few photomicrographs showed the lipid component of the cell membrane, and this for rather short stretches only. This apparent paucity may be the result of a highly irregular surface which can be sectioned transversely only occasionally, and would not be distinguishable in any other plane. Or the scarcity of electron micrographs of lipid components of cell membranes may reflect the real situation. If the latter possibility were real, a corollary is that there is less reactive lipid in the cell membrane than in RER or nuclear membrane, and it is distributed in spots. According to this view, most of the cell surface would not be characterized by a lipid layer, or if there were a lipid layer, the specific lipids would not be as reactive as those of the RER or nuclear membrane.

Also difficult to see was the protein component of the cell membrane. When seen, it consisted of a single thickness of "globular" protein molecules tightly packed as a sheet. It was equivalent, morphologically, to a half-thickness of the intracisternal membrane of the RER, nuclear membrane, or mitochondrial compartmental membranes. Accordingly, the tentative view of the cell membrane is that it is primarily protein in structure, but contains lipids interspersed between the aligned protein molecules, sometimes in patches. It can be assumed, that the cell membrane lipids may not be as reactive as those of the RER or nuclear membrane, or that there may be less of them in proportion to the reduction in thickness of the former as compared with the latter.

The distribution of the lipid component of the Golgi apparatus is extremely varied. It may occur as thin layers continuous with those of the RER, or the lipid may appear as part of a "solid" lipid-containing cord or sheet, in both of which swellings may occur; the latter resemble lipid droplets. The most striking property is that the greater part of the cords and sheets impinge directly on the RNA-rich, wide intervals between the paired cisternal plates of the RER. The possible significance of this morphological relationship is that it offers a path whereby substances synthesized in the RNA-rich spaces between the RER lipid plates (the intercisternal spaces) may be transported directly to the Golgi apparatus. These observations suggest that newly synthesized protein molecules are transferred from the RER via intercisternal pathways to the periphery of the Golgi zone where they may be aggregated as secretory granules. Evidence for such a method of transport and storage of a secretory product is presented later in this volume (Chapters 4 and 6). According to this view, an intermediate stage of transfer via the intracisternal space does not occur.

The occurrence of anastamotic cords, rods, and sheets, nearly all of which appear solid, and of peripherally disposed channels connected with the RER resembles the reconstruction of the Golgi apparatus by Flickinger *(17)* and by Mollenhauer and Morré *(48)*.

Summary

Lipid components of pancreatic acinar and hepatic cells were stained by reactions for the carboxyl ester linkage, the unsaturated aliphatic carbon bonds, as well as aldehyde and ketone bonds. Extreme precautions were taken to prevent displacement, molecular reorganization, and extraction of the lipid components.

The density of staining of the RER is greatest, followed by that of the nuclear membrane, and the cell membrane, which is barely appreciable when identifiable. No support could be obtained for the unit membrane.

In the RER, the lipid layer is thin, discontinuous, and of variable thickness and density. The opposite lipid layers are separated by two layers of closely packed, difficultly stainable protein molecules and probably very little lipid, which together make up the contents of the intracisternal space. Externally the cisternae are clothed in a layer of readily stainable protein which is several times thicker than the lipid layers. The outer protein sheath contains the ribosomes and probably the whole enzymic and nucleotide apparatus for protein synthesis.

The nuclear membrane differs from the RER membrane in that the lipid component seems to be more uniform in thickness and more continuous. Between the two lipid layers are packed protein molecules resembling those of the intracisternal space of the RER. Lipid and protein components of the cell membrane of exocrine pancreatic cells are described on pp. 16–20 and p. 23.

The lipid component of the Golgi apparatus occurs as membranes, "solid" cords, and sheets distributed as highly irregular nets. Local swellings of the latter two structures may appear in sections as "droplets." The density of the lipid reacion ranges from about the minimal in cell membranes to the maximal in the RER. The protein constituent of this pleomorphic structure has been described in Chapter 2 (Vol. I).

It is suggested that protein synthesis takes place in the outer wall of the RER, and that the protein molecules are transferred in the intercisternal spaces to the periphery of the Golgi apparatus, where they are aggregated to form secretory granules.

References

1. Bailey, P. S. (1958). The reaction of ozone with organic compounds. *Chem. Rev.* **58**, 925–1010.
2. Benedetti, E. L., and Emmelot, P. (1968). Structure and function of plasma membranes isolated from liver. *In* "The Membranes" (A. J. Dalton and F. Haguenau, eds.), pp. 33–120. Academic Press, New York.
3. Benson, A. A. (1968). The cell membrane: a lipoprotein monolayer. *In* "Membrane Models and The Formation of Biological Membranes." (L. Bolis and B. A. Pethica, eds.), pp. 190–202. Wiley, New York.
4. Brown, H. D., and Chattopadhyay, S. K. (1971). Organelle-bound enzymes. *In* "Chemistry of the Cell Interface" (H. D. Brown, ed.), Part A, pp. 73–203. Academic Press, New York.
5. Bullivant, S. (1962). Consideration of membranes and associated structures after cryofixation. *Proc. 5th Int. Congr. Electron Microsc.*, Vol. 2, p. R–2. Academic Press, New York.
6. Bullivant, S. (1970). Present status of freezing techniques. *In* "Some Biological Techniques in Electron Microscopy" (D. F. Parsons, ed.), pp. 101–146. Academic Press, New York.
7. Bullivant, S., and Ames, A, III (1966). A simple freeze-fracture replication method for electron microscopy. *J. Cell Biol.* **29**, 435–447.
8. Chapman, D., and Leslie, R. B. (1970). Structure and function of phospholipids in membranes. *In* "Membranes of Mitochondria and Chloroplasts" (E. Racker, ed.), pp. 91–126. Van Nostrand-Reinhold, New York.
9. Coombs, R. R. A., and Lachmann, P. J. (1968). Immunological reactions at the cell surface. *Brit. Med. Bull.* **24**, 113–117.
10. Crane, F. L., Stiles, J. W., Prezbindowski, K. S., Ruzicka, F. J., and Sun, F. F. (1968). The molecular organization of mitochondrial cristae. *In* "Regulatory Functions of Biological Membranes" (J. Järnefelt, ed.), pp. 21–56. Elsevier, Amsterdam.
11. Danielli, J. F. (1968). Phospholipid membranes are necessarily bimolecular. *In* "Molecular Associations in Biology" (B. Pullman, ed.), pp. 529–537. Academic Press, New York.
12. Deuel, H. J. (1951). "The Lipids; their Chemistry and Biochemistry," Vol. 1. Wiley (Interscience) New York.
13. Fawcett, D. W. (1964). Structural and functional variations in the membranes of the cytoplasm. *Proc. 1st Int. Symp. Cell Chem.*, pp. 15–36. Japan Society for Cell Chemistry, Okayama.
14. Finean, J. B. (1967). "Engström-Finean Biological Ultrastructure," 2nd ed. Academic Press, New York.
15. Finean, J. B. (1969). Biophysical contributions to membrane structure. *Quart. Rev. Biophys.* **2**, 1–23.
16. Finean, J. B., Bramley, T. A., and Coleman, R. (1971). Lipid layer in cell membranes. *Nature (London)* **229**, 114.
17. Flickinger, C. J. (1969). Fenestrated cisternae in the Golgi apparatus of the epididymis. *Anat. Rec.* **163**, 39–54.
18. Gersh, I. (1964). A cytochemical method for fixation and staining of lipids in frozen-dried mouse pancreas for study with light and electron microscopes. *Histochemie* **4**, 322–329.
19. Glauert, A. M. (1968). Electron microscopy of lipids and membranes. *J. Roy. Microsc. Soc.* **88**, 49–70.
20. Glauert, A. M., and Lucy, J. A. (1968). Globular micelles and the organization of membrane lipids. *In* "The Membranes" (A. J. Dalton and F. Haguenau, eds.), pp. 1–32. Academic Press, New York.

21. Glauert, A. M., and Lucy, J. A. (1968). Electron microscopy of lipids: effects of pH and fixatives on the appearance of a macromolecular assembly of lipid micelles in negatively stained prparations. *J. Microsc. (Oxf.)* **89**, 1–18.
22. Green, D. E., and Goldberger, R. F. (1967). "Molecular Insights into the Living Process," pp. 222–270. Academic Press, New York.
23. Green, D. E., and Vanderkooi, G. (1970). Structure of the mitochondrial cristael membrane. *In* "Physical Principles of Biological Membranes" (F. Snell, J. Wolken, G. Iverson, and J. Lam. eds.), pp. 287–304. Gordon and Breach, New York.
24. Gurd, F. R. N. (1960). Association of lipides with proteins. Some naturally occurring lipoprotein systems. *In* "Lipide Chemistry" (D. J. Hanahan, ed.), pp. 208–325. Wiley, New York.
25. Kaplan, D. M., and Criddle, R. S. (1971). Membrane structural proteins. *Physiol. Rev.* **51**, 249–272.
26. Kidwai, A. M., Radcliffe, M. A., and Daniel, E. E. (1971). Studies of smooth muscle plasma membrane. I. Isolation and characterization of plasma membrane from rat myometrium. *Biochim. Biophys. Acta* **233**, 538–549.
27. Kiehn, E. D., and Holland, J. J. (1970). Membrane and nonmembrane proteins of mammalian cells. Synthesis, turnover, and size distribution. *Biochemistry* **9**, 1716–1728.
28. Kiehn, E. D., and Holland, J. J. (1970). Membrane and nonmembrane proteins of mammalian cells. Organ, species and tumor specificities. *Biochemistry* **9**, 1729–1738.
29. Koehler, J. K. (1968). The technique and application of freeze-etching in ultrastructure research. *Advan. Biol. Med. Phys.* **12**, 1–84.
30. Korn, E. D. (1966). Structure of biological membranes. *Science* **153**, 1491–1498.
31. Korn, E. D. (1968). Structure and function of the plasma membrane. A biochemical perspective. "Biological Interfaces: Flows and Exchanges" pp. 257–278. New York Heart Association, Little, Brown, Boston, Massachusetts.
32. Korn, E. D. (1969). Cell membranes: structure and synthesis. *Annu. Rev. Biochem.* **38**, 263–288.
33. Korn, E. D. (1971). Composition of an amoeba plasma membrane. *Biochem. Biophys. Res. Commun.* **45**, 90–97.
34. Lenard, J., and Singer, S. J. (1966). Protein conformation in cell membrane preparations as studied by optical rotatory dispersion and circular dichroism. *Proc. Nat. Acad. Sci. U.S.* **56**, 1828–1835.
35. Lison, L. (1960). "Histochimie et Cytochimie Animales," 3rd ed. Gauthier-Villars, Paris.
36. Lovelock, J. E. (1957). The denaturation of lipid-protein complexes as a cause of damage by freezing. *Proc. Roy. Soc. London* **B147**, 427–433.
37. Lovern, J. A. (1959). "The Chemistry of Lipids of Biochemical Significance," 2nd ed. Wiley, New York.
38. Lucy, J. A. (1968). Ultrastructure of membranes: micellar organization. *Brit. Med. Bull.* **24**, 127–129.
39. Lucy, J. A., and Glauert, A. M. (1964). Structure and assembly of macromolecular lipid complexes composed of globular micelles. *J. Mol. Biol.* **8**, 727–748.
40. Luzzati, V., Reiss-Husson, F., and Saludjian, P. (1966). Phase changes in organized lipid and polypeptide structures. *In* "Principles of Biomolecular Organization," (G. E. W. Wolstenholme, and M. O'Connor, eds.), Ciba Foundation Symposium, pp. 69–81. Little, Brown, Boston, Massachusetts.
41. Luzzati, V., Gulik-Krzywicki, T., Tardieu, A., Rivas, E., and Reiss-Husson, F. (1969). Lipids and membranes. *In* "The Molecular Basis of Membrane Function" (D. C. Tosteson, ed.), pp. 79–93. Prentice-Hall, Englewood Cliffs, New Jersey.
42. Maddy, A. H. (1966). The chemical organization of the plasma membrane of animal cells. *Int. Rev. Cytol.* **20**, 1–65.
43. Malhotra, S. K. (1968). Freeze-substitution and freeze-drying in electron microscopy. *In* "Cell Structure and Its Interpretation" (S. M. McGee-Russell and K. F. A. Ross, eds.), pp. 11–21. St. Martin's Press, New York.

References

44. McGee-Russell, S. M., and De Bruijn, W. C. (1968). Image and artifact—comments and experiments on the meaning of the image in the electron microscope. In "Cell Structure and Its Interpretation" (S. M. McGee-Russell and K. F. A. Ross, eds.), pp. 115–133. St. Martin's Press, New York.
45. Meldolesi, J., Jamieson, J. D., and Palade, G. E. (1971). Composition of cellular membranes in the pancreas of the guinea pig. I. Isolation of membrane fractions. *J. Cell Biol.* **49**, 109–129.
46. Meldolesi, J., Jamieson, J. D., and Palade, G. E. (1971). II. Lipids. *J. Cell Biol.* **49**, 130–149.
47. Meldolesi, J., Jamieson, J. D., and Palade, G. E. (1971). III. Enzymatic activities. *J. Cell Biol.* **49**, 150–158.
48. Mollenhauer, H. H., and Morré, D. J. (1966). Golgi apparatus and plant secretion. *Annu. Rev. Plant Physiol.* **17**, 27–46.
49. Moor, H. (1966). Use of freeze-etching in the study of biological ultrastructure. *Int. Rev. Exp. Pathol.* **5**, 179–216.
50. Moretz, R. C., Akers, C. K., and Parsons, D. F. (1969). Use of small angle X-ray diffraction to investigate disordering of membranes during preparation for electron microscopy. I. Osmium tetroxide and potassium permanganate. *Biochim. Biophys. Acta* **193**, 1–11.
51. Moretz, R. C., Akers, C. K., and Parsons, D. F. (1969). Use of small angle X-ray diffraction to investigate disordering of membranes during preparation for electron microscopy. II. Aldehydes. *Biochim. Biophys. Acta* **193**, 12–21.
52. Neufeld, A. (1888). Ueber die Halogenderivate des Phenylhydrazins. *Ann. Chim. Phys.* **248**, 93–99.
53. Noller, C. R. (1957). "Chemistry of Organic Compounds," 2nd ed. Saunders, Philadelphia, Pennsylvania.
54. Parsons, D. F., Williams, G. R., Thompson, W., Wilson, D., and Chance, B. (1967). Improvements in the procedure for purification of mitochondrial outer and inner membrane. Comparison of the outer membrane with smooth endoplasmic reticulum. In "Mitochondrial Structure and Compartmentation" (E. Quagliariello, S. Papa, E. C. Slater, and J. M. Tager, eds.), pp. 29–73. Adriatica Editrice, Bari.
55. Racker, E. (1970). Function and structure of the inner membrane of mitochondria and chloroplasts. In "Membranes of Mitochondria and Chloroplasts" (E. Racker, ed.), pp. 127–171. Van Nostrand-Reinhold, New York.
56. Racker, E. (1972). Introduction to biochemical society symposium on assembly of intracellular structures *Fed. Proc.* **31**, 10–11.
57. Ralston, A. W. (1948). "Fatty Acids and Their Derivatives." Wiley, New York.
58. Robertson, J. D. (1966). The unit membrane and the Danielli-Davson model. In "Intracellular Transport" (K. B. Warren, ed.), pp. 1–31. Academic Press, New York.
59. Rothfield, L., and Romeo, D. (1971). Enzyme reactions in biological membranes. In "Structure and Function of Biological Membranes" (L. I. Rothfield, ed.), pp. 251–284. Academic Press, New York.
60. Rothfield, L., Romeo, D., and Hinckley, A. (1972). Reassembly of purified membrane components. *Fed. Proc.* **31**, 12–20.
61. Seno, S., and Yoshizawa, K. (1960). Electron microscope observations on frozen-dried cells. *J. Biophys. Biochem. Cytol.* **8**, 617–638.
62. Singer, S. J. (1971). The molecular organization of biological membranes. In "Structure and Function of Biological Membranes" (L. I. Rothfield, ed.), pp. 145–222. Academic Press, New York.
63. Singer, S. J., and Nicolson, G. L. (1972). The fluid mosaic model of the structure of cell membranes. *Science* **175**, 720–731.
64. Sjöstrand, F. S. (1953). The ultrastructure of the outer segments of rods and cones of the eye as revealed by the electron microscope. *J. Cell. Comp. Physiol.* **42**, 15–44.
65. Sjöstrand, F. S. (1953). Electron microscopy of mitochondria and cytoplasmic double membranes. *Nature (London)* **171**, 30–32.
66. Sjöstrand, F. S. (1964). Molecular strtucture of cytoplasmic membranes and of mitochondria. *Proc. 1st Int. Symp. Cell. Chem.*, pp. 103–122. Japan Society for Cell Biology, Okayama, Japan.

67. Sjöstrand, F. S. (1968). Ultrastructure and function of cellular membranes. *In* "The Membranes" (A. J. Dalton and F. Haguenau, eds.), pp. 151–210. Academic Press, New York.
68. Sjöstrand, F. S. (1969). Morphological aspects of lipoprotein structures. *In* "Structural and Functional Aspects of Lipoproteins in Living Systems" (E. Tria and A. M. Scanu, eds.), pp. 73–128. Academic Press, New York.
69. Sjöstrand, F. S., and Baker, R. F. (1958). Fixation by freezing-drying for electron microscopy of tissue cells. *J. Ultrastruct. Res.* **1**, 239–246.
70. Sjöstrand, F. S., and Barajas, L. (1968). Effect of modifications in conformation of protein molecules on structure of mitochondrial membranes. *J. Ultrastruct. Res.* **25**, 121–155.
71. Sjöstrand, F. S., and Elfvin, L.-G. (1964). The granular structure of mitochondrial membranes and of cytomembranes as demonstrated in frozen-dried tissue. *J. Ultrastruct. Res.* **10**, 263–292.
72. Stoeckenius, W., and Engelman, D. M. (1969). Current models for the structure of biological membranes. *J. Cell Biol.* **42**, 613–646.
73. Stossel, T. P., Pollard, T. D., Mason, R. J., and Vaughan, M. (1971). Isolation and properties of phagocytic vesicles from polymorphonuclear leukocytes. *J. Clin. Invest.* **50**, 1745–1757.
74. Tria, E., and Barnabei, O. (1969). Lipoproteins and lipopeptides of cell membranes. *In* "Structural and Functional Aspects of Lipoproteins in Living Systems" (E. Tria and A. M. Scanu, eds.), pp. 143–171. Academic Press, New York.
75. Vanderkooi, G., and Green, D. E. (1971). New insights into biological membrane structure. *Bioscience,* **21**, 409–415.
76. Wallach, D. F. H. (1969). The organization of cellular membranes. *Int. Arch. Allergy Appl. Immunol. Suppl.* **36**, 672–681.
77. Wallach, D. F. H. (1969). Cellular membrane alterations in neoplasia: a review and a unifying hypothesis. *Curr. Top. Microbiol. Immunol.* **47**, 152–176.
78. Wallach, D. F. H., and Gordon, A. S. (1968). Lipid-protein interactions in cellular membranes. *In* "Regulatory Functions of Biological Membranes" (J. Järnefelt, ed.), pp. 87–98. Elsevier, Amsterdam.
79. Weiss, L. (1969). The cell periphery. *Int. Rev. Cytol.* **26**, 63–105.
80. Wittcoff, H. (1951). "The Phosphatides." Reinhold, New York.

2

Densitometric Studies of Lipid Membranes in Pancreatic Acinar Cells of the Mouse

Isidore Gersh

An attempt is made in this chapter to interpret some of the cytochemical studies on lipids in the preceding chapter in quantitative terms based on densitometric data, which are derived from the same electron micrographs. The aim was to construct a three-dimensional model (of submicroscopic but not molecular dimensions) of the lipid distribution in the RER, the nuclear membrane, and the Golgi structures. Although densitometric measurements were made of all clearly defined cell membranes, they were too few in number and too variable to be treated statistically as were the lipid components of other cell structures. Other lipid distributions which were not treated quantitatively were the diffuse or background density of the cytoplasm not directly related to a cytoplasmic structure, the mitochondrial lipid, and the nuclear lipid.

The chief difficulties that had to be circumvented in the quantitative studies were that the exact section thickness was unknown and that the exact plane of section through a structure could not be ascertained by simple inspection of an electron micrographic image. To these should be added certain difficulties of lesser importance. At the submicroscopic, almost molecular, level, the cut surfaces cannot be assumed to be smooth—small pieces may be chipped out, or raised portions may be left standing, only to fall back, on drying, out of its morphological context; or fragments floating on the water surface of the sectioning tray may settle on the section in some unnatural position. In addition, one must assume that the chemical reaction which identifies lipids must be incomplete. Ideally, the aim of the procedure was to achieve maximum staining of lipids consistent with an adequate infiltration of the stained specimen with plastic monomer. Were the infiltration complete, it would be impossible to infiltrate the specimen with plastic monomers, and hence, to make sections suitable for this kind of study. It was assumed that the more reactive sites of lipids were more completely stained than the less reactive ones. The term "more reactive sites" is used in this work to include all sites which reacted with the chemical reagents used as stains either because of their chemical properties or because of steric accessibility. The term, therefore, cannot be used to characterize chemically the nature of the reacted or unreacted lipids. Like all other reactions studied, the ones used for the identification of lipids cause some shrinkage or molecular displacement. Finally, it is certain that the compression and stretching, which every section undergoes during microtomy, and variations in the thickness of a single section must add to all the other uncertainties at the molecular level.

Despite this almost forbidding account of the difficulties inherent in any quantitative analysis of any single electron microscope image, or part thereof, progress

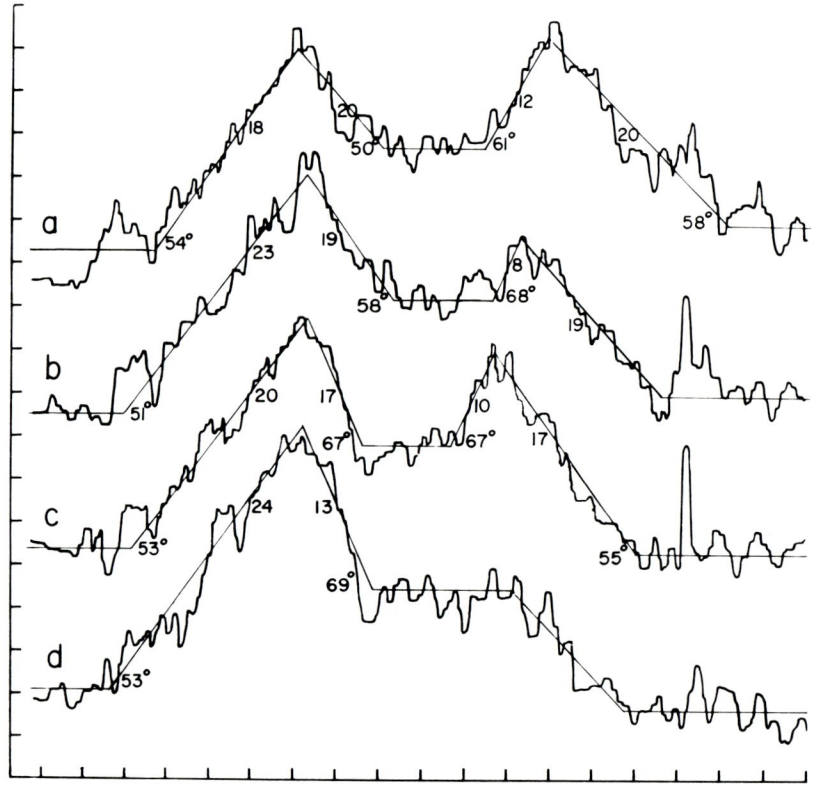

FIG. 2.1. Examples of densitometric curves at high magnification to show the nature of the tracings passing through the RER; linear magnification X 50. The successive curves are parallel to the base line on the electron micrograph and each other and are made along lines 1.5 mm apart. This and other sites analyzed are indicated on the electron micrograph in Chapter 1, Fig. 1.3 (this volume). This series was made at site 4. The right half of the lowest curve was discarded. The numbers along the thin lines represent $\frac{1}{2}w_1$ or $\frac{1}{2}W_r$; angles are given in degrees.

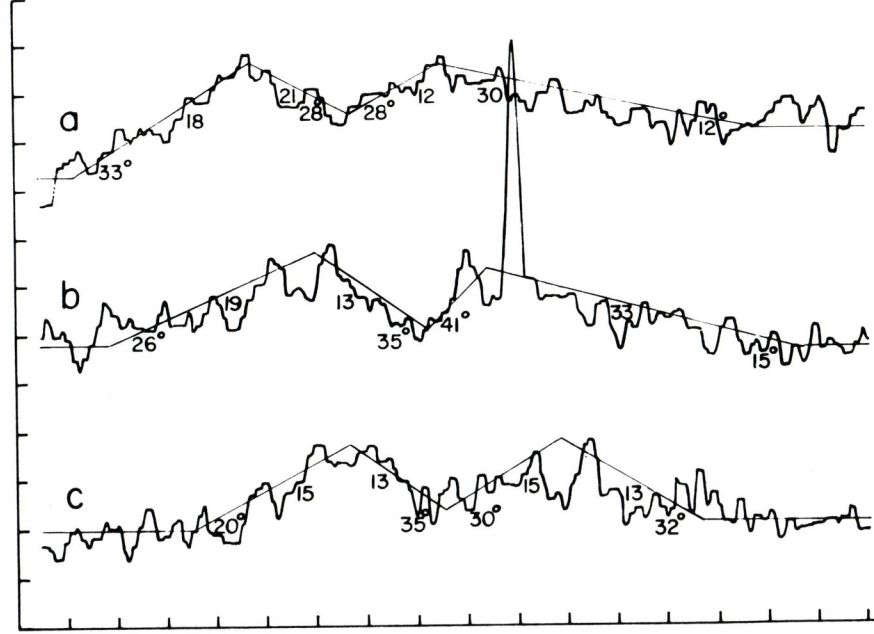

Fig. 2.2. Examples of densitometric curves through nuclear membranes at the same magnification as in Fig. 2.1 for comparison of the height of curves in the two sets of figures. These tracings were made from Chapter 1, Fig. 1.4 (Vol. I) at site 3 and were made along lines 1.5 mm apart. The numbers along the lines represent $\tfrac{1}{2}w_l$ or $\tfrac{1}{2}W_r$; angles are given in degrees.

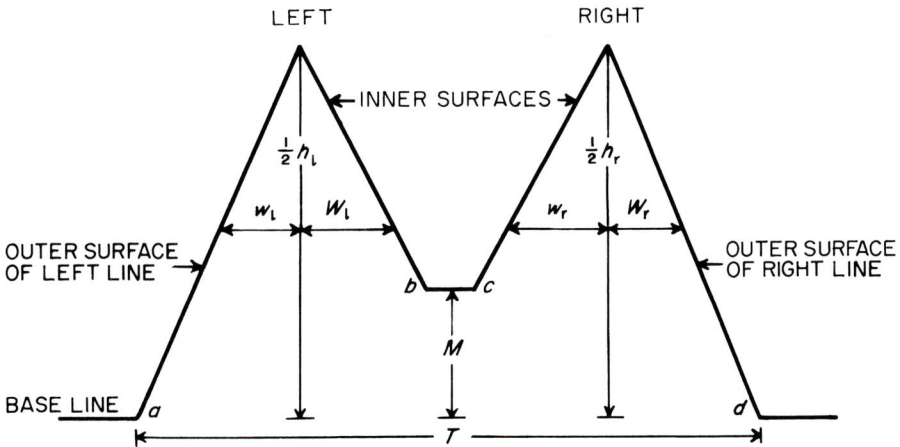

Fig. 2.3. Diagram to illustrate kinds of data obtained from densitometric tracings through lipid components of RER and nuclear membrane. See pp. 34–35 for symbols.

Methods

Selected electron microscope photographic plates were prepared for densitometry by marking certain sites with lines. These were then recorded photographically on the electron micrographs. Typical sites selected for densitometry are shown in Chapter 1, Figs. 1.2–1.4 (this volume). The sites were selected to favor the areas of electron micrographs where the plane of section through a portion of a membrane could be as nearly transverse as possible, and the regions where the walls of the RER were not too closely apposed. A series of densitometric curves was run adjacent to and parallel to each line from a marked starting point and at intervals of 0.75 mm (or a multiple thereof) with a Joyce–Loebl automatic recording microdensitometer, Model MK IIIC. Some typical tracings are shown in Figs. 2.1 and 2.2; see also Fig. 2.7.

Results and Discussion

THE ROUGH ENDOPLASMIC RETICULUM (RER)

In order to make possible statistical treatment of these tracings, they were simplified in the following way:

1. Sites were selected because of their relative freedom from overlay or underlay, and because the structure seemed to be at a favorable angle.
2. Some fortuitous irregularities in the density tracings were ignored as they were judged to be irrelevant. These were replaced, by inspection, by straight lines from which certain data could be read off and tabulated (see Figs. 2.1 and 2.2). This simplification was performed on all densitometric recordings at one sitting, and none was altered or eliminated in subsequent treatment of the data.
3. Despite the discontinuities of the dense lines on the electron microscopic plates and variations in their thickness and density, adjacent lines of a RER cisternal pair were treated as uniform and complete, except when the path of the light beam passed directly through a gap. This part of the record was not used (for example, the right side of curve d in Fig. 2.1).
4. Each adjacent pair of plates (or lines) of RER was treated as a pair whose surfaces were parallel to each other.

From these simplified curves certain data could be read off and tabulated according to the following scheme (Fig. 2.3), which applied particularly to the RER, the nuclear membrane, and part of the Golgi complex. The notations are defined as follows:

a and d = angles of external faces of RER plates
b and c = angles of internal faces of paired RER plates
l = left plate
r = right plate

Results and Discussion

h_l = peak deflection of left plate
h_r = peak deflection of right plate
w_l = width of left half of left plate at $\tfrac{1}{2} h_l$
W_l = width of right half of left plate at $\tfrac{1}{2} h_l$
w_r = width of left half of right plate at $\tfrac{1}{2} h_r$
W_r = width of right half of right plate at $\tfrac{1}{2} h_r$
T = distance between outer faces of a pair of RER plates
M = height of deflection between the plates

The angles a, b, c, and d were recorded in degrees, all other measurements were made in millimeters. The optical magnification of the image was × 22.5, and the mechanical enlargement along the line (left–right) was × 50 or × 10.

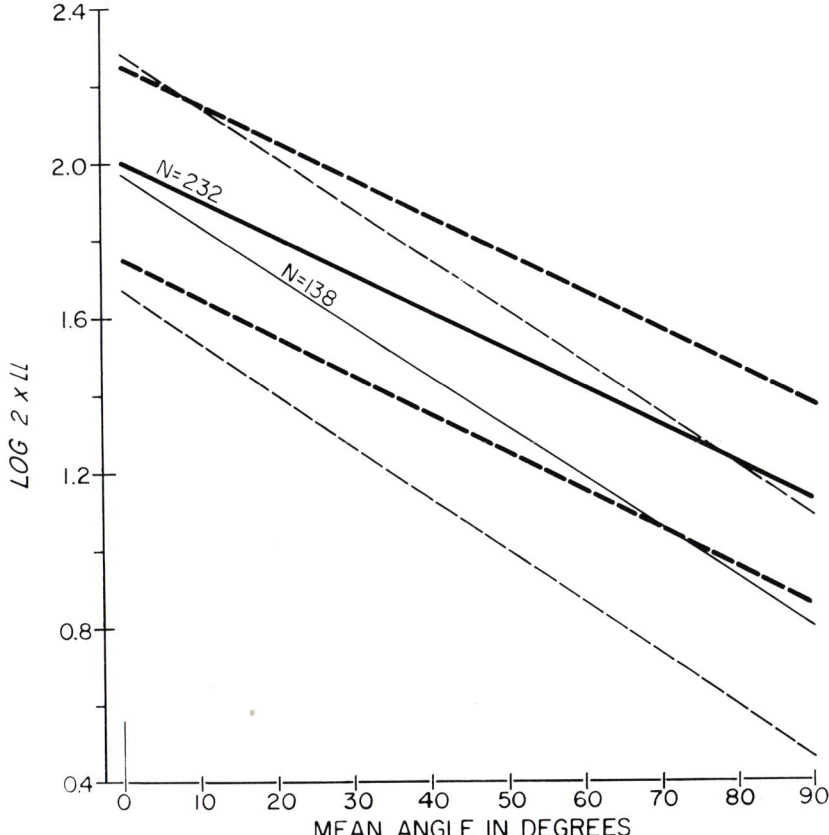

FIG. 2.4. Regression line (solid thick line) with 95% confidence limits (thick dash lines) of log of thickness of lipid component of one membrane of a pair which together enclose a cisternal space. The X axis is log $2w_l$; the Y axis is the outer angles $(a + d)/2$. $Y = 1.1296$ at an angle of 90°. Regression line (solid thin line) with 95% confidence limits (thin dash lines) of log of thickness of a lipid component of nuclear membranes. The X axis is log $2w_l$; the Y axis is the outer angles $(a + d)/2$. $Y = 0.7961$ at an angle of 90°. The values at 90° are derived from the regression lines and are outside the 95% confidence limits of each other.

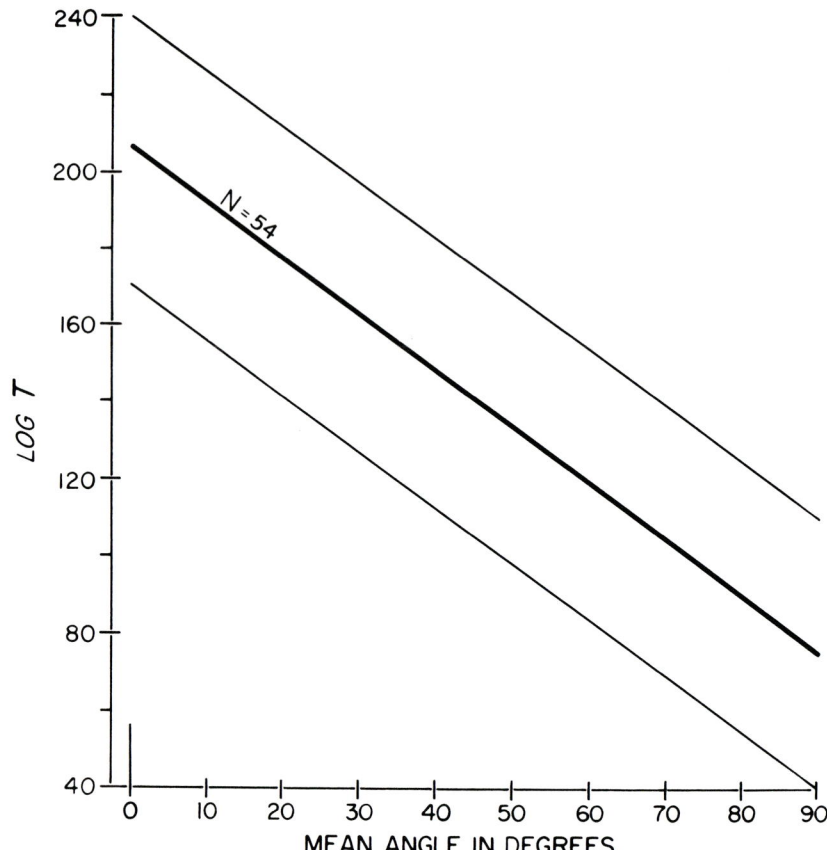

Fig. 2.5. Regression line with 95% confidence limits of T versus mean angle $(a + d)/2$ of the densitometric tracings.

It was found that when the angle a or d (in degrees) was plotted against log $2w_1$ (or log $2W_r$), a straight line relationship was obtained. The regression line derived from these data when extended to 90° gives the statistical thickness of each plate, which comes out at 77 Å (Fig. 2.4, solid thick line). A similar plot of the inner surfaces of both lines resulted in a statistically identical value. A similar plot of the two outer angles a and d against T in millimeters (Fig. 2.5), when plotted and extended to 90° as a regression line, gave the outside statistical distance between the two lines when these were supposedly parallel to each other and cut strictly transversely. This amounted to 431 Å.

The values given above may be used qualitatively to show that some lipid material is present between the lipid plates. These values are graphically represented in Fig. 2.6 as vertical plates cut transversely, with the plates at an angle of 90°. If both are tilted, with the outside distance the same and the plates parallel, three kinds of arrangements may be considered. In (A), the tilt is such that the plates

Results and Discussion

do not overlap in a section 600 Å thick, when viewed from above. In (B), they overlap somewhat, and in (C) they overlap considerably. Densitometric curves (Aa, Bb, and Cc) from such specimens, with the space between them (IS) and outside of them (OS) free of absorbing material, would give the curves shown in solid lines below (Aa, Bb, and Cc). It is clear that none of these curves resembles those from actual specimens (see Fig. 2.1). The addition of density (to an amount equal to 20% of the density of the lipid of the plates) to the outside of the plates abolishes the plateaus (dotted lines in Aa, Bb, and Cc). The addition of density (to an amount equal to 20% of the density of the lipid of the plates) between the plates changes the shape of the curve so that the plateaus are replaced by peaks, and the level between the peaks (M) is raised (dashed lines in Aa, Bb, and Cc). The shape of the curves with dashed lines in Aa and Bb approximates that actually obtained. Thus, it is clear that some lipid-stained dense

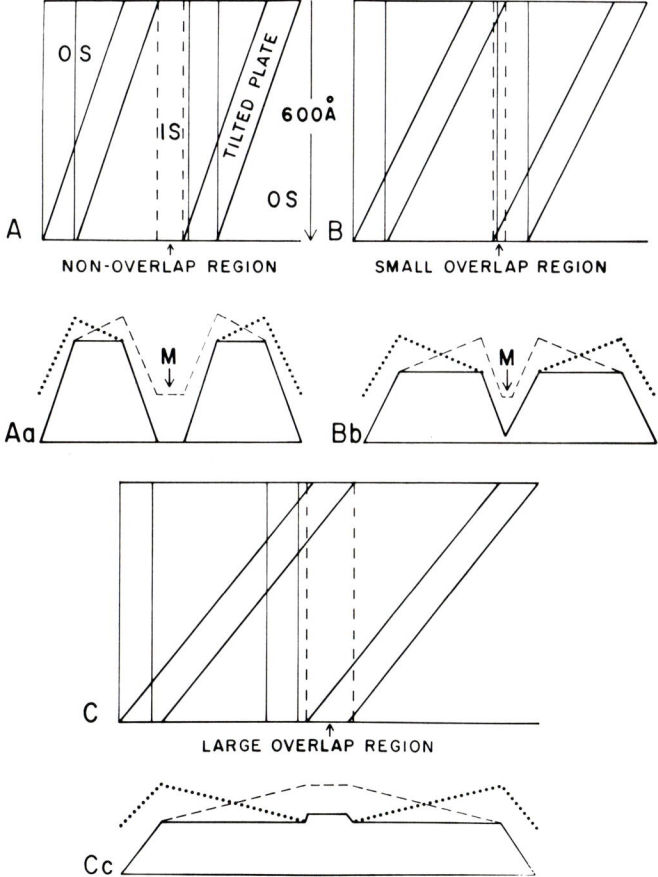

FIG. 2.6. Models of lipid component of RER based on statistical analysis of densitometric curves. See text for discussion.

material must be present between the plates. Whether the relative amount is equal to 20% or some other value of the lipid density of the plates is not known.

It should be stated that control densitometric recordings were made of RER in sections of unstained specimens. These showed no significant deviations from a flat record (i.e., there were no significant peaks).

In summary, a model of the lipid component of the simplified plates of the RER of pancreatic acinar cells, based on statistical analyses of densitometric data, shows that each lipid component (line) is approximately 77 Å thick, and the pairs are separated by a gap of about 277 Å. The gap (or intracisternal space) is much less dense than that of the lipid plates, indicating differences in the amount of reactive lipid in the two sites. It should be recalled that, in fact, the lipid component is not uniformly dense or thick, and that there are discontinuities in each plate. It should be stated also that the meaning of an exact value for the intracisternal space after freeze-drying and postfixation in IIPH is not clear, because the value may vary with the nature of the postfixation. Presumably, the value would be influenced by the nature of the polypeptide distortion caused by the denaturing agent used for postfixation and staining.

Nuclear Membranes

The lipid component of the nuclear membranes was studied in the same way, and a regression line to 90° with 95% confidence limits could be drawn in order to obtain a comparable value for the thickness of the lipid component (Fig. 2.4, solid thin line). Statistically, the regression value for both parts of the lipid portion of the nuclear membrane are indistinguishable, and all the values were pooled to yield an estimated thickness of 36 Å. The regression line of the pooled values is very probably significantly different from the regression line of all RER values.

A comparison was made between the mean heights of densitometric curves of RER and nuclear membranes. The results are presented in arbitrary, but comparable, optical density units. The mean height of RER curves was 32 compared to 23 for nuclear membrane.

In summary, the statistical analysis of densitometric curves of nuclear membrane show that the lipid component of the nuclear membrane is 36 Å thick, and relatively uniform, compared to the lipid component of the RER. The reactive components of the lipid are approximately two-thirds as dense as those in the RER. The same qualifications on the meaning of the exact value of the thickness of the lipid portion of the nuclear membrane apply here as they do to the exact values of the membranes of the RER.

Golgi Apparatus

The lipid component of the Golgi complex is pleomorphic. When densitometric curves are made across RER and the Golgi complex on the same plate, it is clear that the shapes of the curves differ in most instances (Fig. 2.7).

Model density curves were made of solid spheres (C, D), of thin- or thick-

Results and Discussion

walled hollow balls (A, B), and of sections through them at various levels (Fig. 2.8). As these did not resemble most densitometric curves of parts of the Golgi complex, by trial and error, a model was constructed whose density curves could be adjusted to match the densitometric tracings. It consists of a section (with plane parallel surfaces) through an irregularly branched, three-dimensional, chiefly solid net or reticulum, whose walls may appear thicker or thinner in spots (E). The height of the density curve would depend on the density of the material, while the width of deflection would depend on the thickness of the solid at that point and also on its geometry. This pleomorphic model is more satisfactory than a model (F) based on a simple trapezoidal structure in section. The pattern of densitometric curves through most parts of the Golgi complex resembles that of the variable model, but some parts appear to resemble sections of solid cords or sheets with slits or relatively clear lumina running through them. A minority resemble the thin plates of the RER or nuclear membrane, especially where they join and become continuous with the former.

Sometimes, the Golgi structures contain droplets somewhat deformed through sectioning, and frequently these are enclosed by a separated layer of the same density. They may correspond with a protein surface on secretory granules which is continuous with a branch of the Golgi mini-reticulum as illustrated in Chapter 2, Figs. 2.11, 2.12, 2.18, and 2.19 (Vol. I).

The optical density of the lipid component of the Golgi complex is as variable as its shape in sections. The height of the densitometry curve through parts of the

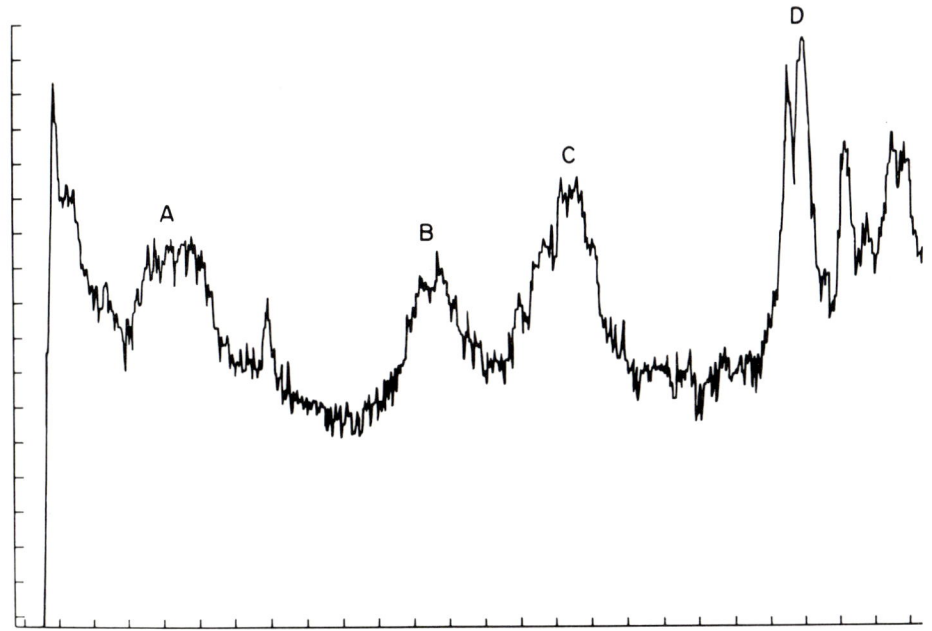

Fig. 2.7. Examples of densitometric curve through RER and Golgi complex; linear magnification × 10. (A), (B), and (C) are sites of Golgi structures, while (D) passes through RER.

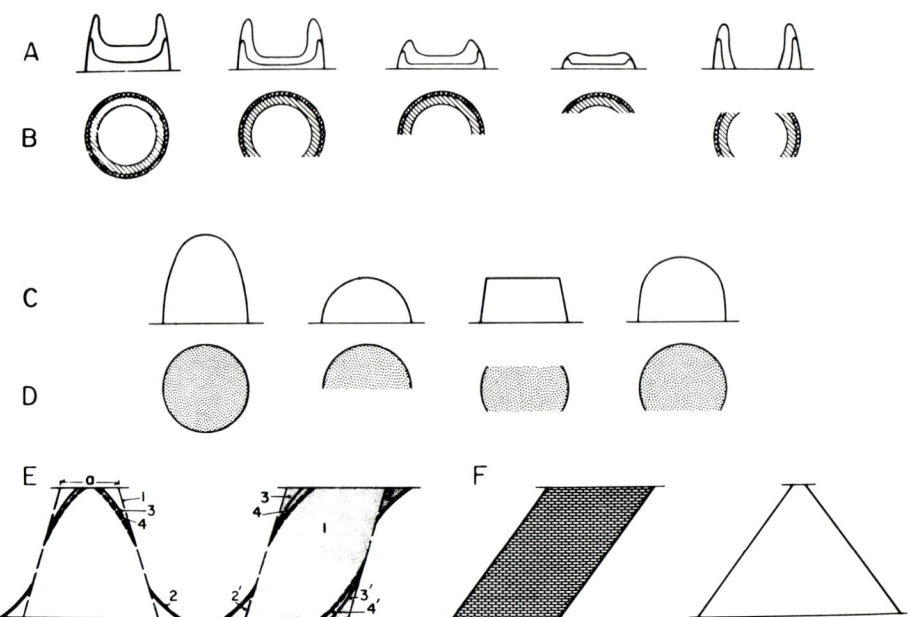

FIG. 2.8. Densitometric representations of various models. Solid sphere (D) and various kinds of sections through the solid (coarse stippling); thin-walled (double hatched) and thick-walled (single hatched and double hatched) hollow balls (B); (F) part of plane with parallel surfaces (brick pattern); (E) part of plane with parallel surfaces (1) distorted to varying degrees (2, 3, and 4). The densitometric curves of model (E) resemble most closely the majority of sections through the strands of the Golgi complex.

The densitometric curves for the solid sphere (D) and various sections through it are given in (C). The densitometric curves for the thin- and thick-walled hollow ball (B) and of various sections through them are given in (A), where the upper line is for the thick-walled ball and the lower line for the thin-walled ball. The densitometric curve for the trapezoid (F) is given in the figure to its right. This solid figure (E) (fine stippling) represents three cases: (1) a trapezoid, (2) the trapezoid deformed by displacing 3 and 4 to 2 and 3′ and 4′ to 2′, and (3) the trapezoid deformed by displacing 3 to 2 and 3′ to 2′. The corresponding densitometric curves are to the left: the trapezoid is outlined by the thin dashed line; the heavily deformed body by the heavy solid (2) and the thin solid (3) line. The width of the plateau (a) is a function of the degree of deformation if the thickness between the plane parallel surfaces and the width of the body are kept constant.

Golgi complex may vary from something just above the base line to a level as high as that of the lipid component of the RER. The mean height of the densitometric curves in arbitrary but comparable optical density units was 43 ($N = 36$), as compared with that of RER on the same plates which was 48 ($N = 23$). Although the mean height of the densitometry tracings of the two structures is nearly the same, their fundamental difference is clear from the nonidentity of the class frequency distributions. These are given in the histogram (Fig. 2.9).

In summary, the lipid component of the Golgi complex is a highly pleomorphic mini-reticulum whose lipid content is very variable. While most of the net appears

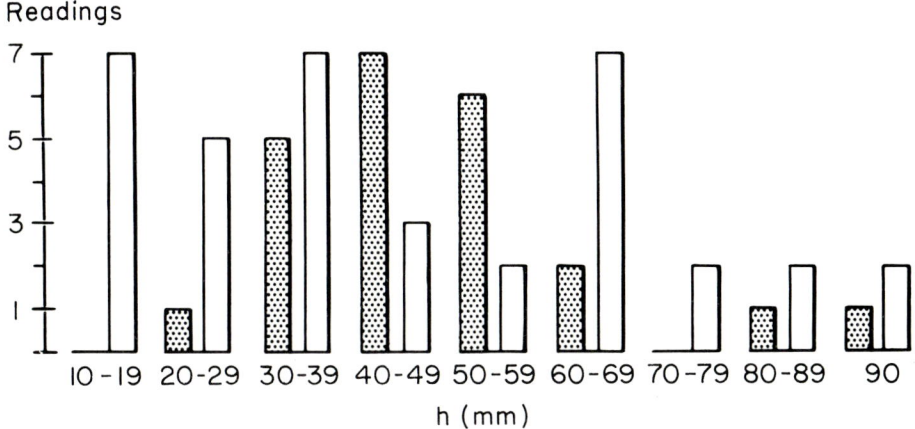

Fig. 2.9. Histogram of height of densitometry tracings (h) through lipid components of RER and Golgi complex in relative density units. RER, stippled column; Golgi complex, open column.

solid, some of the cords appear to have a lumen, and some minor tubelike connections are made with the RER. The general picture corresponds closely with that of the Golgi apparatus as described by others in certain plant cells. By far the greater part of the surface of the net is in direct contact with the intercisternal spaces between the adjacent pairs of RER walls, where ribosomes and soluble RNA are concentrated.

Intranuclear lipid is visible in sites rich in deoxyribonucleoprotein (chromatin) and ribonucleoprotein (nucleoli). In addition, there is an appreciable diffuse (unresolvable) background density.

In hepatic cells, the pattern of lipid distribution in the RER, nuclear membrane, and nucleus is similar to that of the pancreatic acinar cells. By contrast, the lipid pattern did not make possible a positive identification of the cell membrane or the Golgi apparatus, possibly because of the more prominent background density in hepatic cells.

Summary

Statistically considered, the lipid component of each membrane of the RER of the pancreatic acinar cell is approximately 77 Å. They are separated by a gap, the intracisternal space, which in the regions studied measured about 277 Å. There is much less reactive lipid in the intracisternal space than in the walls enclosing it. The lipid component of the nuclear membranes is more uniform and about 36 Å thick. The amount of reactive lipid of Golgi structures varies enormously, and matches their pleomorphism. Models were devised for the RER and the Golgi apparatus based on the qualitative treatment of the densitometric data.

3

Cytochemical Study of Lipid, Protein, and Nucleic Acids of Mitochondria in Exocrine Pancreatic and Hepatic Cells of the Young Adult Mouse

Isidore Gersh

Perhaps nowhere else in the animal cell is compartmentation as clearly manifest as in mitochondria. The almost universal occurrence in them of outer and inner "membranes" and of matrix, as well as the specific activities of each part, are visible and functional demonstrations of this. Yet there is this fundamental paradox: the morphological aspects of this aggregation are apparently related chiefly to lipid components of the membranes, while functional aspects are apparently related to proteins. The possible relations of these major components to each other and to structural proteins have been important topics considered by numerous researchers and are being studied in many laboratories. Some of the pertinent articles are referred to in Chapter 2 (Vol. I) and Chapter 1 (this volume), and additional general references limited to mitochondria are cited here *(5–7, 11, 12, 21–23)*. Of consuming interest also, is the occurrence and possible functions of RNA and DNA in mitochondria. The brief history of knowledge of mitochondrial nucleic acids has been reviewed so well that it is unnecessary to document specifically the references to original studies *(1, 2, 14–16)*.

This chapter is based on observations on the distribution of lipid and protein in mitochondria of hepatic and exocrine pancreatic cells of adult mice, as well as of nucleic acids in mitochondria of embryonic cells. The two major components are integrated to form membranes which separate protein-rich compartments (which also contain lipids). The pattern of membranes and compartments is remarkable in its uniformity; so much so that the current concept of the structure of mitochondria (cristae, matrix, inner membrane, outer envelope) requires some modification. Nucleic acids are not prominent in mitochondria of most adult cells and were studied mostly in the cells of chick embryonic somites and liver, where they are more abundant.

Methods

Most specimens were frozen ultrarapidly and dried *in vacuo* at $-40°C$ or lower (as described in Chapter 2, Vol. I). Some were dried *in vacuo* by the flowing-gas freeze-drying method described by Jensen *(10)*. Details of subsequent treatments are given below or in appropriate references.

Controls

Some freeze-dried specimens were infiltrated *in vacuo* with alcohol as vapor (which was condensed on the specimens) or as a fluid. Some of these were then extracted continuously for 1 day with a mixture of chloroform–methanol in a Soxhlet apparatus *(8, 9)*. All were embedded in Durcupan. Electron micrographs of ultrathin sections served as unstained zero controls for the primarily protein basis of cell structure. Other freeze-dried specimens were heated *in vacuo* for about 18 hours at 90°C and were then infiltrated with alcohol as above, to serve as another kind of unstained zero control.

Metallic Salts in Alcoholic Solution as Stains for Proteins

Freeze-dried specimens postfixed in alcohol as above were stained with alcoholic platinum tetrabromide *(8, 9)*. Excess stain was washed out in alcohol and specimens were embedded in Durcupan. Electron micrographs of ultrathin sections showed the sites predominantly of protein in cells. Some ultrathin sections were stained with uranyl acetate to enhance contrast. Alcoholic platinum tetrabromide is one of many cross-linking metal salts which have similar staining properties.

Organometallic Vapors *in Vacuo* as Stains for Proteins

Freeze-dried specimens were stained *in vacuo* with a variety of organometallic vapors and embedded in Durcupan directly or after postfixation in alcohol. Mercury(II) hexafluoroacetylacetonate (Hg hfac) is typical of this series of cross-linking reagents (Chapter 2, Vol. I). Some ultrathin sections of Durcupan-embedded specimens were stained with uranyl acetate.

Organic Vapors *in Vacuo* as Stains for Proteins

These include difluorodinitrobenzene (FFDNB) and difluorodinitrodiphenylsulfone (FFSulfone), which react chiefly with free NH_2 and other groups of proteins and tend to cross-link them (see Chapter 2, Vol. I). Some ultrathin sections of Durcupan-embedded blocks were stained with uranyl acetate.

ORGANIC VAPORS *in Vacuo* AT LOW TEMPERATURE AS STAINS FOR LIPID

This includes a series of primary amines of which the most favorable seemed to be diiodophenylhydrazine (IIPH) (this volume, Chapter 1). Some ultrathin sections of Durcupan-embedded blocks were stained with uranyl acetate to see the relationships between lipids and protein.

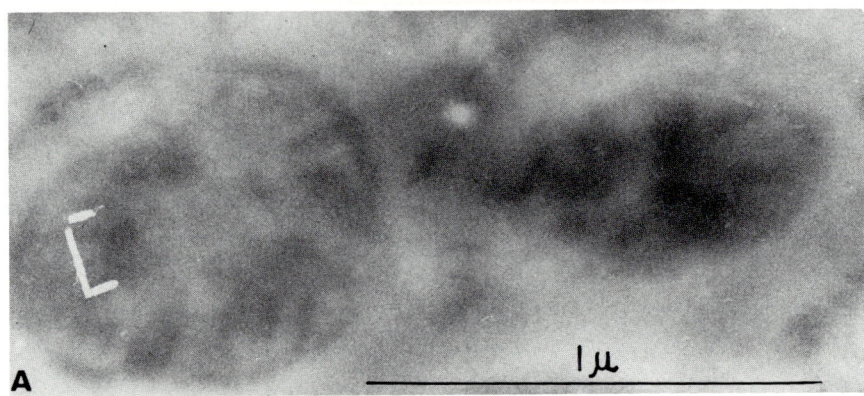

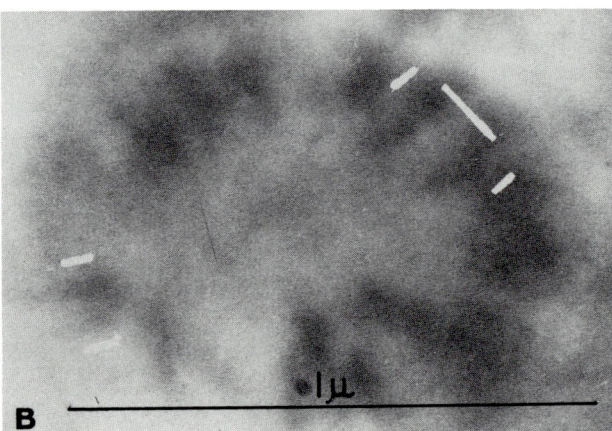

FIG. 3.1(A–C). Lipid test (IIPH) on mitochondria of three different specimens of hepatic cells of young adult mouse. The lipid is distributed in lumps which, from their measurements, must extend over several compartments. Lipid is also present between lumps as unresolvable background density. In many mitochondria the lipid is uniformly distributed. The white marks are sites where densitometric measurements were made. A, × 58,000; B and C, × 70,000.

Methods

ORGANIC VAPOR *in Vacuo* FOR BI- AND POLYVALENT
CATIONS AS WELL AS FOR PROTEINS

The use of the reagent heptafluorodimethyloctanedione (HEFDOD) is described in Chapter 2 (Vol. I) and Chapter 9 (this volume). Some ultrathin sections were stained with uranyl acetate.

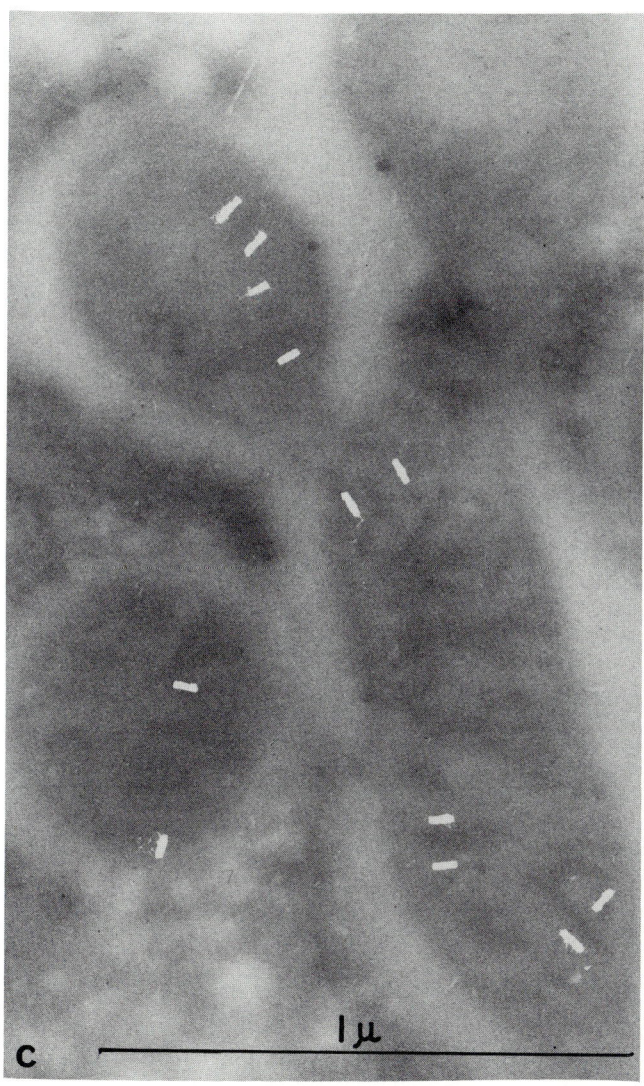

FIG. 3.1(C).

POSTFIXATION BY Hg hfac VAPORS AND ALCOHOL, FOLLOWED BY AQUEOUS GALLOCYANIN-CHROMALUM FOR NUCLEIC ACIDS

See Vol. I, Chapter 3 for greater details.

NOTE ON MITOCHONDRIAL SHRINKAGE

In many electron micrographs of freeze-dried preparations, mitochondria appear shrunken. There seemed to be three variables which could be manipulated to control this artifact: (1) temperature of drying, (2) the degree of vacuum, and (3) the humidity. Drying at temperatures down to $-78°C$, and in an all glass system (without joints) with a pressure of 10^{-5} mm (as judged with a McLeod gauge) did not reduce mitochondrial shrinkage. In reexamining previously reported instances of the successful use of freeze-drying for electron microscopy, the most common factor remaining seemed to be that after drying, specimens were exposed to vapors of osmium tetroxide in room air at a pressure of ½ to 1 atmosphere *(3, 13, 20)*. This suggested that there were enough water molecules in the air to be absorbed by the dry specimens. One consequence of this was the

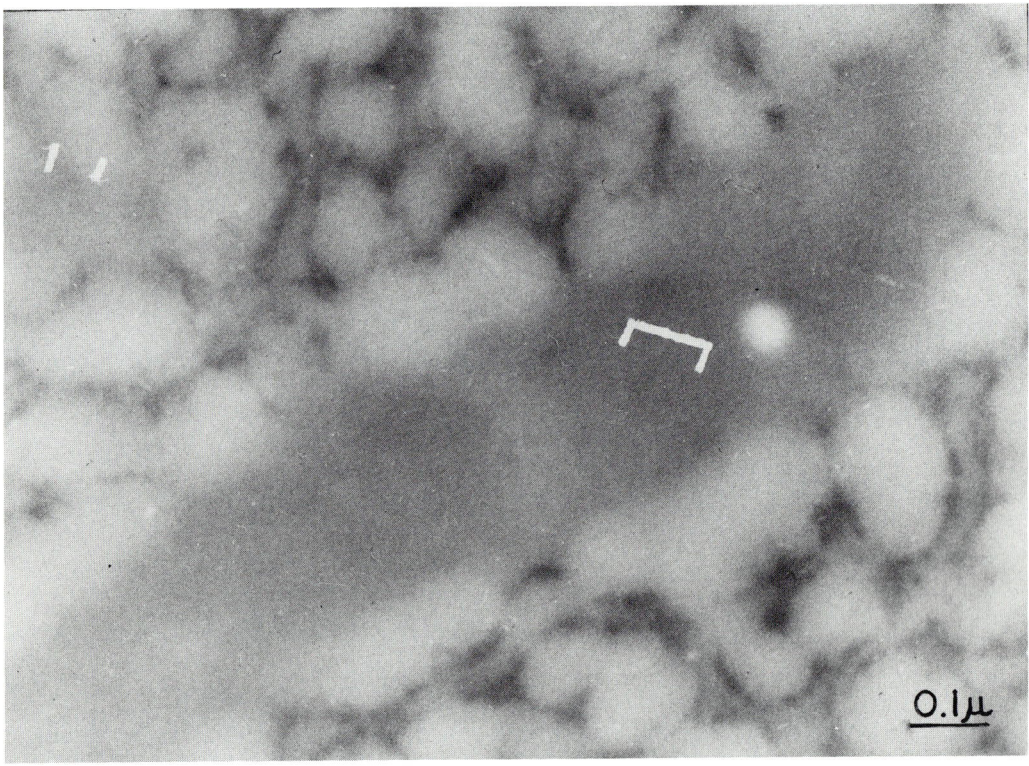

FIG. 3.2. Lipid test (IIPH) on mitochondrion of exocrine pancreatic cell. Lipid is distributed uniformly in the mitochondrion. Lipid component of the RER has more contrast. The white marks are sites where densitometric measurements were made. × 100,000.

swelling to its original size of mitochondria, which had shrunk during the antecedent drying. It seemed plausible to reconstruct the process in this way: when mitochondria are dried beyond a certain point, sufficient bound water is removed so that intermolecular forces become effective and are sufficiently strong to cause a closer aggregation of protein molecules and a consequent shrinkage of the whole mitochondrion away from the surrounding cytoplasm. Partly to compensate for this process, water vapor was introduced during the exposure of freeze-dried specimens to vapor reagents, with inconsistent results. More constant were the results achieved with the aid of Dr. G. L. Rossi, when constant leaks of ambient air were introduced into the vacuum line which reduced the vacuum to a range of 0.1–0.85 mm Hg. This was not altogether successful in reducing mitochondrial shrinkage. The final solution to this technical problem was to adapt the moving gas–freeze-drying method developed by Jensen *(10)*. The apparatus used for this purpose and detailed procedures will be published later in a cytochemical paper concerned with DNA in meiosis.

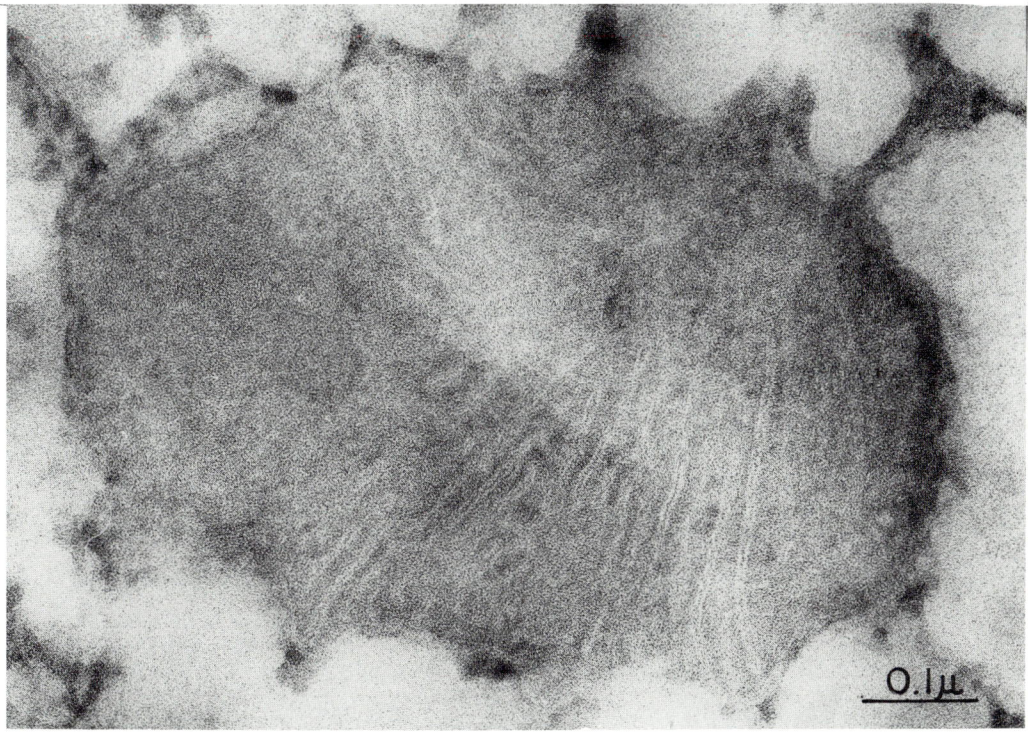

Fig. 3.3. Mitochondrion of exocrine pancreatic cell to show orderly arrangement of compartments. The walls of the compartments are pale, the contents are dark. The pale band is split by a thin, dense stripe. Stained for lipid (IIPH), and section stained for protein (uranyl acetate). × 157,000.

48   3. Lipid, Protein, and Nucleic Acids of Mitochondria

Observations

PROTEIN

Mitochondria appear rather homogeneous (with granular but unresolvable structure) in sections of unstained control specimens (Chapter 3, Fig. 3.28, Vol. I), whether extracted or not, as well as in sections of specimens stained for protein by metallic salts in solution (see Vol. I, Chapter 2, Figs. 2.15 and 2.17), or by organometallic (Vol. I, Chapter 3, Fig. 3.36) or organic vapors (Vol. I, Chapter 3, Fig. 3.32). No membranes are visible in the mitochondria of hepatic or exocrine pancreatic cells.

LIPID

Hepatic mitochondria are characterized by irregular, broad densities which do not correspond with cristae or membranes (Figs. 3.1A–C). The denser regions, as measured with a densitometer, range from 190 to 615 Å in sections through their thinnest regions, with a mean value of 337 Å. This is nearly three times the thickness of the pale stripe described in the next paragraph. Pancreatic cell mitochondria appear rather uniform (Fig. 3.2).

PROTEIN AND LIPID

The appearance changes dramatically after lipid-stained sections are stained on the grid with uranyl acetate. In such preparations, mitochondria are characterized by alternating pale and dark bands (i.e., plates, when considered in three dimensions), of approximately equal thickness, which are transverse to the organoid. The pale bands are continuous with a similarly pale structure which encloses the mitochondrion. The dark bands or plates are completely separated from each other and from the rest of the cytoplasm by the pale sheets or membranes. The latter may be con-

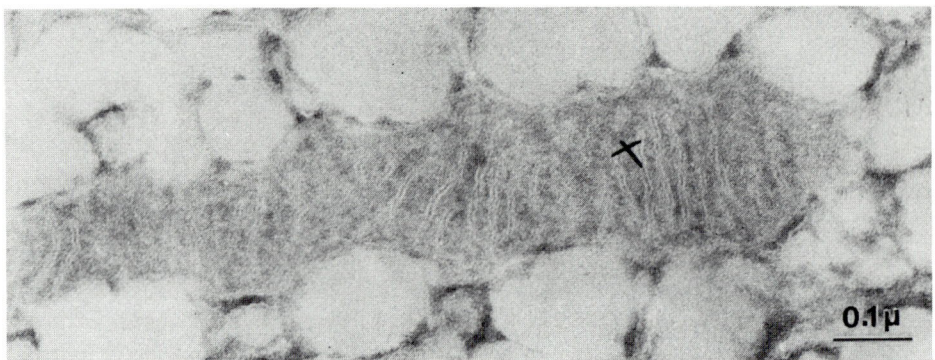

FIG. 3.4. Mitochondrion from a specimen of exocrine pancreatic cell treated as in Fig. 3.3. × 100,000.

sidered as walls of compartments, the contents of which constitute the dark plates. The contents are irregular in density. The pale membranes vary somewhat in thickness (Figs. 3.3 and 3.4).

This concept is represented diagrammatically in Fig. 3.10A.

When sections pass through mitochondria at angles which differ only slightly from the longitudinal, the pale and dense bands appear to have bizarre shapes.

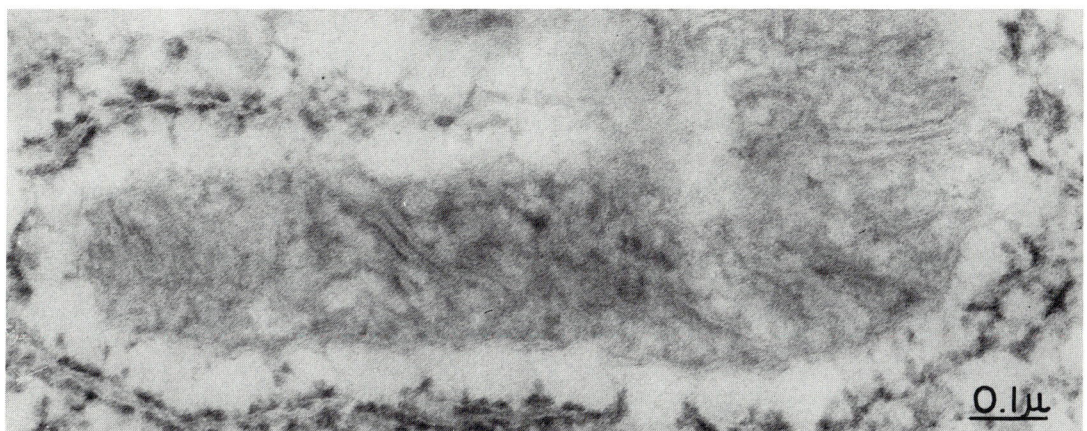

Fig. 3.5 ▲   ▼ Fig. 3.6.

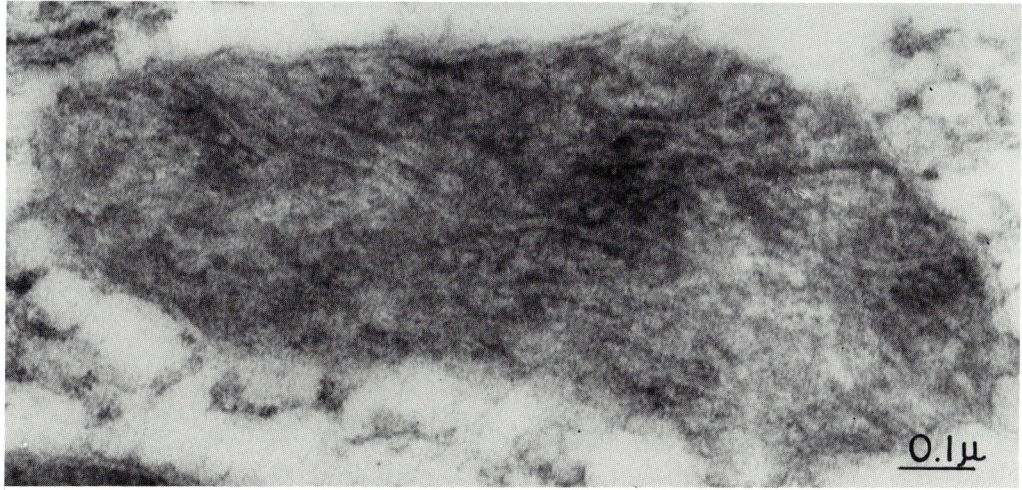

Figs. 3.5 and 3.6. Mitochondrion of a hepatic cell (Fig. 3.5) and of an exocrine pancreatic cell (Fig. 3.6), stained for lipid (IIPH), and section stained for protein (uranyl acetate). Both mitochondria are shrunken. As a consequence of the distortions induced by shrinkage and bending, the pale compartment walls and the darker staining contents of the compartments lose their regular, orderly arrangement, and assume bizarre patterns. × 100,000.

These become fantastically complicated when, in addition, mitochondria are bent, or distorted through shrinkage (Figs. 3.5 and 3.6).

The pale band is split approximately equally by a thin sheet, densely stained with uranium, which is about 15 Å thick (see Fig. 3.7). This thin dense sheet is continuous with another which splits the membrane enclosing the entire mitochondrion. The outer half of this membrane is the only complete and continuous membrane of the organoid. All others form enclosures or compartments of surprisingly uniform dimensions, which are only slightly affected by the method of postfixation.

In the liver, the pale band, after IIPH treatment, has a mean value of 118 Å (ranging from 97 to 152 Å), and, after HEFDOD treatment, has a mean value of 144 Å (ranging from 121 to 178 Å). In the pancreas, the corresponding values are 107 Å (86–126 Å) and 134. The dense stripe (or thin sheet as referred to above) in the middle of the pale band is also rather constant. In the liver, its thickness has a mean value of 13 Å (10–20 Å) after treatment with IIPH, and

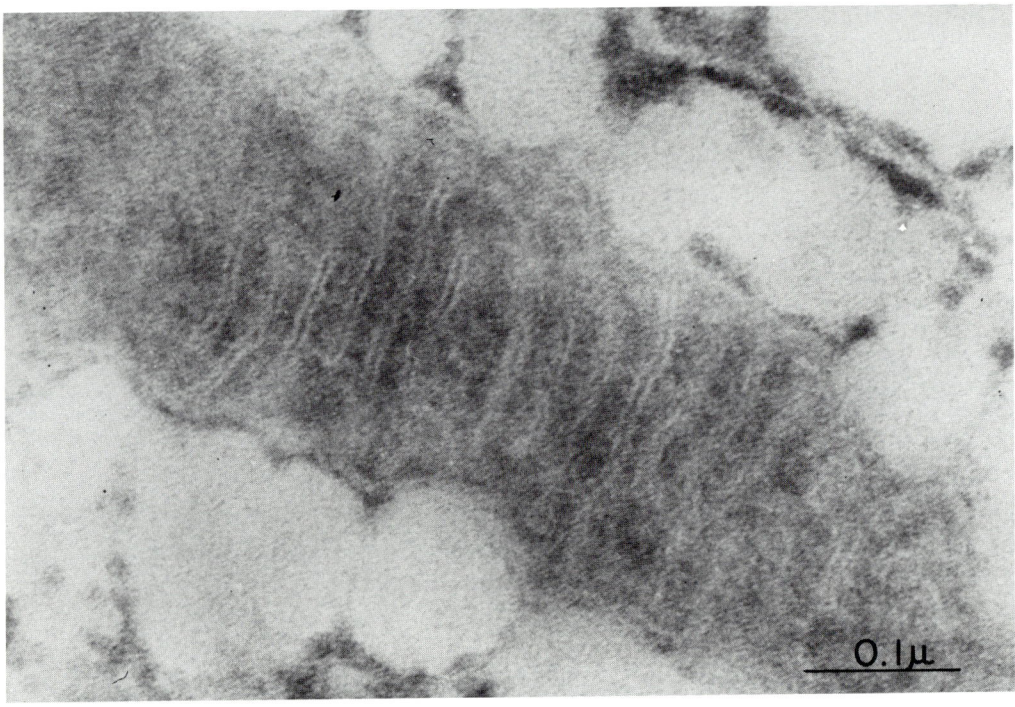

Fig. 3.7. Another mitochondrion, from pancreas, treated as in Figs. 3.3 and 3.4. The thin dense stripe in the middle of the pale band is not of uniform thickness. The irregularities in thickness are associated with the pale "globules," which are tightly packed in layers on either side of the thin, dense stripe. The outside of each pale globule is regarded as the hydrophilic portion of the protein molecule; the inside as the lipophilic portion. The thin, dense stripe is about 15 Å thick, and is regarded as the hydrophilic portions of protein molecules (globules) on opposite sides, where they impinge on each other. × 200,000.

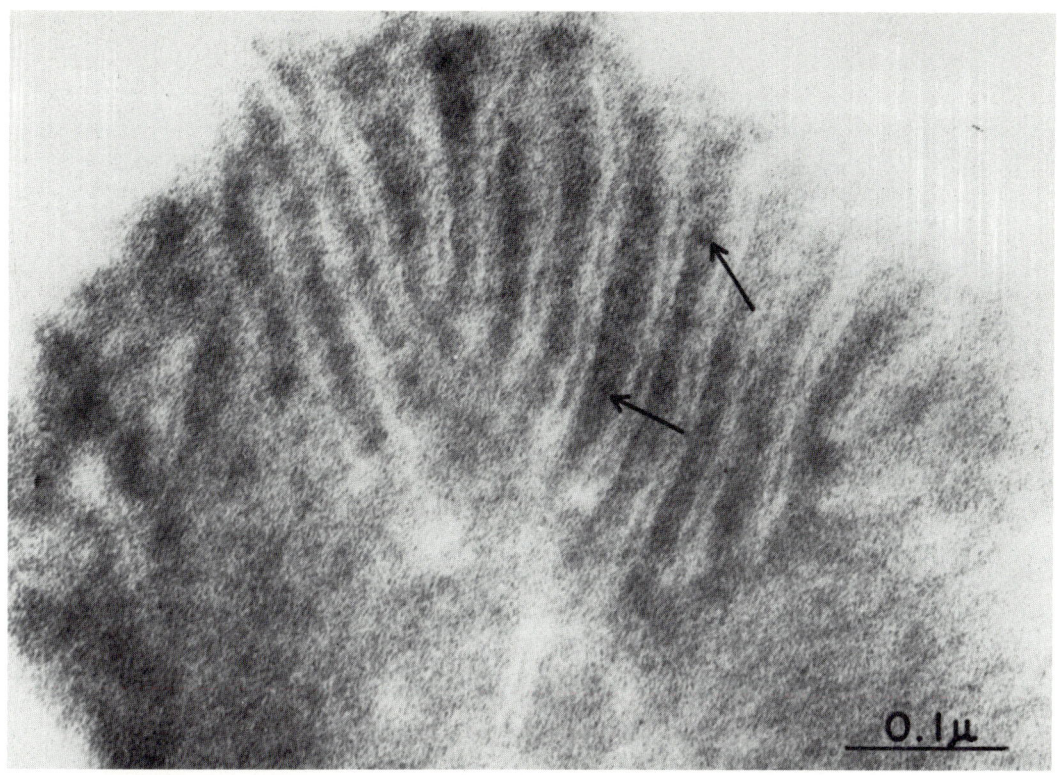

Fig. 3.8. ▲   ▼ Fig. 3.9.

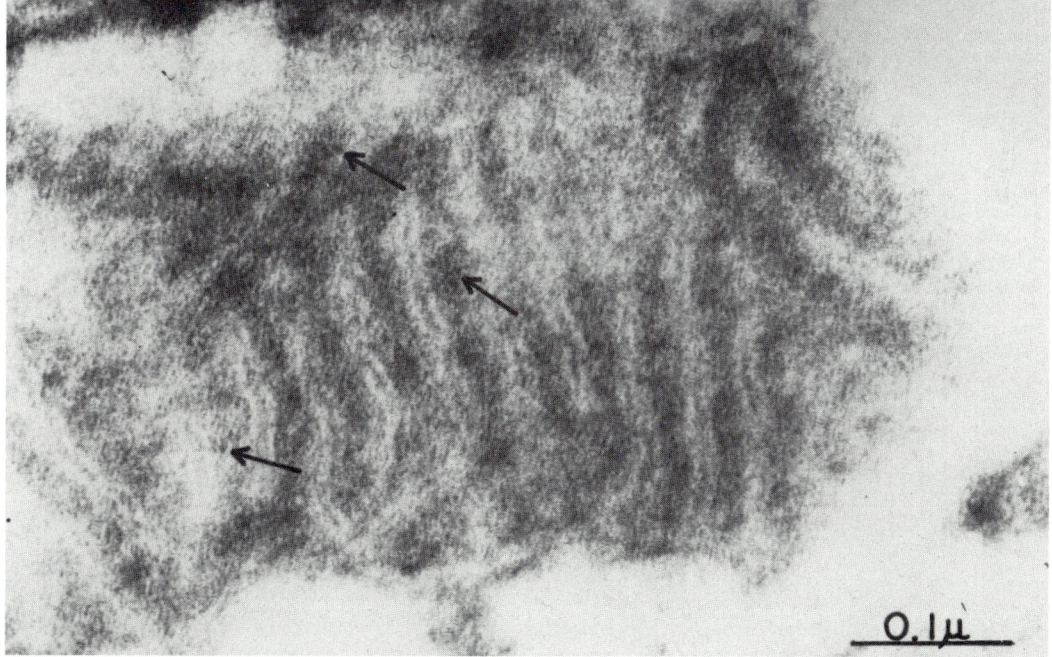

Figs. 3.8 and 3.9. Mitochondrion from pancreas treated as in Figs. 3.3 and 3.4. The contents of the compartments are dense, and contain small and large granules. The latter are marked by arrows. × 215,000. G2 prints, defined in Preface.

17 Å (11–24 Å) after treatment with HEFDOD. In the pancreas, the dense stripe measured 18 Å in thickness (11–28 Å) after treatment with IIPH.

Essentially the same results may be observed after postfixation with certain vapor reagents when the sections are stained with uranyl acetate. In addition to IIPH and HEFDOD mentioned above, other vapors include those of Ca acac, Co acac, Zr hfac, and the successive exposure of the same preparation to vapors of Hg hfac and FFDNB. The exact details of the methods of preparation are given in Chapter 2 (Vol. I).

On closer examination of even thinner sections the thin dense band appears irregular in thickness and the membrane on either side of this band is observed to consist of linear (in sections) or planar pale regions about 50 Å in diameter, separated by thin, faintly stained lines (Figs. 3.8 and 3.9). Where these small globules impinge on each other from opposite sides of the pale band, the area is thicker and forms the single dense thin line visible at lower magnifications. The surface of the thin, dense line is irregular, the somewhat scalloped appearance conforming with the impinging surface of the rounded contours of the globules. The globules have some stainable content, which varies with the kind of vapor postfixation; on the whole, there is more stainable material in these globules than in globules of

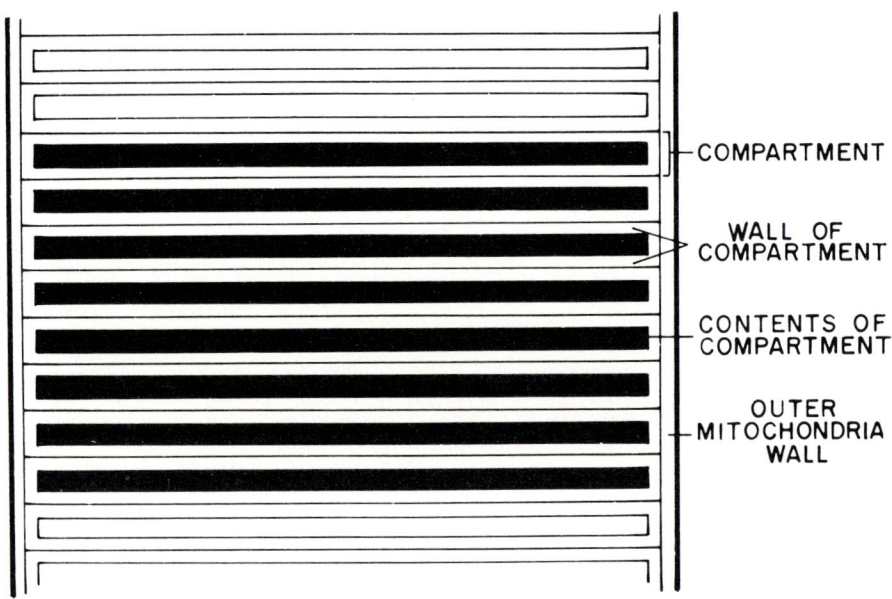

Fig. 3.10. (A) Diagram of a model of a longitudinal section of a mitochondrion. The mitochondrion consists of regularly disposed stacks of compartments, each completely enclosed by membranes (pale). The outer membrane (also pale) is the only continuous membrane. The contents of each compartment are dark.

Observations

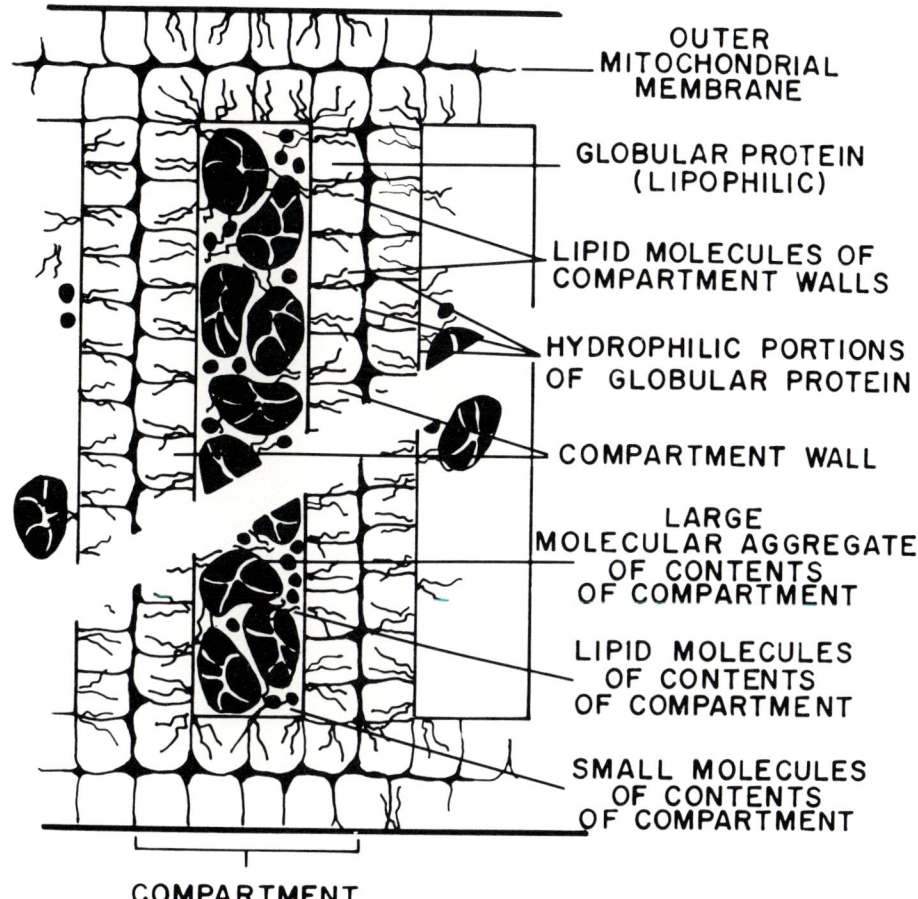

Fig. 3.10.(B) Diagram of a model of one compartment with adjacent structures to show the molecular basis of its organization. The membranes are formed by planar layers of "globular" molecules. Associated with them are phospholipid molecules which are so disposed that their lipophilic portions extend and penetrate within and between the globular molecules, while their hydrophilic ends are associated with the hydrophilic outer portions of the globular molecules. The contents of the compartment consist of (a) hydrophilic proteins which vary in size from 10 and 20 to 50 Å, the latter being aggregates of smaller granules, and (b) phospholipid molecules showing no particular orientation. There may also be some unresolvable background structure.

the RER or nuclear membrane. This description is very nearly like that of Sjöstrand and Barajas (21), and is represented in diagrammatic form in Fig. 3.10B.

The irregularities in density of the material enclosed in each compartment are also resolved in electron micrographs of exceedingly thin sections at higher magnification into small granules which range from 10 and 20 Å to about 50 Å (Figs. 3.8 and 3.9, arrows). The latter appear to be made of still smaller gran-

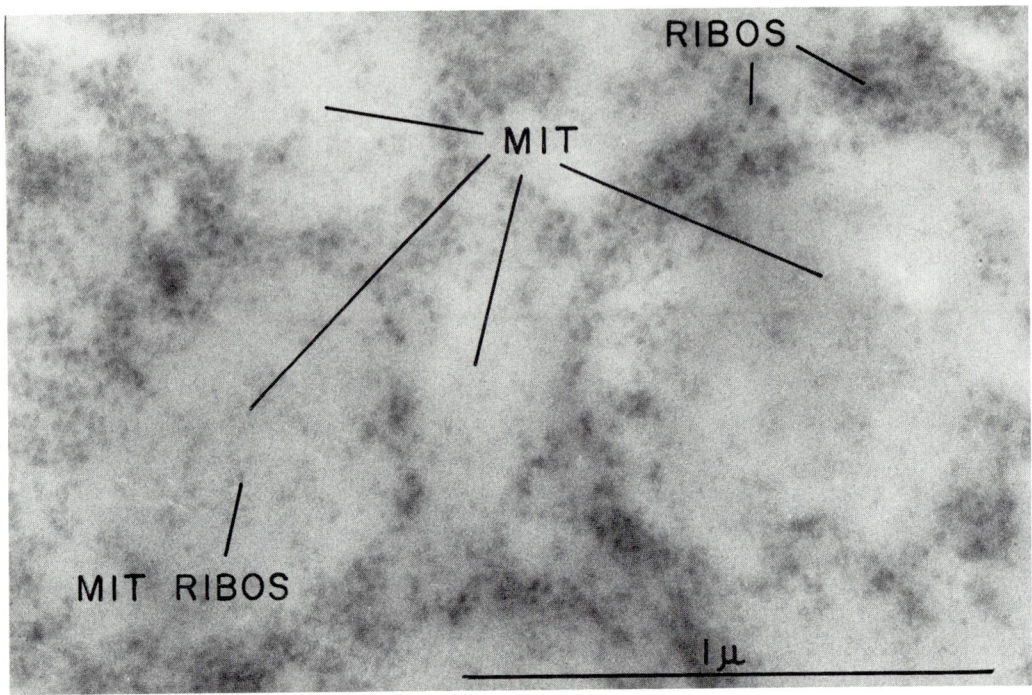

Fig. 3.11. Mitochondria of liver cells of early chick embryo (6-day incubation), postfixed with vapors of Hg hfac, and stained for nucleic acids with gallocyanin–chromalum. Ribosomes (RIBOS) are tightly packed between the mitochondria (MIT). Within the mitochondria are numerous RNA granules (MIT RIBOS). These are smaller than ribosomes, and are sometimes aggregated. × 70,000.

ules. There seems to be a homogeneous (unresolvable) material between the granules.

Nucleic Acids

The material was prepared with Drs. J. W. Lash and R. J. Alperin in connection with other studies, and consisted of cells of somites of 17 and 18 somite stage, and hepatic cells of 6-day-old chick embryos. Mitochondrial ribosomes are very numerous in these rapidly dividing cells (Figs. 3.11 and 3.12). They are smaller than cytoplasmic ribosomes (of the RER), and their density is somewhat reduced by digestion with RNase, but not with DNase, prior to staining with gallocyanin–chromalum. They are less numerous in mitochondria of adult cells, where they are often seen singly and widely separated from the few other mitochondrial ribosomes, which may be present in the same section of the mitochondrion. They appear sometimes to be strung out along segments of an exceedingly fine line, which may represent portions of messenger RNA (Fig. 3.13). It is ap-

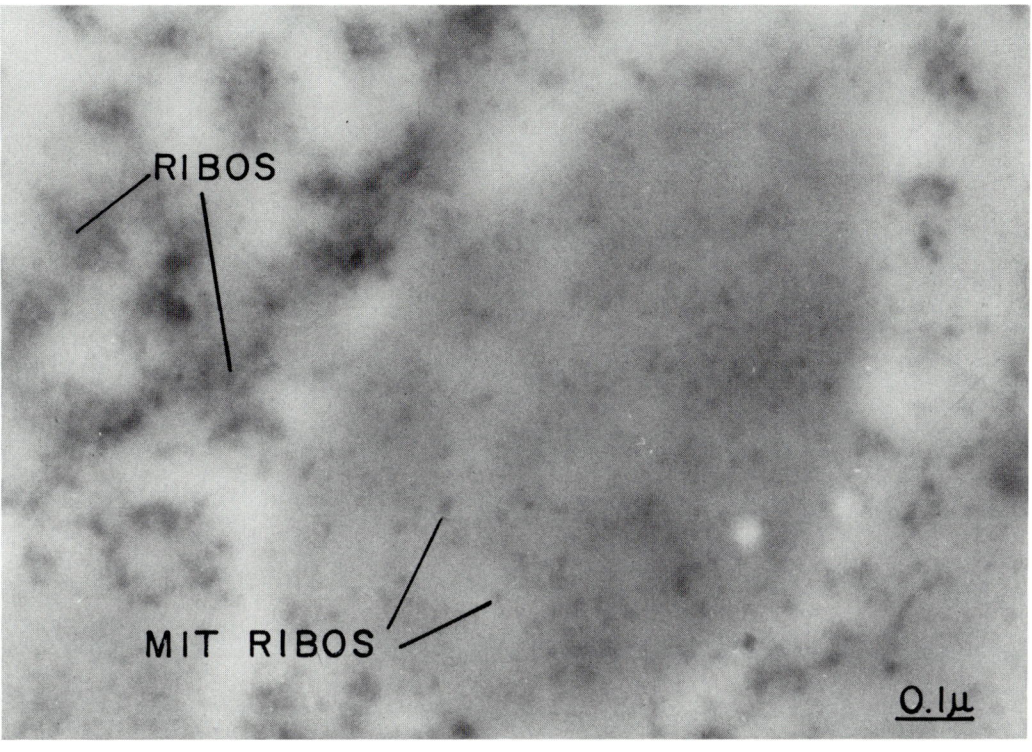

Fig. 3.12. Mitochondria of another chick embryo liver of the same age as in previous figure, prepared in the same way to stain nucleic acids. Extramitochondrial ribosomes (RIBOS) are aggregated, sometimes in relation to RER. The smaller intramitochondrial RNA ribosomes (MIT RIBOS) are scattered, or loosely aggregated. × 100,000.

preciably thinner than the short linear segments of stainable material which are 40–50 Å in diameter (Fig. 3.14). Such thicker strands were not visible after treatment of the specimen with DNase prior to staining, but were present after treatment with RNase.

Discussion

The observations may be summarized briefly: (1) Fixation and staining of mitochondria for proteins and for bi- and polyvalent cations shows that proteins are distributed apparently uniformly. (2) Fixation and staining of mitochondria for lipids reveals that they are broadly distributed in these organelles, sometimes more or less uniformly, sometimes in irregular (rather large) shapes and sizes, which do not correspond with cristae. (3) Fixation of mitochondria for lipids and

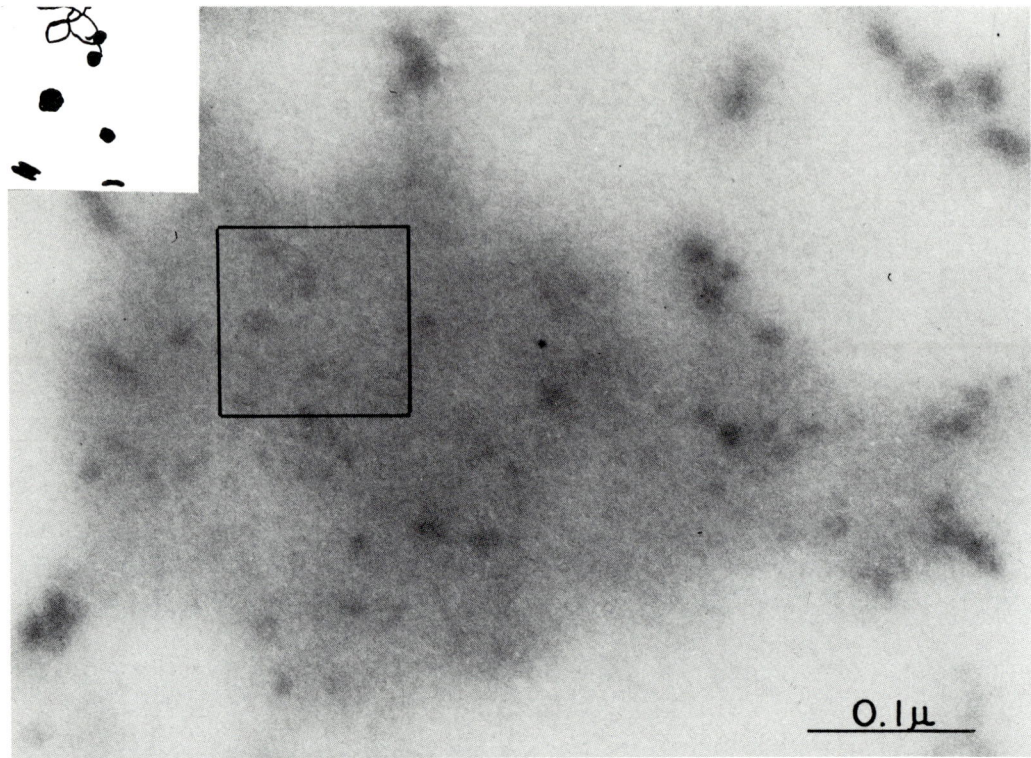

Fig. 3.13. Mitochondrion of the same preparation as in Fig. 3.12 stained for nucleic acids. In these regions, the mitochondrial ribosomes may appear to be strung along a twisted, extremely fine filament. This is indicated in the insert traced from the electron micrograph. × 214,000.

proteins with uranyl acetate used in section staining for proteins, reveals a highly structured organization of membranes enclosing compartments containing deeply staining proteins as well as lipids. (4) At very high resolution, the membrane is resolved into (protein) globules or protein molecules, with which lipids are associated. (5) In the same kinds of preparations, the compartments contain granules of about 10–20 Å to 50 Å in diameter, together with lipid. (6) DNA and RNA are visible in mitochondria, the former as short segments, the latter as mitochondrial ribosomes and perhaps as very fine filaments. Points (3) through (5) are more or less in accord with descriptions in the literature of mitochondrial structure in material fixed by freeze-drying, by freeze-substitution, as well as by immersion in aqueous fixatives *(3, 4, 13, 21)*.

There seems to be a contradiction between the observations summarized in (1) and (2) above, and those in (3), but this contradiction is more apparent than

Discussion

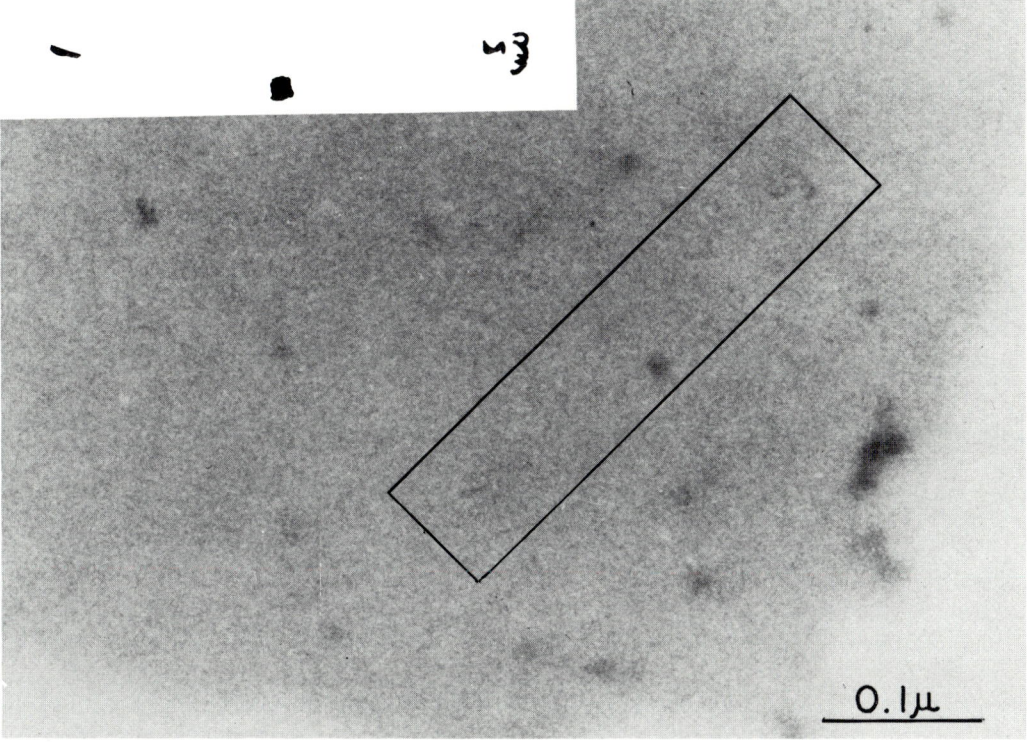

Fig. 3.14. Portion of a mitochondrion from the same preparation as in Fig. 3.12, stained for nucleic acids. Mitochondrial ribosomes. Between them are lines which probably represent DNA. These are shown in the insert traced from the electron micrograph. × 214,000.

real. Mitochondria in unstained, control specimens appear "homogeneous" whether lipid is retained or extracted, because the lipid component does not contribute as much to the mass as the protein component, and in any case seems to be distributed uniformly or as large masses. If lipid is extracted and the proteins are denatured and stained, it is possible to conceive of lipophilic proteins being denatured to such a degree that they become hydrophilic. Under such circumstances, the hydrophilic proteins thus stained by platinum tetrabromide, uranyl acetate, or other metal salts in solution would be indistinguishable from previously undenatured hydrophilic proteins in reactivity, and the protein structure of mitochondria would indeed appear homogeneous.

The lipid masses are broader than the membranes, and may extend across compartments. This probably means that lipid molecules are not confined to membranes, but are present also among the protein molecules of the compartments.

The structure of mitochondria of liver and exocrine pancreatic cells is remarkably consistent. Mitochondria could be considered essentially as stacks of compartments. (See diagram, in Fig. 3.10A.) Each compartment consists of a wall or membrane about 50 Å thick (white in diagram), which completely encloses a dense platelike granular protein material about 100 Å thick (black in the diagram). Each unit extends transversely across the mitochondrion. Adjacent units are separated by a very thin densely stained layer. Enclosing the stacks is a continuous membrane also about 50 Å thick which constitutes the outer surface of the mitochondrion. It is impossible to ascertain whether the compartments are completely enclosed, or whether their contents might not be continuous through some holes or gaps in their membranes between adjacent compartments. If the gaps exist, they are not prominent. Whether the basic repeat plan occurs in mitochondria of all other cells is not known. Until a detailed study of mitochondrial structure in a wide range of cells is completed, the general discussion which follows should be considered to be tentative.

My present hypothesis on the molecular basis of mitochondrial structure may be presented with reference to the diagram in Fig. 3.10B. Protein is present within the membrane as well as within the compartment. The former is more or less regularly disposed as an organized sheet of globular (structural?) molecules and is normally lipophilic. The latter is restricted to the compartment, where it may be functionally organized into the three major integrated catalytic systems, and is normally hydrophilic. Lipid is present in both membranes and compartments. There is no informatinn as to how lipid is related to proteins in mitochondra—as lipids, as lipoproteins, or as proteolipids—although there is abundant biochemical information on separate fractions; nor is there any reason to exclude the possibility that some of the membrane proteins may exercise enzymic functions, or that some of the compartment contents may also be structural in nature.

Adjacent membranes are separated by a very thin, densely stained protein sheet, which is hydrophilic. It is possible that this sheet is real, and consists of outstretched protein molecules corresponding with those called for by the classical Davson–Danielli model. It seems more probable that the thin sheet represents the hydrophilic, densely stained outer parts of the largely lipophilic molecules, possibly distorted during cross-linking by the vapor reagents, and compressed as a result of rearrangements accompanying denaturation. This is the view favored also by Sjöstrand *(17, 18, 19)*. The dimensions of the globular proteins are perhaps not very significant at this time, since they may reflect changes in the ionic environment, degree of cross-linking and consequent distortions, denaturation, etc. Added to these difficulties are distortions which could follow when sections 300–400 Å thick, or less, are stretched on the water trough where they are subject to high surface tension forces. For these reasons the quantitative aspects of mitochondrial molecules are not stressed. The measurements given are considered to be only approximate, and are replacable by others when some technique can be developed to measure mitochondrial molecules *in situ*, or in purified preparations of aggregates subjected to minimal alteration, as was done, for example, for ribosomes and tRNA.

According to this hypothesis, structural protein may be confined mainly to membranes, but some may also be present among the compartment proteins. "Matrix" is restricted to packets of compartment contents with their associated lipids. Lipids and lipophilic proteins probably would be oriented in all membranes. The compartment contents and associated lipids may also be organized as functional assemblies. In the formation and persistence of mitochondria, there would thus seem to be a high value on self-assembly of like molecules in membranes, and of unlike but functionally related catalytic molecules, aggregated in small packets, which are the contents of compartments.

Another corollary of this concept of mitochondrial organization is that enzymic activity is virtually planar. Substrate may have little or no opportunity to diffuse within the narrow, molecularly crowded confines of the compartment, especially if the proteins and lipids are very highly organized and related by various forces not only to each other, but also to the structural molecules of the membrane. Opportunities for free diffusion would seem to be very limited.

A final corollary is that this concept leaves no room for cristae and matrix as currently conceived. According to the hypothesis presented above, mitochondrial cristae would be the equivalent of the compartment membranes whenever the former are complete, transverse walls, as is frequently observed in skeletal or cardiac muscle mitochondria after the use of aqueous fixatives. Whenever the septa are incomplete or the gaps between them are large and irregular as in hepatic mitochondria fixed in aqueous fluids, it is possible that the missing compartment membranes were not preserved completely. When they disintegrate, their remnants would be added to the now disorganized and disaggregated assemblies already present in the compartments. The total mixture of displaced membrane and compartment proteins with their associated lipids would accordingly become what is known as matrix. It is not known exactly why the compartment membranes in mitochondria of some cells should resist fixation and persist, while they are destroyed partially or completely in mitochondria of other cells. This is a general problem without a solution in fixation for light microscopy also. It is also not known how tubular structures are formed in the mitochondria of certain protozoa and certain endocrine cells of mammals.

It is necessary to reemphasize that the findings reported in this chapter, and the hypothesis on which they are based, are restricted to studies on mitochondria of hepatic and pancreatic exocrine cells. It is essential to study mitochondria in many other cell types, as it was important in another context to investigate the DNA molecular pattern of nuclei of many cell types (see Vol. I, Chapter 8). In the course of other studies reported in this work on many cell types (Vol. I, Chapters 7–11), mitochondria have been observed either unstained or after staining for proteins or for nucleic acids. It is perhaps suggestive that the basic structure of mitochondria in all cells appeared homogeneous as it did in this study when protein alone was stained in mitochondria of hepatic and exocrine pancreatic cells. Proof for the generality of a uniform, single, mitochondrial compartmentation plan in cells requires that the special fixation and staining procedures, such as some of those used in the research reported here, be applied to see whether the postulated structure of mitochondria actually exists in these cell varieties. Also required is a study of mitochondria of some selected cell types in different species, and in different pathological and physiological states. Material for such studies in normal and stimulated cells is at hand and will be prepared for study in due course.

When examined at higher resolutions mitochondrial membranes (after certain procedures) appear as aligned protein globules (with some lipid) and the contents of the compartments appear as protein granules (with some lipid). The former are considered as primarily hydrophobic, the latter as hydrophilic. The relationship of these two classes of protein molecules to enzymic activity is not known. The various possibilities are: (1) Structural (nonenzymic) proteins are confined to membranes and are the exclusive elements in them. (2) Some or all of the structural proteins are also enzymic in nature. (3) The proteins within the compartment are the sole enzymic proteins in the mitochondria. (4) The activity of proteins contained within the compartment is limited to one variety of enzymic activity (as, for example, Krebs cycle) which is connected with other related enzymic activities which are located in compartment membranes (as, for example, electron transfer and phosphorylation). The morphological appearance of granules is not sufficiently differentiated at present to distinguish between these possibilities. For example, the globular proteins and lipid of outer and compartment membranes are indistinguishable in appearance despite their alleged widely different enzymic content. It thus becomes essential to demonstrate enzymic activities morphologically. This could be achieved by high resolution cytochemical studies or by a program involving parallel morphological and biochemical studies of suitably fixed and stained preparations beginning with mitochondria *in situ* and continuing through all stages of preparation of submitochondrial particulates. Perhaps such program might bring order to the frequently contradictory reports of activity of mitochondrial particles of uncertain reality.

The observations on mitochondrial nucleic acids reported in this chapter serve only to identify their presence. Nucleic acids are especially notable in mitochondria, which are multiplying rapidly. Several authors have claimed they have demonstrated DNA filaments, individually or in aggregates, in sections of mitochondria. A precondition for their demonstration appears to have been a fixation which results in a pale background or large spaces in the region. One may assume that in the pale regions proteins were partially or wholly removed or displaced, and that the residue was then precipitated as networks of filaments. It seems highly probable that the DNA double helix could not survive such poor fixation, or that, if it did, it could not be distinguished from reprecipitated filaments. The use of the nucleases may not aid in the identification of the filaments because of the uncertainties introduced by poor fixation. This is not intended to cast doubt on the existence of mitochondrial DNA, which has been amply proved by various biochemical methods.

I have been unable to reconstruct the course and relations of the DNA in sections of mitochondria, or the relations of mitochondrial ribosomes to each other and to the filament suggestive of messenger RNA. The very process of ultrathin sectioning of mitochondria imposes restrictions on the study of nucleic acids which seem to limit the usefulness of this method of study.

Summary

Studies on the distribution of lipid and protein in mitochondria of exocrine pancreatic and hepatic cells of the mouse were the basis for an analysis of the structure of these mitochondria. The basic repeating structures of mitochondria are stacked disklike, apparently closed compartments about 200 to 250 Å thick. Each compartment (as well as the outer mitochondrial membrane) consists of walls which are about 50 Å thick. The cavity enclosed by them is about 100 Å thick, and is stuffed with a rather granular material. The walls are composed of "globular" proteins which behave as lipophilic molecules with a hydrophilic outer portion. Interspersed among them are lipids. The granules within each compartment are readily stained and small, and may be aggregated. They behave like hydrophilic protein molecules. Some lipid is present between the protein molecules. There is no evidence to support the unit membrane hypothesis. Possible relations of the structural components mentioned above to function are discussed.

Scarce ribosomal particles are visible in sections of mitochondria in cells of the young adult mouse. They are very numerous in hepatic and chondrogenic cells of embryos. Short lengths of DNA are also detectable in mitochondria of embryonic cells. The relations of the nucleic acids to the basic structures of mitochondria were not elucidated.

References

1. Attardi, G., and Attardi, B. (1970). The informational role of mitochondrial DNA. *In* "Problems in Biology: RNA in Development" (E. W. Hanly, ed.), pp. 245–283. Univ. of Utah Press, Salt Lake City, Utah.
2. Borst, P., and Kroon, A. M. (1969). Mitochondrial DNA: Physicochemical properties, replication, and genetic function. *Int. Rev. Cytol.* **26**, 107–190.
3. Bullivant, S. (1970). Present status of freezing techniques. *In* "Some Biological Techniques in Electron Microscopy" (D. F. Parsons, ed.), pp. 101–146. Academic Press, New York.
4. Chase, W. H. (1959). Fine structure of rat adipose tissue. *J. Ultrastruct. Res.* **2**, 283–287.
5. De-Thé, G. (1968). Ultrastructural cytochemistry of the cellular membranes. *In* "The Membranes" (A. J. Dalton and F. Haguenau, eds.), pp. 121–150. Academic Press, New York.
6. Ernster, L., and Kuylenstierna, B. (1970). Outer membrane of mitochondria. *In* "Membranes of Mitochondria and Chloroplasts" (E. Racker, ed.), pp. 172–212. Van Nostrand-Reinhold, New York.
7. Fessenden-Raden, J. M., and Racker, E. (1971). Structural and functional organization of mitochondrial membranes. *In* "Structure and Function of Biological Membranes" (L. I. Rothfield, ed.), pp. 401–459. Academic Press, New York.
8. Gersh, I., Isenberg, I., Stephenson, J. L., and Bondareff, W. (1957). Submicroscopic structure of frozen-dried liver specifically stained for electron microscopy. I. Technical. *Anat. Rec.* **128**, 91–111.
9. Gersh, I., Isenberg, I., Bondareff, W., and Stephenson, J. L. (1957). II. Biological. *Anat. Rec.* **128**, 149–169.
10. Jensen, W. A. (1962). "Botanical Histochemistry: *Principles and Practice*," pp. 100–127. Freeman, San Francisco, California.
11. Lehninger, A. L. (1964). "The Mitochondrion." Benjamin, New York.
12. Lehninger, A. L. (1970). "Biochemistry," pp. 395–416. Worth, New York.
13. Malhotra, S. K. (1968). Freeze-substitution and freeze-drying in electron microscopy. *In* "Cell Structure and Its Interpretation" (S. M. McGee-Russell and K. F. A. Ross, eds.), pp. 11–21. St. Martin's Press, New York.

14. Nass, M. M. K. (1969). Mitochondrial DNA: advances, problems, and goals. *Science* **165**, 25–35.
15. Rabinowitz, M., and Swift, H. (1970). Mitochondrial nucleic acids and their relation to the biogenesis of mitochondria. *Physiol. Rev.* **50**, 376–427.
16. Schatz, G. (1970). Biogenesis of mitochondria. *In* "Membranes of Mitochondria and Chloroplasts" (E. Racker, ed.), pp. 251–314. Van Nostrand-Reinhold, New York.
17. Sjöstrand, F. S. (1968). Molecular structure and function of cellular membranes. *In* "Regulatory Functions of Biological Membranes" (J. Järnefelt, ed.), pp. 1–20. Elsevier, Amsterdam.
18. Sjöstrand, F. S. (1968). Ultrastructure and function of cellular membranes. *In* "The Membranes" (A. J. Dalton and F. Haguenau, eds.), pp. 151–210. Academic Press, New York.
19. Sjöstrand, F. S. (1969). Morphological aspects of lipoprotein structures. *In* "Structural and Functional Aspects of Lipoproteins in Living Systems" (E. Tria and A. M. Scanu, eds.), pp. 73–128. Academic Press, New York.
20. Sjöstrand, F. S., and Baker, R. F. (1958). Fixation by freezing-drying for electron microscopy of tissue cells. *J. Ultrastr. Res.* **1**, 239–246.
21. Sjöstrand, F. S., and Barajas, L. (1968). Effect of modifications in conformation of protein molecules on structure of mitochondrial membranes. *J. Ultrastruct. Res.* **25**, 121–155.
22. Stoeckenius, W. (1970). Electron microscopy of mitochondrial and model membranes. *In* "Membranes of Mitochondria and Chloroplasts" (E. Racker, ed.), pp. 53–90. Van Nostrand-Reinhold, New York.
23. van Deenen, L. L. M. (1968). Membrane lipids. *In* "Regulatory Functions of Biological Membranes" (J. Järnefelt, ed.), pp. 72–86. Elsevier, Amsterdam.

4

Possible Precursor Granules in Fibroblasts

Giovanni L. Rossi, Isidore Gersh, and Zelma Molnar

This chapter deals with possible precursors in cytoplasm which may give rise in fibroblasts to tropocollagen or precursors of tropocollagen. Chapter 6 (this volume) deals with the same topic in cartilage and bone cells. The main thesis is that certain identical granules occur in fibroblasts of tail tendon, in chondrocytes of epiphyseal plate, and in osteoblasts of the parietal bone and their derivatives; the granules were absent from all other cell types studied. These observations suggest that the granules may be precursors of tropocollagen. Other features will also be presented and analyzed—the morphological relations of the granules to ribosomes and the endoplasmic reticulum, cell processes, the cell surface, and the extracelluar fibers.

Materials and Methods

PROCEDURE FOR STAINING POSSIBLE PRECURSOR GRANULES OF TROPOCOLLAGEN

Rat tail tendons were prepared from animals of the following ages (in days): 2, 3, 4, 6, 8, 10, 12, and 14. The rats were bled when decapitated, and transferred to a moist box where the tail was grasped firmly by fine forceps and pulled away from the body. As the tail vertebrae and skin separated, continued gentle pulling removed the distal part of the tail tendons, leaving bare tendon fibers extending from the proximal stump. They were severed from the animal, and cut quickly with a thin razor blade. Pieces, 0.2 mm thick or less, were transferred to small aluminum foil squares, and frozen ultrarapidly by immersion in freon 13 cooled with liquid nitrogen to a temperature of approximately $-180°C$.

In general, the tendon bundles of younger rats were finer than those of older rats. However, there was a great overlapping, so that some tendon specimens of younger rats were thicker than some from older rats. This can be attributed to the small size of the specimens frozen and to the fact that tendon bundles were not selected, since all which yielded to the steady severing pull with the forceps were taken. Hence, specimens or preparations of tendons of different age animals are chronological descriptions of the specimens and have no developmental significance.

The methods involve ultrarapid freezing and drying at a low temperature, followed by postfixation and staining *in vacuo* with vapors of cross-linking reagents in the absence of fluid water. This procedure was designed (1) to freeze specimens so rapidly that ice crystals do not form; (2) to dry at a low enough temperature so as to favor sublimation of water from the frozen state and thus avoid diffusion; (3) to cross-link proteins and their smaller units in order to reduce diffusion and extraction in subsequent steps, and to add mass to the reactive sites; (4) to harden the specimens; and (5) to infiltrate them more thoroughly and uniformly to improve the sectioning quality of the polymerized tissue block. The vapors used are difluorodinitrobenzene (FFDNB), difluorodinitrodiphenylsulfone (FFSulfone), and mercury(II) hexafluoroacetylacetonate (Hg hfac). Sometimes a mixture of FFDNB and FFSulfone was used. The rationale and technical details are given in detail in Chapter 2 (Vol. I). All the reagents combine with proteins; some combine more extensively than others with nucleic acids.

PROCEDURE FOR STAINING NUCLEIC ACIDS

Specimens treated with any of these vapors may be stained with gallocyanin–chromalum (GCA) for study of nucleic acids. The rationale and technical details are given in Chapter 3 (Vol. I).

For submicroscopic studies, sections were cut with diamond knives, floated on water, mounted on grids, and examined with an Hitachi HU-11A electron microscope operating at an accelerating voltage of 50 kv. Electron micrographs were enlarged × 2 in printing.

For study with the light microscope, thicker sections about $1/2$ μ were cut, mounted on coverslips, stained with toluidine blue as described in this volume (Chapter 5), and mounted in Permount. Two particular sets of observations were made with the light microscope, the first, concerned with the cytoplasmic granules, and, the second, with cytoplasmic basophilia.

Observations

OBSERVATIONS WITH THE ELECTRON MICROSCOPE

After postfixation and staining with FFSulfone, the cells and their processes are clearly distinguishable in electron micrographs, while the extracellular components are barely so (Figs. 4.1–4.3). Extracellular collagen is, however, present and readily demonstrable after section staining with uranyl acetate (Fig. 4.4). The nuclei contain denser bodies of high contrast, which constitute the chromatin. The

Observations

nucleoplasm fills the remainder of the nucleus (Fig. 4.1) except for the nucleous (see Fig. 4.14, lower cell). The cytoplasm contains numerous densely staining areas which are primarily the rough endoplasmic reticulum (RER). These extend into many processes (Figs. 4.1 and 4.3).

Granules of microscopic size (about 0.5 to 1.0 μ in diameter) are visible in the cytoplasm (Figs. 4.1–4.3, GRAN III). They will be referred to as third-order granules. They may be spherical or oval in sections, and are of high contrast. There seems to be no preferred site for the granules, and they have been observed

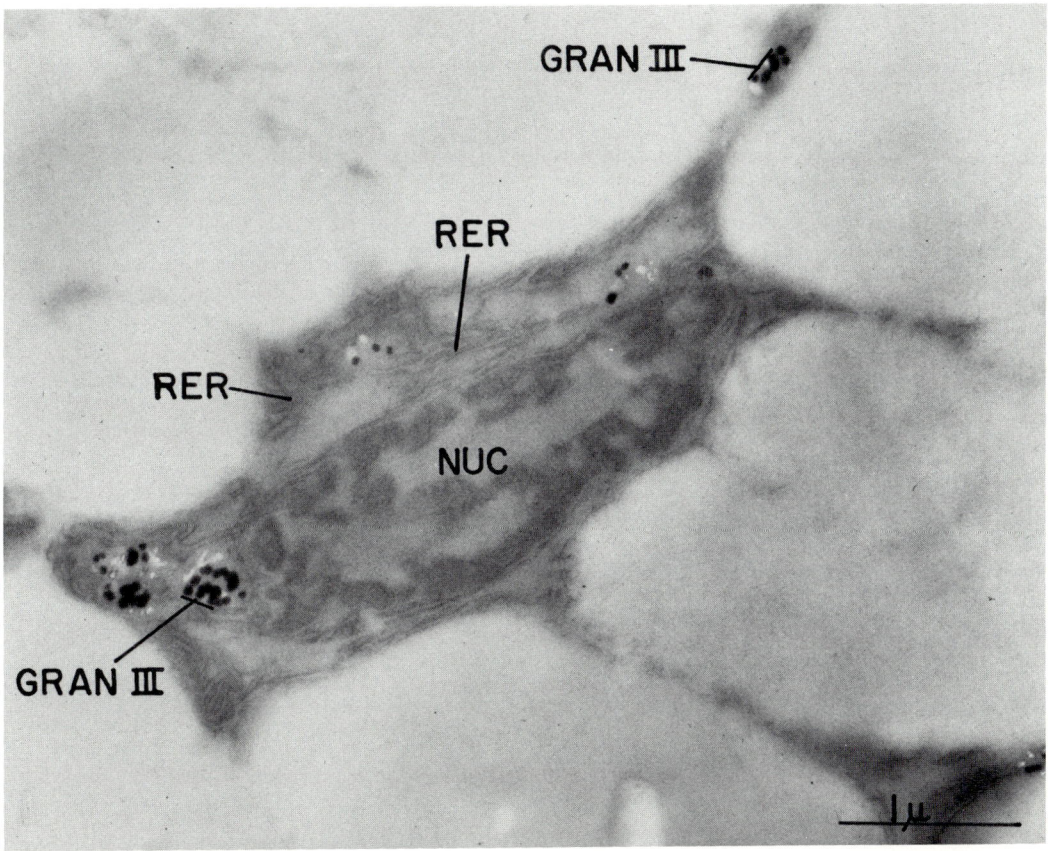

Fig. 4.1. Tail tendon, 10-day-old rat, postfixed and stained *in vacuo* with vapors of difluorodinitrodiphenylsulfone (FFSulfone). The nucleus (NUC) contains small chromatin masses situated chiefly at the periphery of the nucleus. The cytoplasmic endoplasmic reticulum (RER) is clearly marked in the cell body and in some cell processes, but cannot be detected in others. Third-order granules (GRAN III) are present in the cell body as well as in cell process. Sometimes the second-order granules have fallen from the larger granule, leaving a clear space, but most of the second-order granules persist. The extracellular collagen fibers are unstained and do not appear. $\times$ 24,000.

in the juxtanuclear cytoplasm, as well as in the cell processes. They may be located deep in the cytoplasm (Fig. 4.1) or superficially (Figs. 4.5–4.8). The latter may be separated from the extracellular connective tissue by a thin layer of cytoplasm (Fig. 4.5) or the cytoplasm may be so attenuated as to be inappreciable (Fig. 4.8). There are also examples of possible intermediate positions (Figs. 4.6 and 4.7).

The third-order granules contain a variable number of smaller granules, each about 1000 Å in diameter and of high contrast. There may be one to twelve or more per section through a third-order granule. These will be referred to as sec-

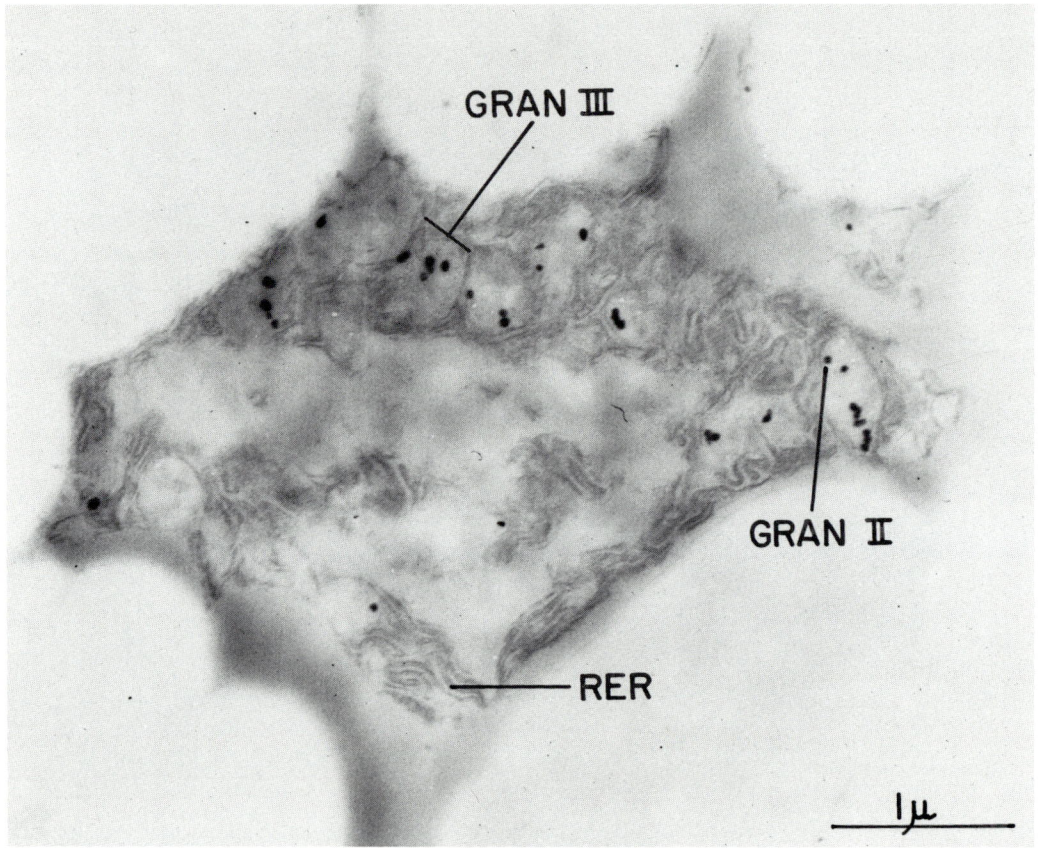

Fig. 4.2.

Figs. 4.2 and 4.3. Same specimen as in Fig. 4.1; it shows the marked variations which occur in the third-order granules (GRAN III) in their size and shape, and in their content of second-order granules (GRAN II). Organized RER may be absent from large stretches of cytoplasm. None of the second- or third-order granules occurs in the space between the RER membranes. × 24,000.

Observations

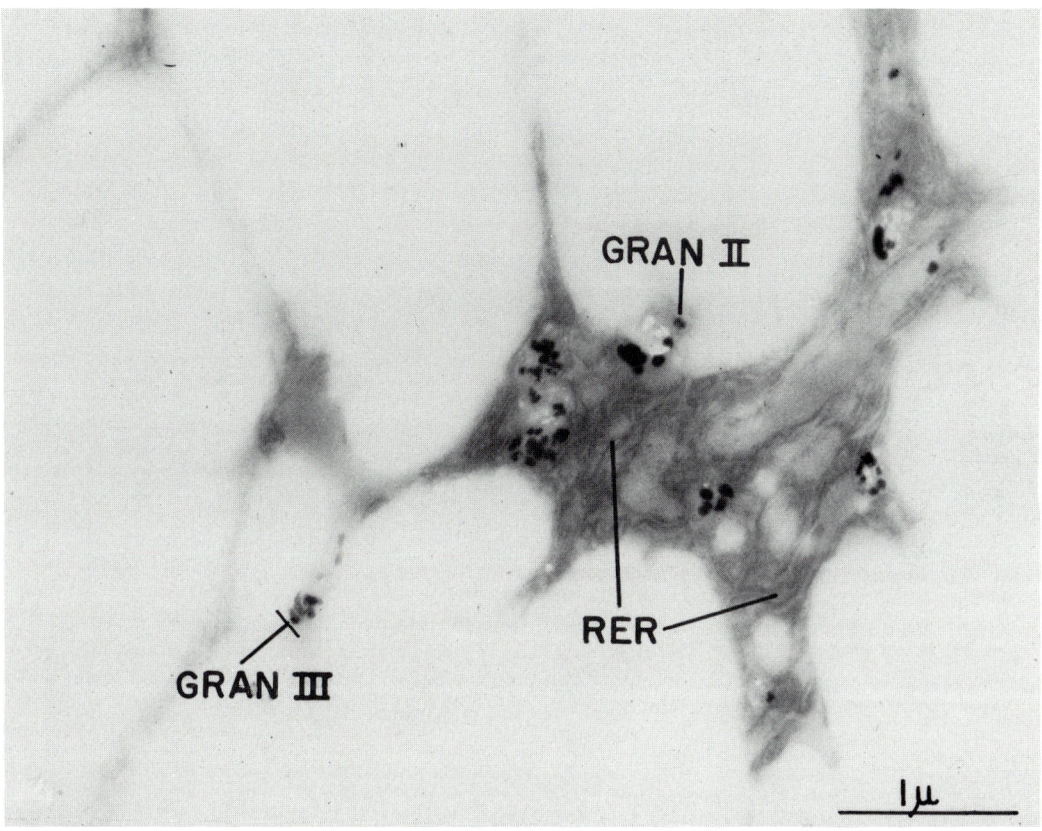

Fig. 4.3.

ond-order granules. Their outer surface is not smooth, but its irregular, like a mulberry (Figs. 4.5–4.8). The irregular surface reflects the fact that each second-order granule comprises a cluster of numerous tightly packed smaller granules about 80 Å in diameter. Each one of these ultimate or first-order granules appears to consist of an outer shell of high contrast surrounding a paler center (Fig. 4.9). The second-order granules are presumably hard, for they frequently fall out of the sections, leaving neat holes.

As there are relatively few free amino groups in the extracellular collagen, the reaction there would be expected to be weak, and in prints the areas rich in collagen may even appear blank (Figs. 4.1–4.3). However, when ultrathin sections are counterstained with uranyl acetate, the collagen fibers are stained well, with a uniform periodicity (Fig. 4.4).

The third-order granules lie outside the cisternae of the RER, which are consistently narrow in specimens postfixed with vapors of FFSulfone. The granules

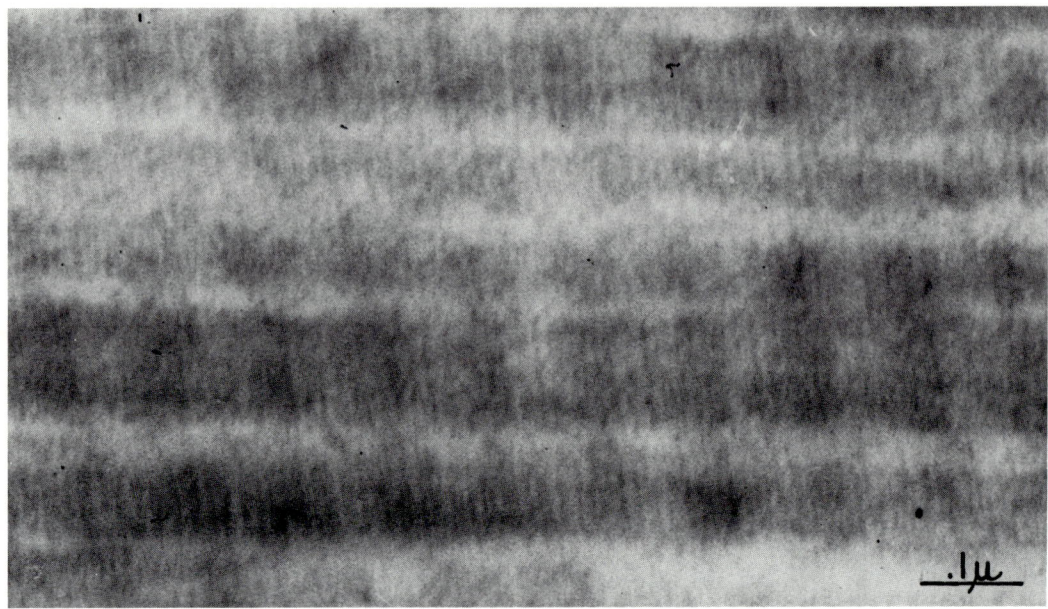

Fig. 4.4. Same specimen as in Figs. 4.1–4.3, to show the periodicity of collagen fibers sectioned longitudinally. Though these are barely or not at all visible in electron micrographs of preparations stained only with vapors of FFSulfone (as in Figs. 4.1–4.3), they are clearly banded after the sections are stained with uranyl acetate. × 142,000.

were never observed within the cisternae of the RER. Even after more selective staining of the RER with GCA, it seems extremely unlikely that the third-order granules could ever be included within the cisternae of the RER.

In the nucleus, DNA is distributed as clumps which may be large, as in Fig. 4.10, cell A, or fine as in cell B. In most cells, the DNA clumps are related to the nuclear membrane, with numerous gaps between them. The nucleolus is of intermediate density (Fig. 4.14).

In the cytoplasm, the RER becomes apparent primarily because the ribosomes have been stained with GCA. The cisternae vary markedly in density and in their pattern (Figs. 4.10–4.12, 4.14, and 4.15), and extend into some cell processes (see Figs. 4.13 and 4.15 for higher magnifications). In general, the picture conforms with the general description of the RER in preparations fixed and stained with FFSulfone alone. In other cell processes, the RNA occurs in small groupings as in Fig. 4.16. In still other cell processes, RNA is absent or minimal (Figs. 4.11 and 4.14). In no case, was the RER ever of a dimension large enough to accommodate a third-order granule.

When ultrathin sections of specimens postfixed with vapors of FFSulfone are stained with uranyl acetate, the density patterns are somewhat different from those

Discussion

stained by FFSulfone alone or by Hg hfac and GCA (Fig. 4.17). In the nucleus, the region occupied by the chromatin is of very high contrast, and the dense walls of the nucleoplasm observed in the preparations stained with GCA are delineated by their high contrast. At the same time, the cisternae of the RER appear as structured elements subdivided by walls into numerous compartments. The cytoplasmic granules are stained, and are clearly excluded from the intracisternal space of the RER. Staining of sections with uranyl acetate seems, then, to emphasize certain nuclear and cytoplasmic subdivisions not otherwise readily apparent.

These subdivisions appear even more pronounced in preparations of freeze-dried specimens postfixed and stained with an alcoholic solution of platinum tetrabromide (Figs. 4.18 and 4.19). The cytoplasmic granules are not preserved by this method.

Two further general observations should be made, and these apply to all preparations studied by the methods described in this report: (1) No cytoplasmic granules have ever been observed extracellularly, and (2) all cell surfaces, whether of the cell body near the nucleus or of its processes, are sharply outlined.

OBSERVATIONS WITH THE LIGHT MICROSCOPE

In the thick sections (about $1/2$ μ), stained with toluidine blue and observed with the light microscope, the fibroblasts and their processes are intensely stained and display their variegated shapes with great beauty (Figs. 4.20 and 4.21). The nuclei are smooth and clear, outlined by numerous chromatin clumps adherent to the nuclear surface. The largest single structure is the nucleolus with its associated chromatin. The only other stainable structures are the fine chromatin particles scattered throughout the (unstained) nucleoplasm.

The cytoplasmic granules, which correspond with the third-order granules described above, are scattered around the scant cytoplasm surrounding the nucleus, and extend into the cell processes (Figs. 4.20 and 4.21.) Sometimes the cell process is so fine and/or unstained (or weekly stained) that it is not visible in, or photographable with, the light microscope; but the course of the process can be traced by the progression of granules. Many or most visible processes contain no granules.

Cytoplasm stained in these preparations is rarely uniform. Irregularities may be discerned in the cytoplasm of nearly all cell bodies, and are especially notable in many cell processes (Figs. 4.22–4.24). The irregularities reflect, in large part, aggregates of ribosomes, whether or not they are associated with the RER. The pattern is more crowded than that in chondrocytes (this volume, Chapter 5).

Discussion

A more complete consideration of the observations reported above will be presented in Chapter 6 (this volume). In anticipation of the results, two general comments may be made: (1) the particular cytoplasmic granules of fibroblasts are

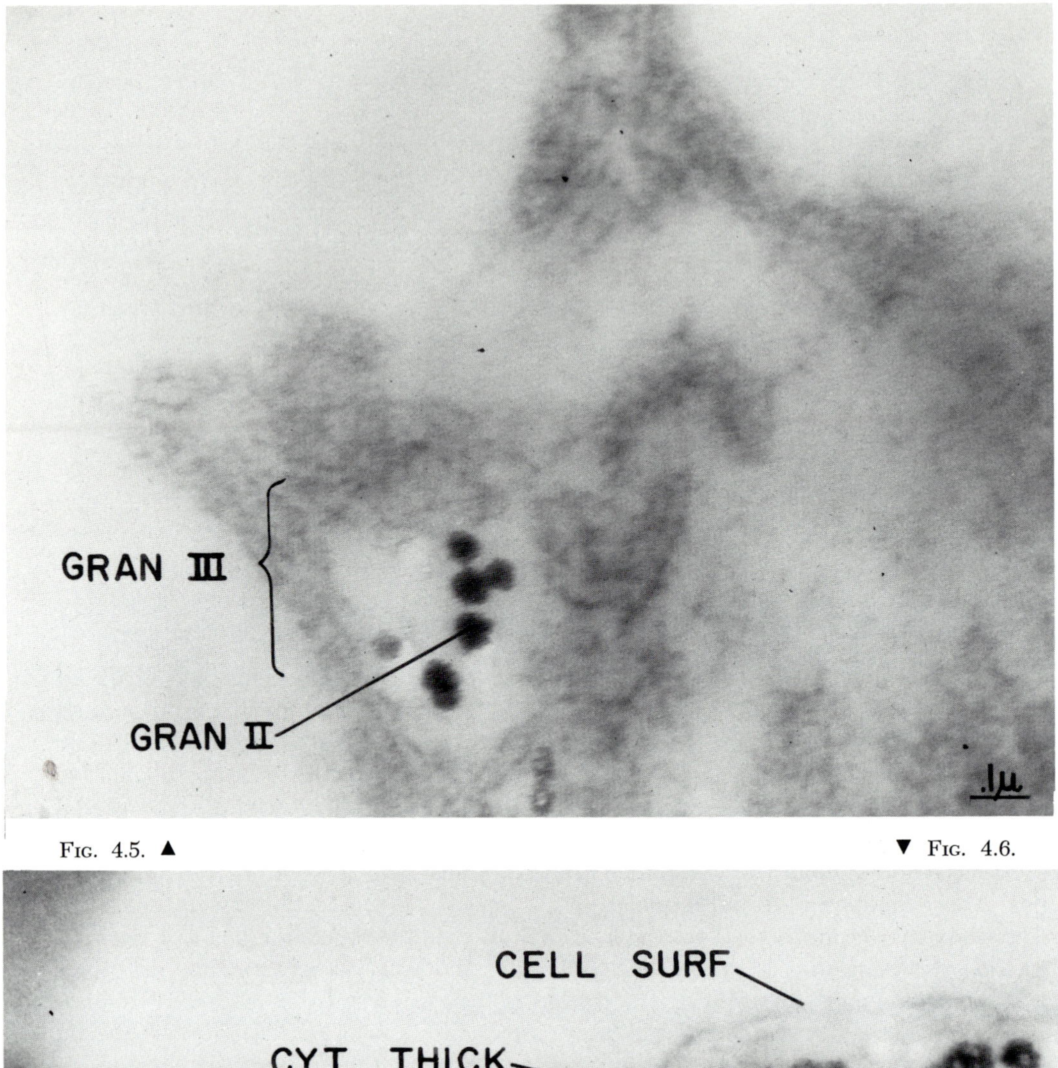

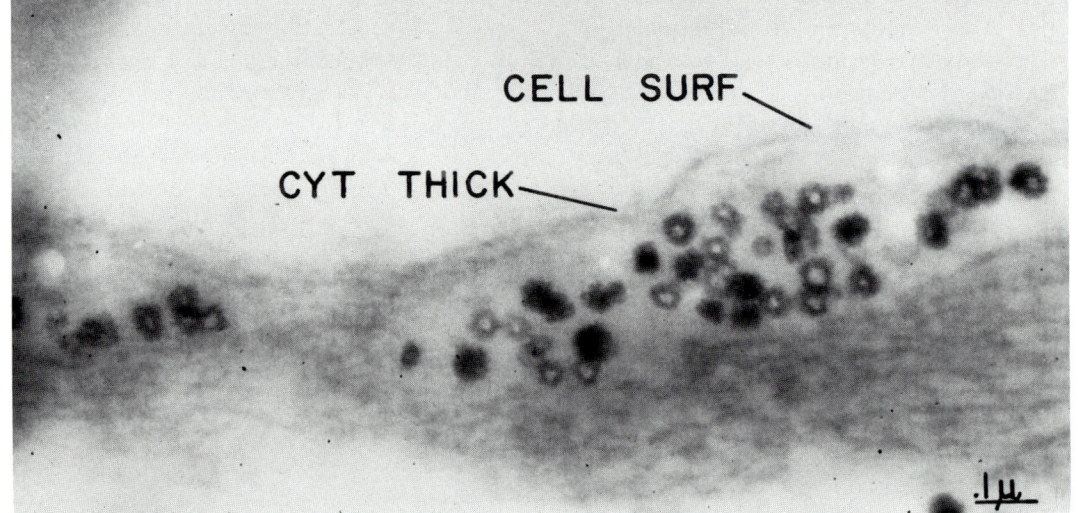

FIGS. 4.5–4.8. Tail tendon, 4-day-old rat, postfixed and stained *in vacuo* only with vapors of FFSulfone. The third-order granules (GRAN III) are progressively closer to the edge of the cell (CELL SURF), and as seen in Fig. 4.8 bulge extracellularly. Though there is a layer of cytoplasm (CYT THICK) between the third-order granule and the extracellular space in Figs. 4.5–4.7, it is inappreciable in Fig. 4.8. Note especially in Figs. 4.5 and 4.6 that at this high magnification, the surface of the second-order granule is irregular and mulberrylike. × 82,000.

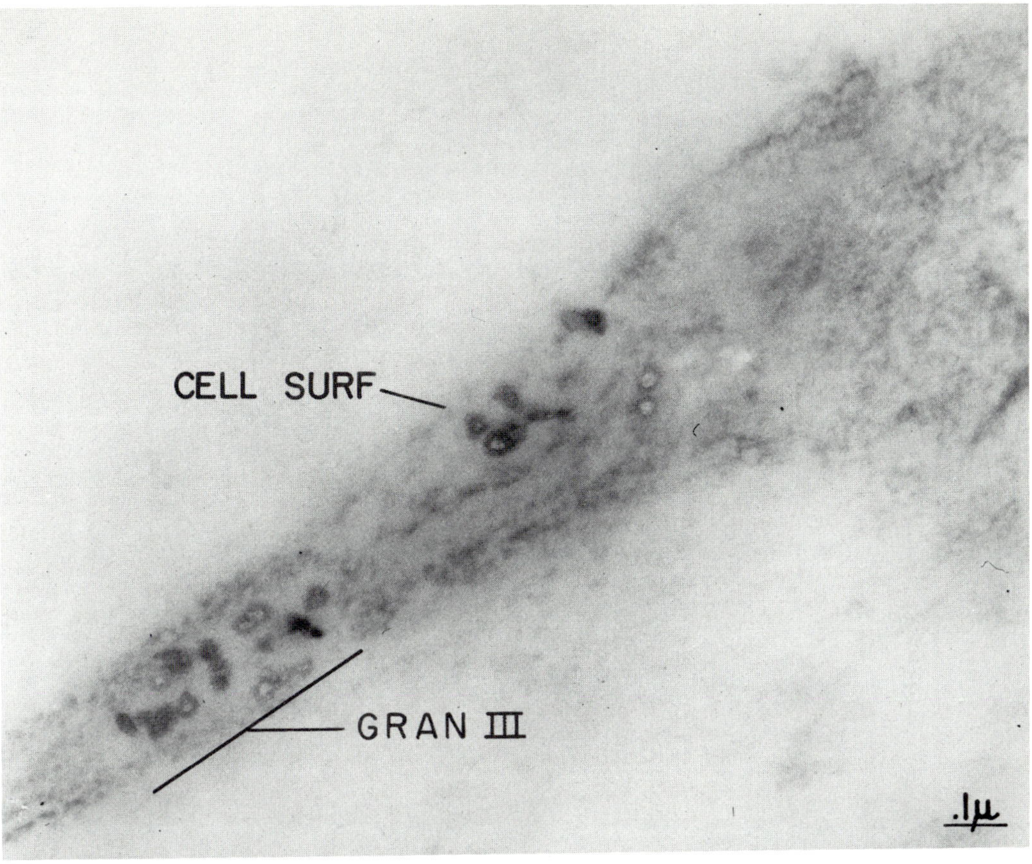

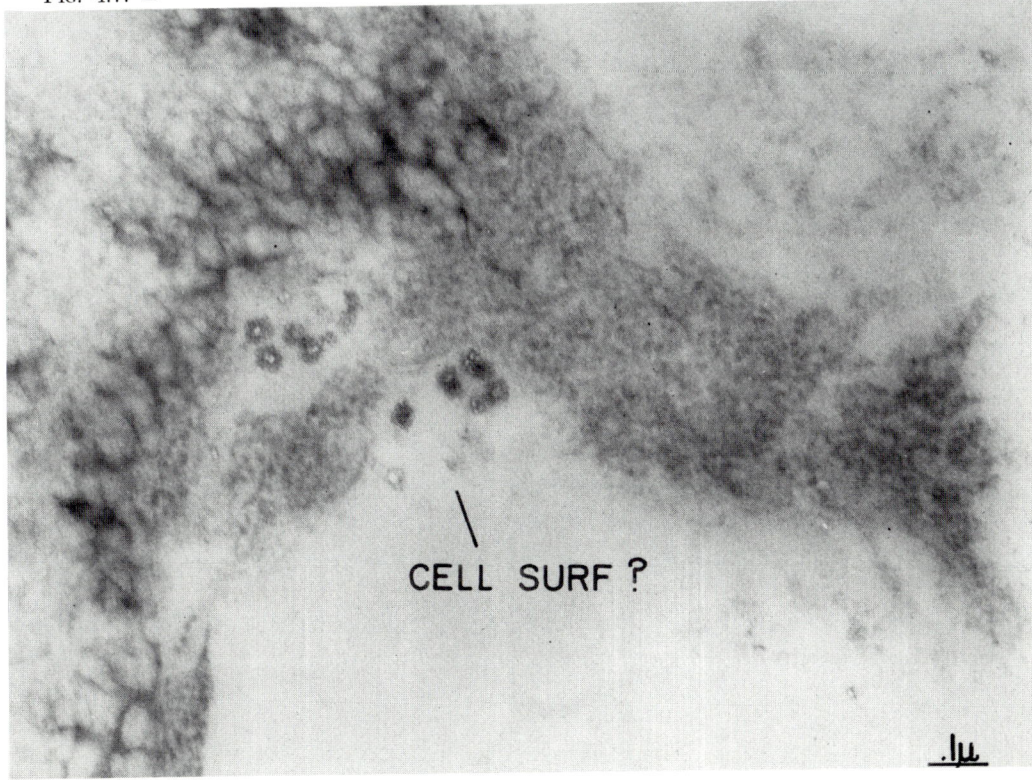

Fig. 4.7.   ▲   ▼ Fig. 4.8.

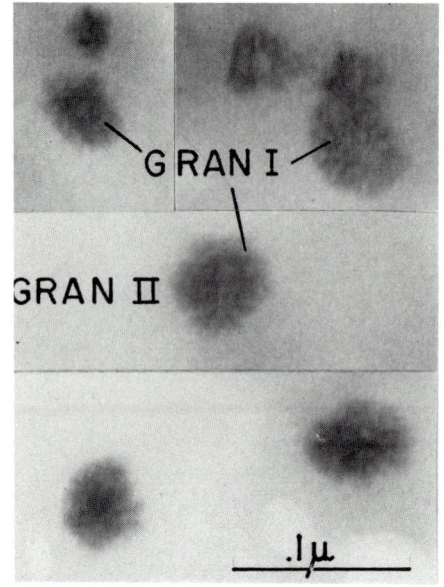

Fig. 4.9. Same specimen as in Figs. 4.5–4.8. Separate second-order granules at higher magnification show that they comprise a cluster of smaller units of about 80 Å, the first-order granules. These consist of pale cores enclosed by high contrast regions. The irregularity in the outline of the second-order granule (GRAN II) noted in Figs. 4.5 and 4.6 is referable to these first-order granules (GRAN I) at their surface. × 240,000.

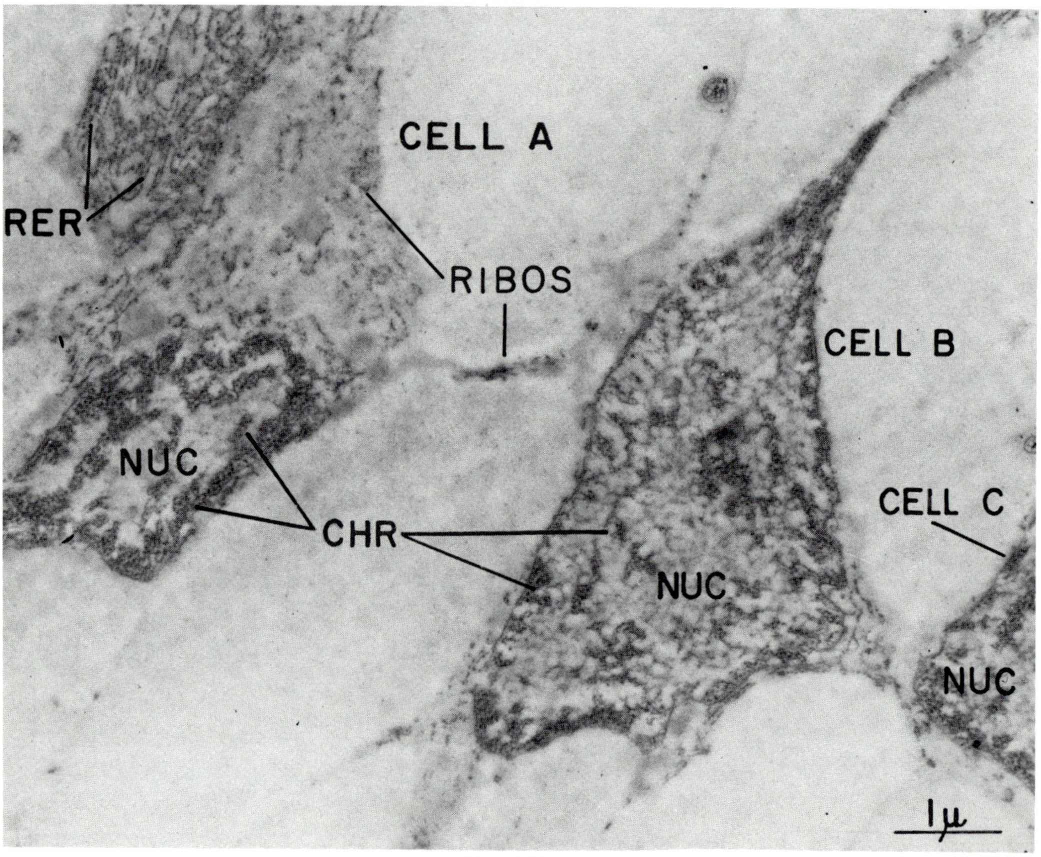

Fig. 4.10.

Discussion 73

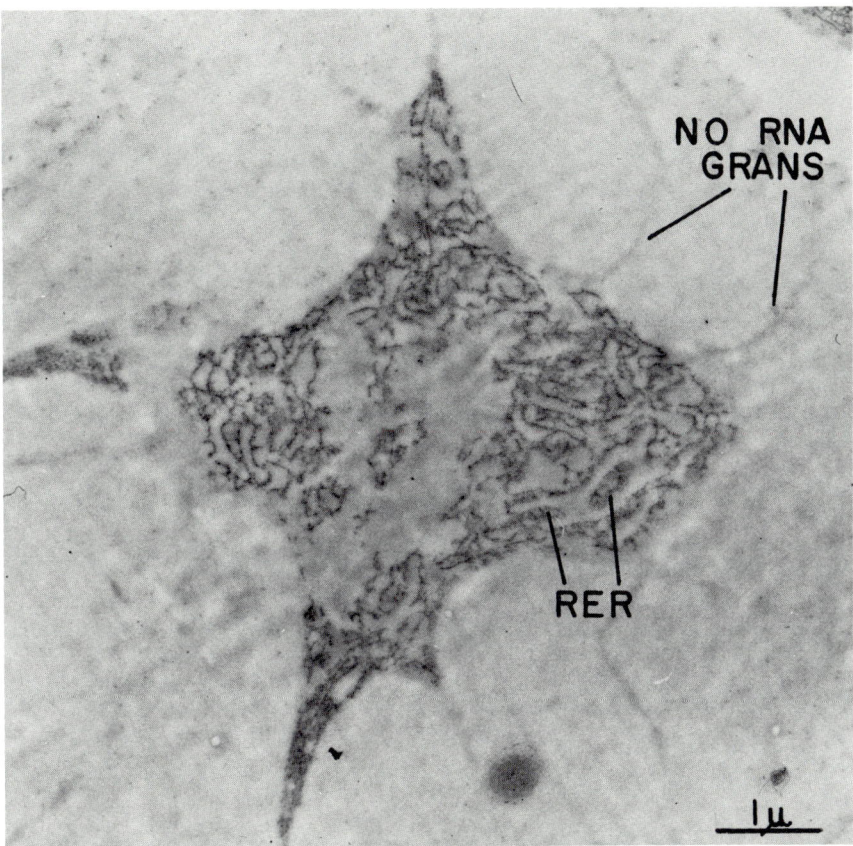

Fig. 4.11.

Figs. 4.10 and 4.11. Tail tendon, 8-day-old rat, postfixed and stained *in vacuo* by mercury (II) hexafluoroacetylacetonate (Hg hfac) followed by gallocyanin–chromalum (GCA) to stain nucleic acids. DNA is distributed in the nucleus in small clumps (CHR). The amount of RNA in the cytoplasm varies greatly in amount and in distribution, but is confined largely to the moderately dispersed RER system. RNA granules are present in some cell processes (RIBOS), but absent from others (NO RNA GRANS). The granules appearing in Figs. 4.1–4.3 and 4.5–4.9 were dissolved during the staining in the aqueous gallocyanin–chromalum. Collagen fibers are unstained. × 14,200.

identical with those in chondrocytes and osteoblasts (as well as osteocytes and osteoclasts), and have not been observed in any other kind of cell studied (see Vol. I, Chapters 8–11), and (2) these granules are different from the PAS-positive granules described in fibroblasts (6) and osteoblasts (11) and from other, as yet unidentified granules in cartilage cells (12). The granules differ morphologically from those described by Fitton Jackson in osteoblasts and other cells (5). These findings have led us to suggest that the granules may be possible precursor granules of tropocollagen.

The possible precursor granules we have described seem to be soluble in water and in many alcoholic solutions. Bondareff (3) was unable to preserve these granules in freeze-dried preparations, despite testing about 200 postfixative solutions. Such granules have not been observed in tissue fibroblasts (13) or cultured cells (10), perhaps for the same reason. Though preserved in 95% alcohol alone and by a concentrated solution of ferric chloride in chondrocytes, they generally require cross-linking reagents in the absence of water for satisfactory preservation. In fibroblasts, the PAS-positive granules differ from the possible tropocollagen granules in that they are less numerous and more readily preserved by postfixation, without the need for cross-linking reagents. In addition, in an experiment with sections of osteoblasts as reported in the next chapter, the possible precursor granules did not correspond with the PAS-positive granules characteristic of these cells.

The possible precursor granules have two components: (1) first-order granules about 80 Å in diameter which have high contrast and are aggregated into clusters about 1000 Å in diameter, which constitute the second-order granules, and (2) a less dense, usually homogeneous matrix, which is of very much lower contrast, in which the latter are embedded. The totality constitutes the third-order granule which is microscopically visible and can be photographed with the light microscope after suitable staining. The first-order granule consists of a very thin, very dense and reactive peripheral region which surrounds a less dense core. Interpretations of the basis for this appearance of the first-order granules are given in Chapter 6.

In the cytoplasm, the possible protropocollagen granules have been observed in the cell body and cell processes of fibroblasts (and also chondrocytes and osteoblasts) but never in the intracisternal spaces of the RER. These are narrow, and rather uniform, and free of the marked distention which has attracted the attention of all other investigators of these cells. The mechanism which provokes the release of the possible precursor granules from the cell is unknown. The actual release by the cell seems to take place by a gradual attenuation of the cytoplasm lying between it and the extracellular space. (See later in this volume, p. 144 for a further discussion of this topic.)

The surface of the fibroblast, including all its processes, is sharply delineated, unmarred by the deposition or accumulation of any homogeneous or fibrillar material. This is contrary to many reports in the literature, which have led their authors to suggest that collagen is somehow formed extracellularly with the surface

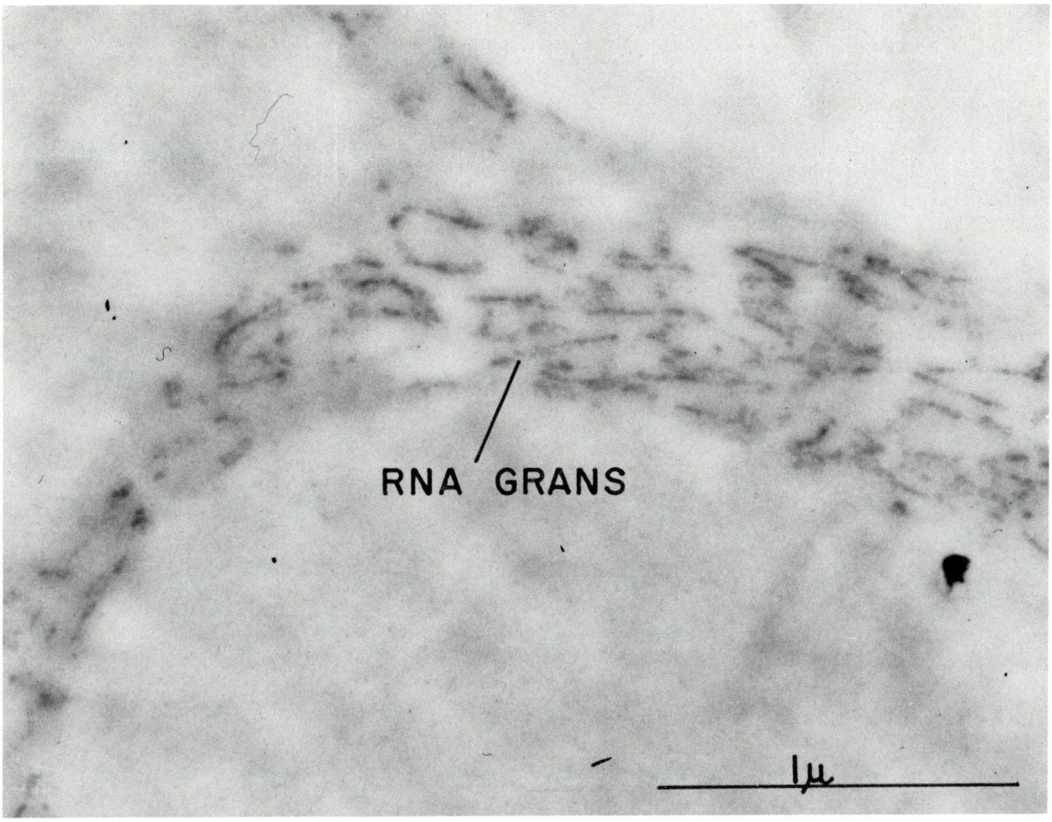

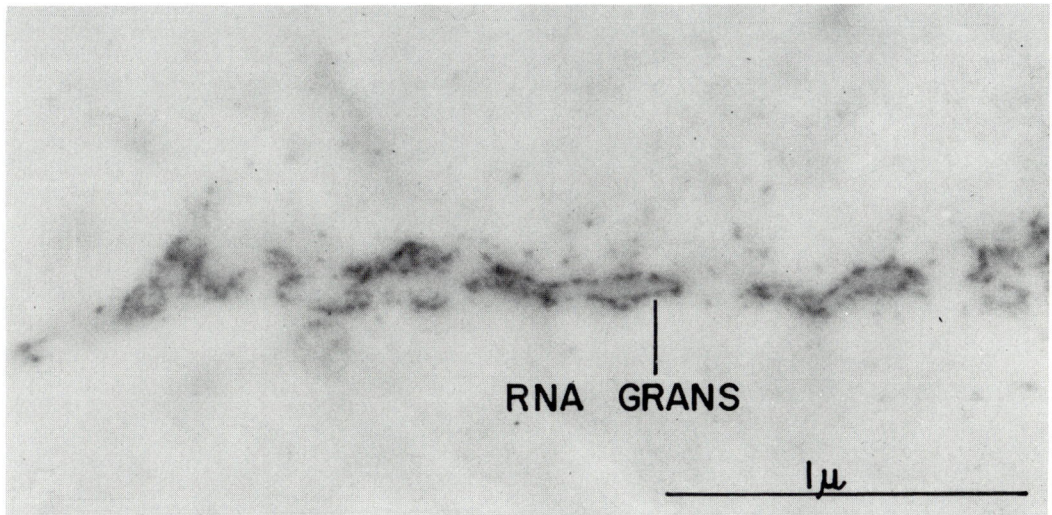

Fig. 4.12. ▲   ▼ Fig. 4.13.

Figs. 4.12 and 4.13. Same preparation as in Figs. 4.10 and 4.11, showing cell processes containing RNA distributed in part in columns or chains. × 48,000.

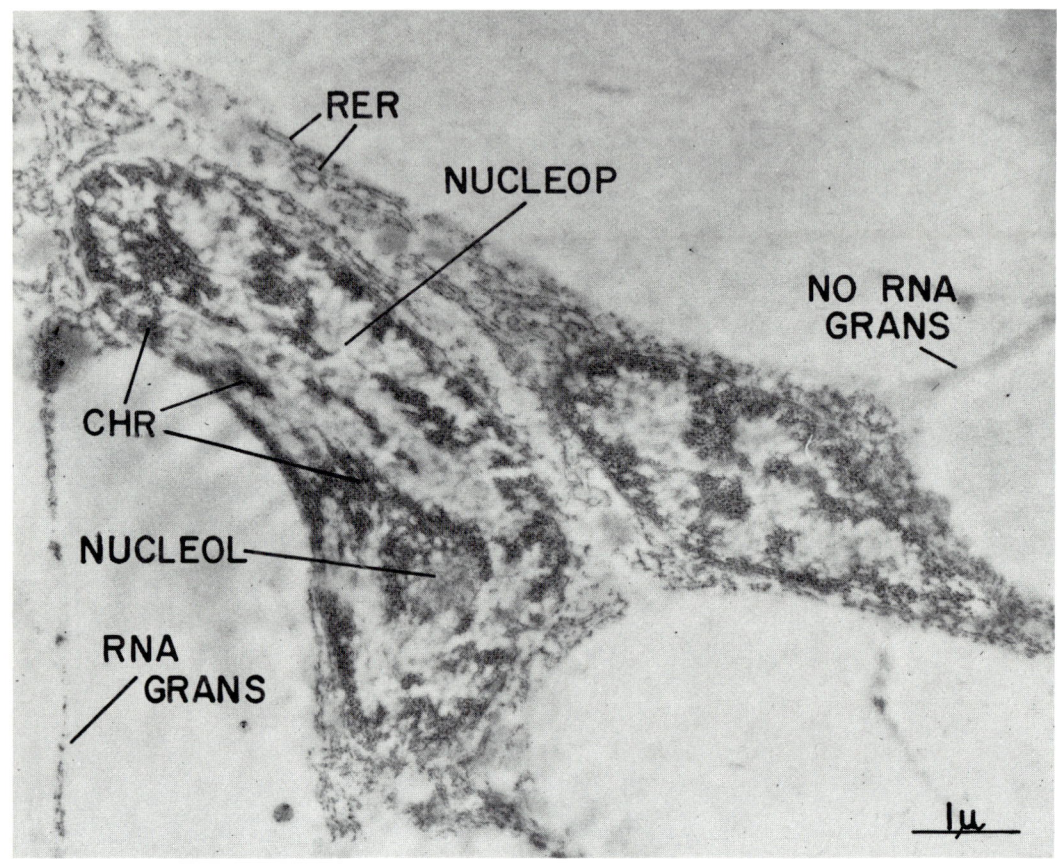

Fig. 4.14.

Figs. 4.14–4.16. Tail tendon, 14-day-old rat, postfixed and stained *in vacuo* with vapors of Hg hfac followed by GCA to stain nucleic acids. DNA is distributed in the nucleus as small clumps (CHR), some of which are related to the nucleolus (NUCLEOL) (Fig. 4.14). RNA occurs in the nucleolus, as well as in a series of interconnected walls in the nucleoplasm (NUCLEOP). Cytoplasmic RNA is distributed chiefly as RER, and to a lesser extent as isolated aggregates or columns of ribosomes. The latter are sometimes more prominent in cell processes. Figures 4.14 and 4.15, × 14,200; Fig. 4.16, × 48,000.

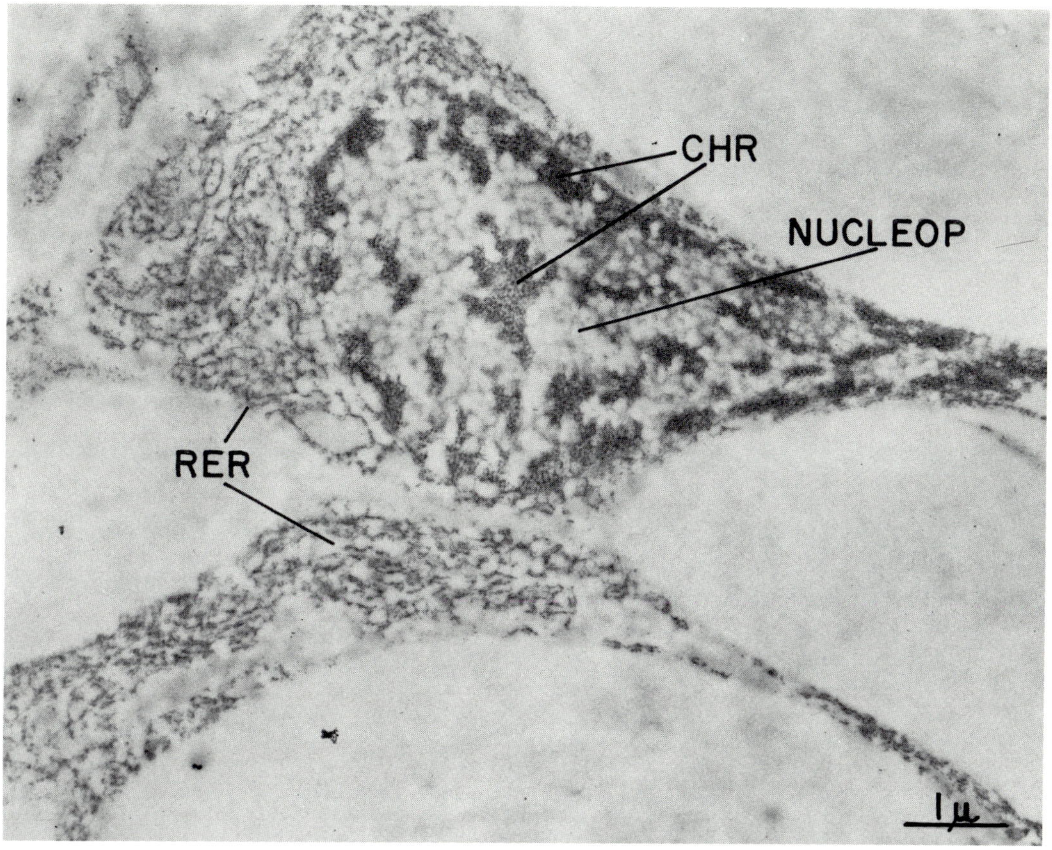

Discussion

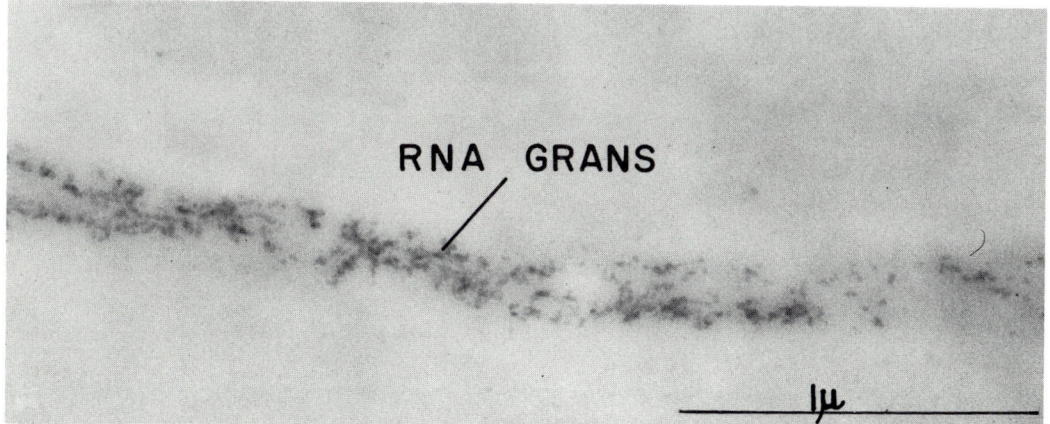

Fig. 4.15.

Fig. 4.16.

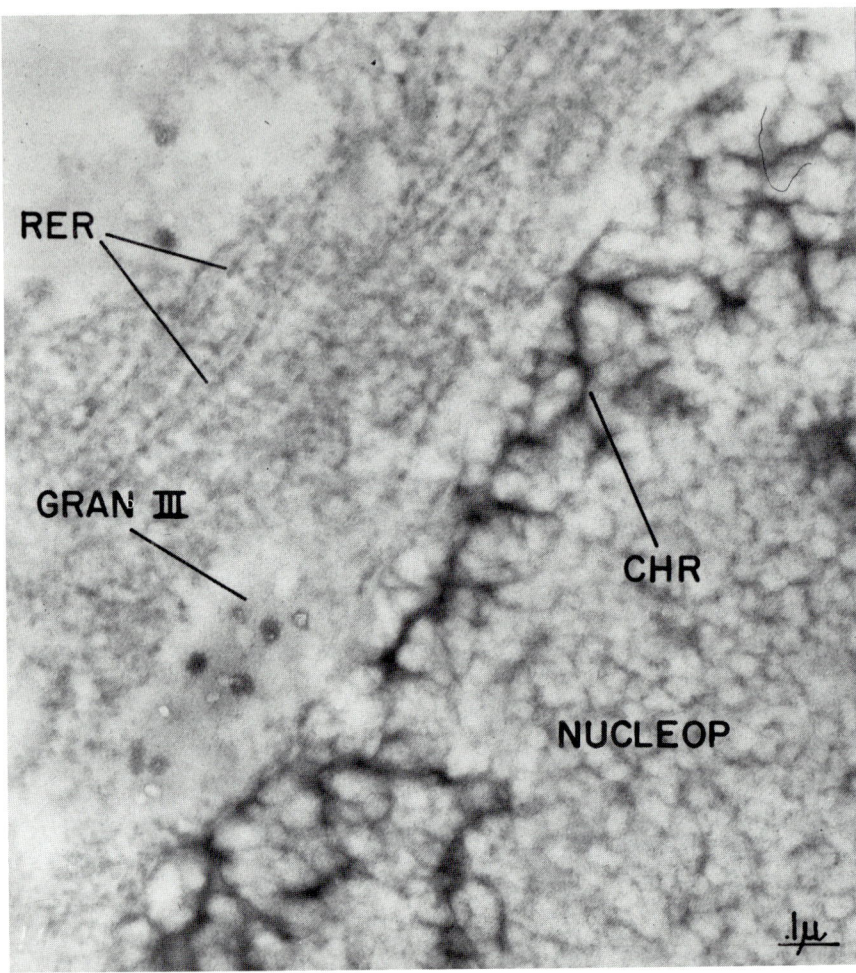

Fig. 4.17. Tail tendon, 14-day-old rat, postfixed and stained with vapors of FFSulfone. Sections were stained with uranyl acetate. The two-phase system illustrated here appears more clearly in Figs. 4.18 and 4.19. The third-order granule (GRAN III) is preserved. CHR, chromatin; NUCLEOP, nucleoplasm. × 82,000.

Figs. 4.18 and 4.19. Tail tendon, 8-day-old rat, postfixed in alcoholic platinum tetrabromide. Both nucleus and cytoplasm consist of numerous submicroscopic pseudovacuoles and their walls. In the nucleus (NUC), the walls are thickened and show more contrast where the chromatin masses occur (CHR). In the cytoplasm, some walls are in pairs which are thicker and have more contrast than others. These correspond with the RER. In both nucleus and cytoplasm, the dense walls appear to be nongranular. Third-order granules are soluble in the solution of platinum tetrabromide and hence are not visible. Extracellular materials are barely visible. Figures 4.18 and 4.19, × 48,000.

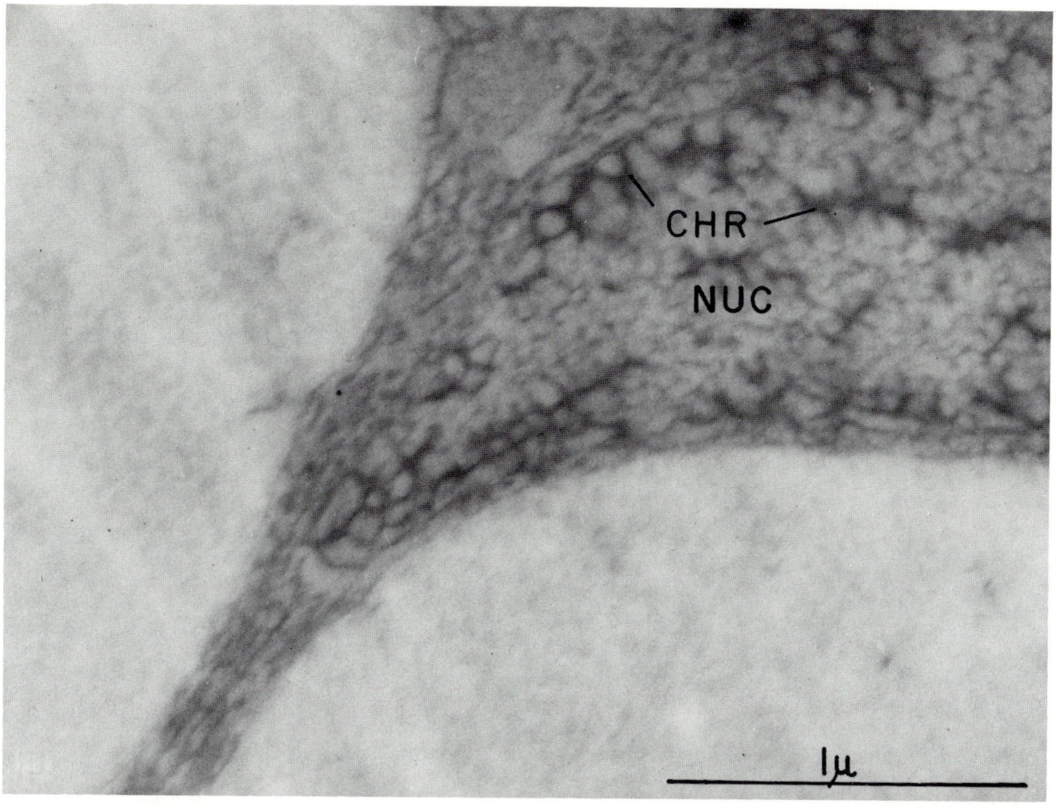

FIG. 4.18. ▲

▼ FIG. 4.19.

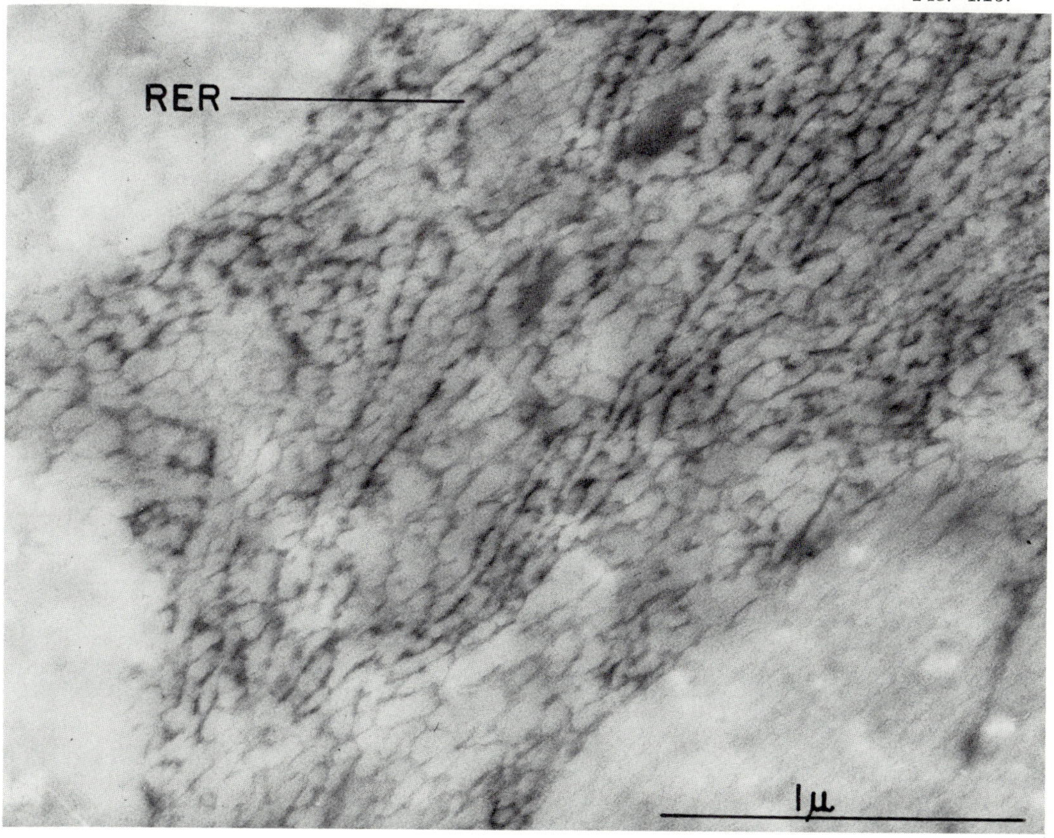

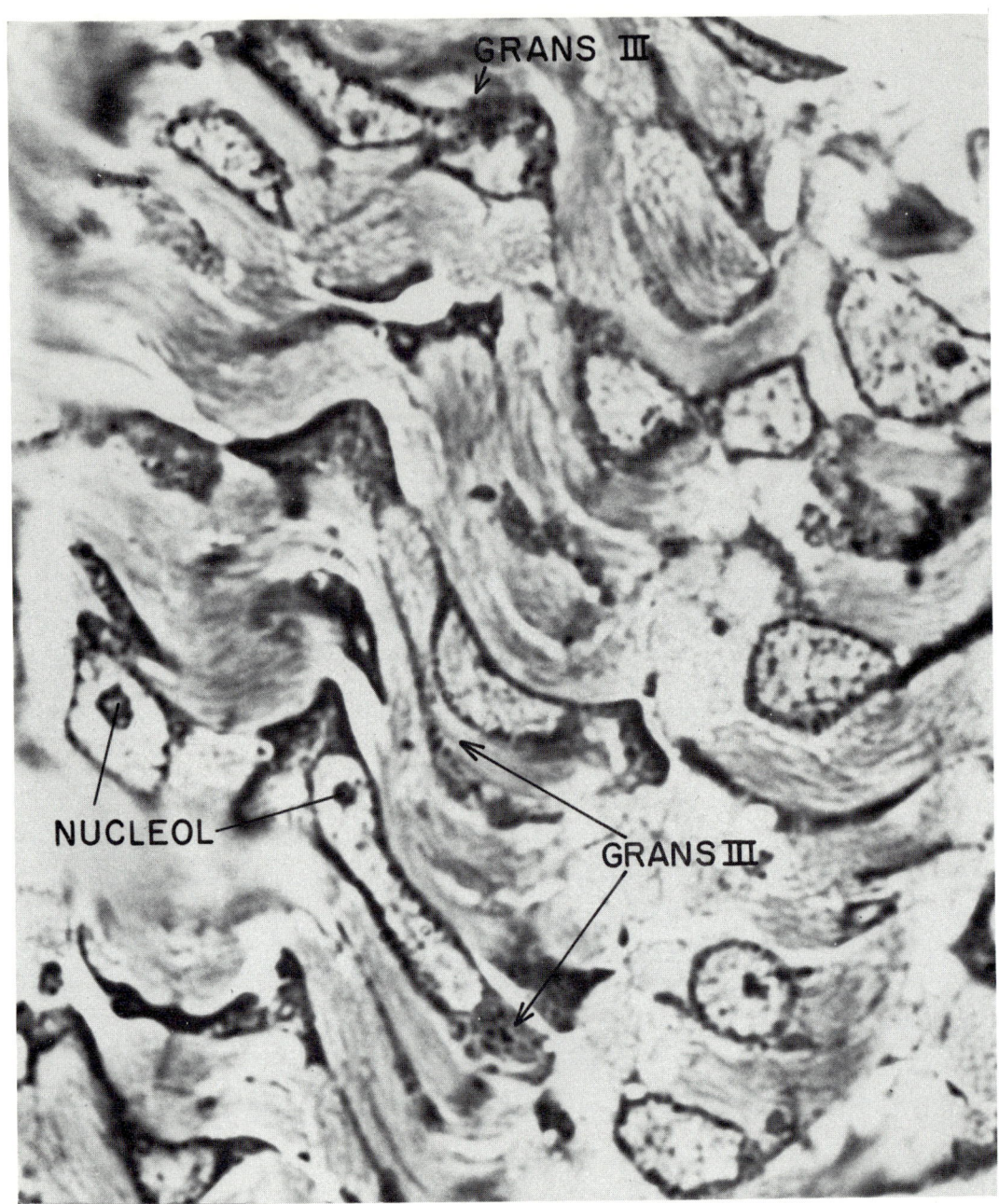

Fig. 4.20.

Figs. 4.20 and 4.21. Photomicrographs of sections of tail tendon of 2- and 10-day-old rats prepared by freezing and drying, cross-linking *in vacuo* with vapors of FFSulfone, embedding in plastic, and staining of the sections with toluidine blue. Cytoplasmic granules, some of which are marked by arrows, correspond with the third-order granules observed in the electron microscope. The granules appear in the perinuclear cytoplasm (Fig. 4.20) as well as in cell processes (Figs. 4.20 and 4.21). In Fig. 4.20, the granules extend chiefly into cell processes. Abbreviations as given earlier. Figure 4.20, 2-day-old rat; Fig. 4.21, 10-day-old rat. $\times$ 3200.

Discussion

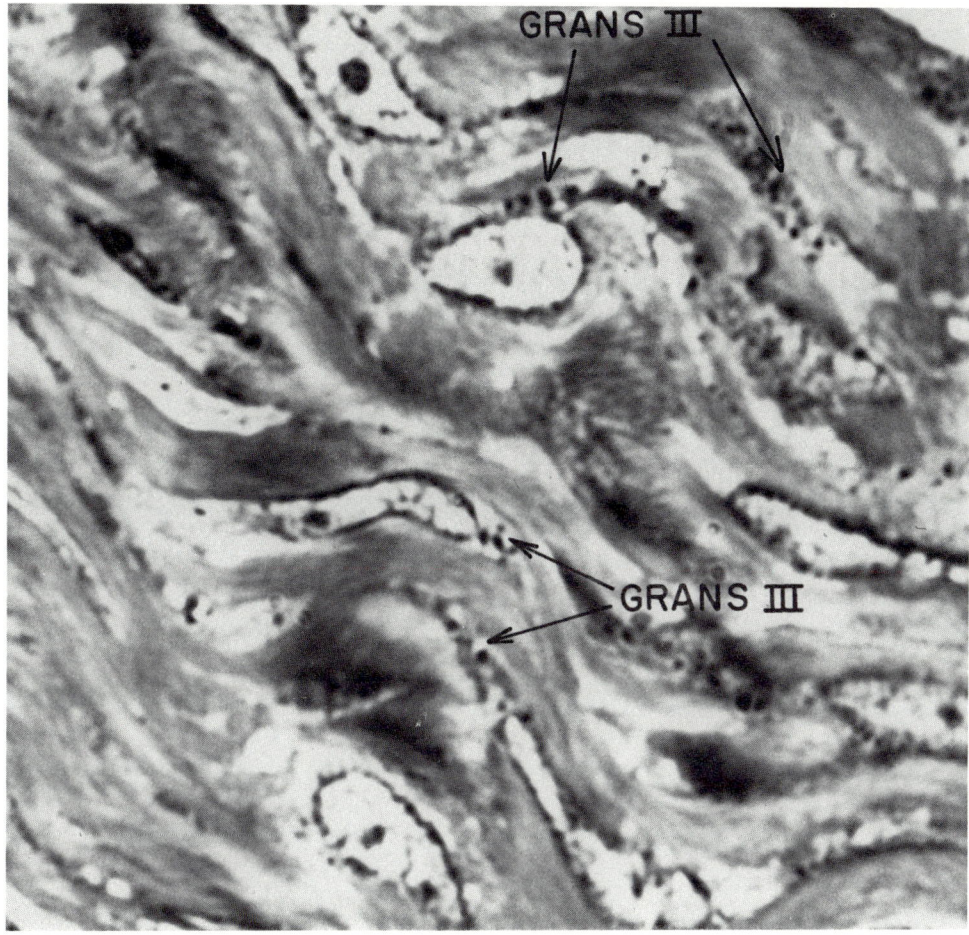

Fig. 4.21.

acting as a kind of template. Fibrils have not been observed in the cytoplasm. This also is contrary to the reported descriptions of fibrils below the surface of cells or in vacuoles, which have led their authors to suggest that fibrils are formed alone or in conjunction with some ground substance component at these sites and then extruded extracellularly. The voluminous literature on the origin of collagen fibers has been adequately reviewed in recent years *(1, 5, 9, 15, 16, 17)*.

In a series of reports *(2, 4)*, followed with two elegant reviews *(8, 14)*, Prockop and his co-workers suggested that the formation of a complex between collagen and a mucopolysaccharide is not required for the formation of extracellular matrix, that the collagen molecules pass directly from the cytoplasm to the extracellular matrix, and that the collagen is not packaged in any membranous structure (such as the Golgi apparatus) prior to secretion.

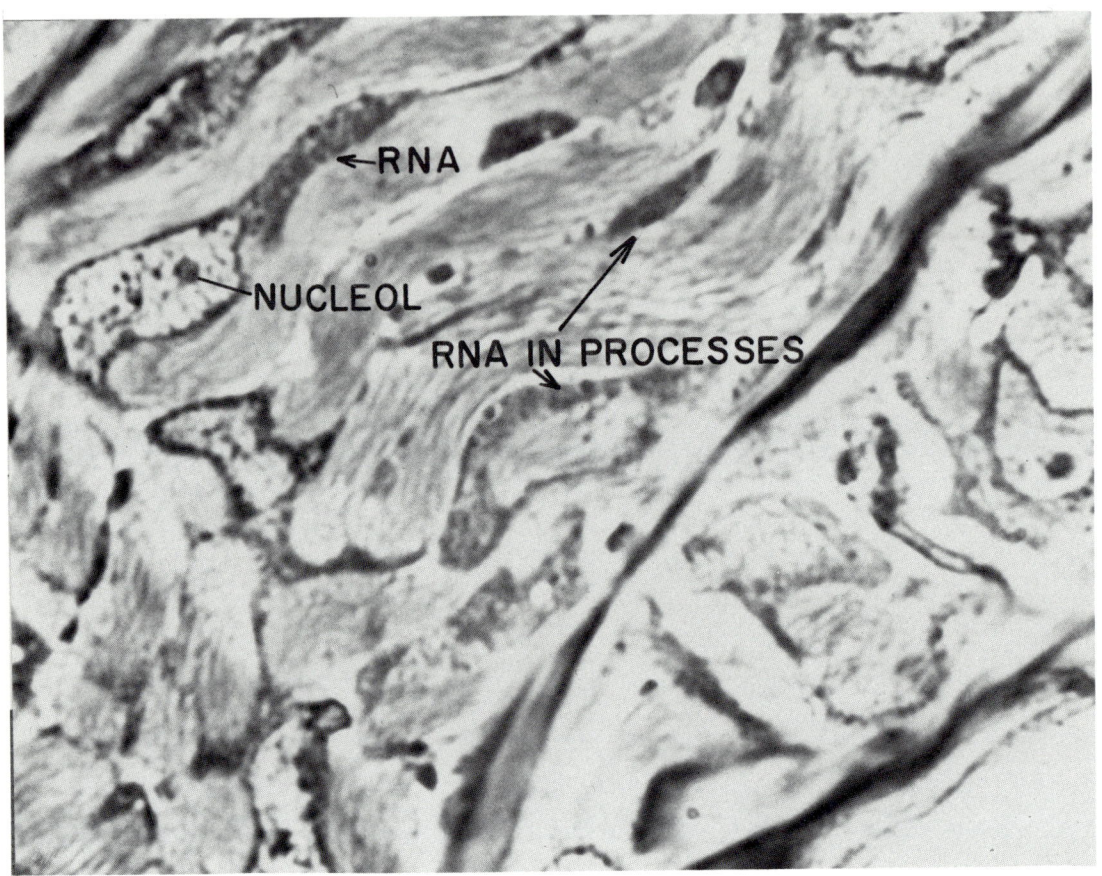

Fig. 4.22.

Figs. 4.22–4.24. Photomicrographs of sections of specimens prepared from tail tendon of 2- and 4-day-old rats as for Figs. 4.20 and 4.21, to show basophilic cytoplasm in cell body close to the nucleus and extending out into cell processes. Many cell processes contain no notable RNA, as observed with the electron microscope, and are invisible in this preparation. Abbreviations as given earlier. Figure 4.22, 2-day-old rat, ×3200; Figs. 4.23 and 4.24, 4-day-old rat, ×2720.

Discussion

83

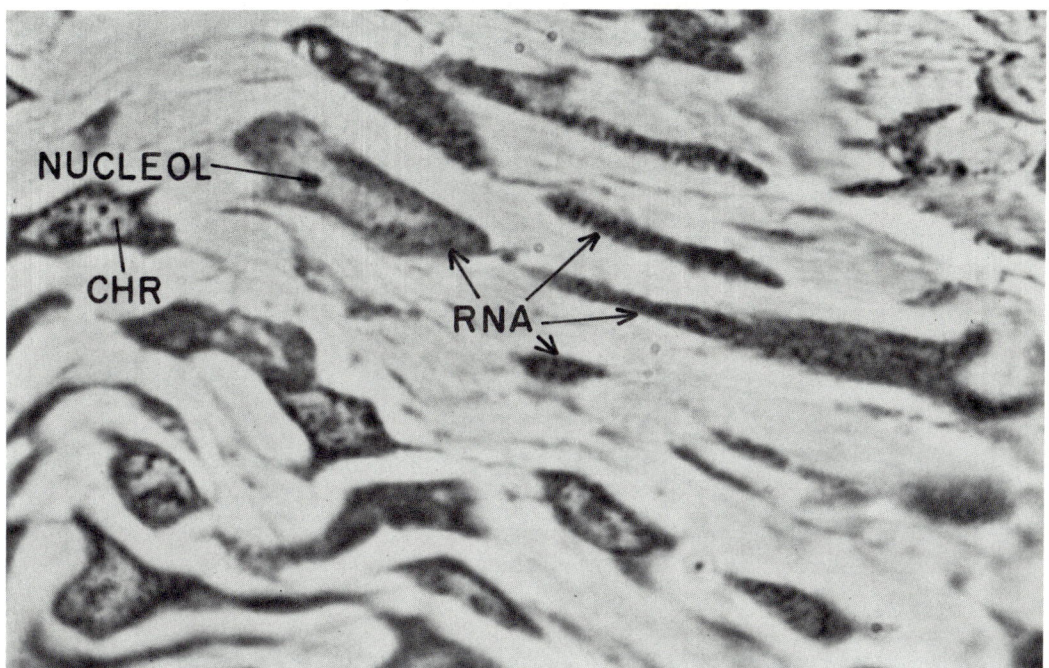

Fig. 4.23.

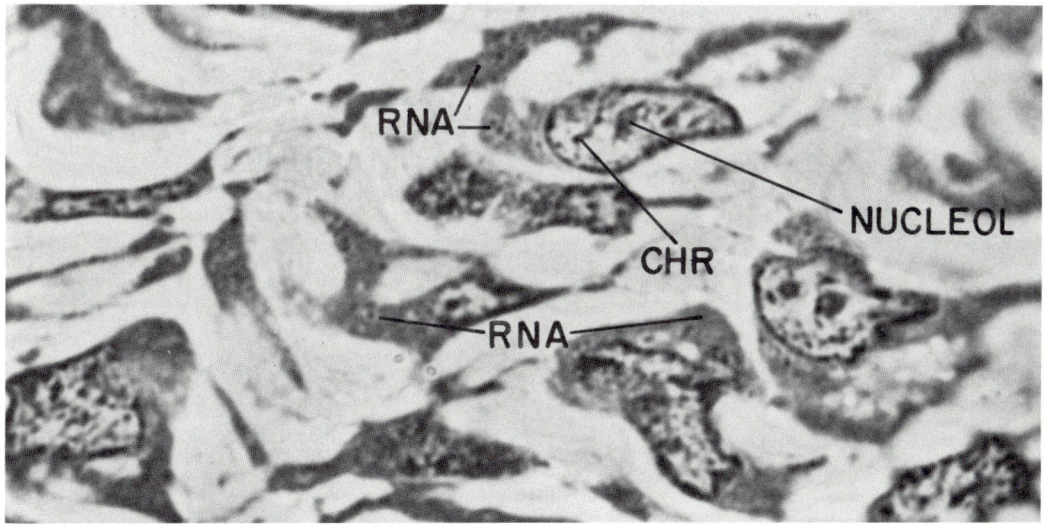

Fig. 4.24.

Nuclear and cytoplasmic compartments are emphasized by staining ultrathin sections with uranyl acetate. These structures are the same as those visible in other cells in frozen-dried specimens postfixed in 95% alcohol or in alcoholic platinum tetrabromide solutions (7). They are similar to those described in Vol. I (Chapters 2 and 5) in other cells, and like them, contain protein which happens to be unstained or very lightly stained by all the reagents mentioned in this chapter. The protein is stained after postfixation and staining with vapors of Hg hfac and FFDNB followed by section staining with uranyl acetate (see this volume, Chapter 6, Fig. 6.10A). This evidence supports the belief that pseudovacuoles are real. A more extensive discussion of the reality of pseudovacuoles is given in the first two chapters of this work.

Summary

Vapor reagents were used *in vacuo* to postfix and stain frozen-dried rat tail tendon by cross-linking reactive groups of proteins and their smaller units. These are difluorodinitrobenzene, difluorodinitrodiphenylsulfone, and mercury(II) hexafluoroacetylacetonate. Cytoplasmic granules of microscopic dimensions are stained. They have two components: (1) a homogeneous matrix of low contrast which encloses (2) second-order granules of high contrast which are about 1000 Å in diameter. The latter are clusters of smaller first-order units about 80 Å in diameter, each of which consists of a dense material enclosing a less dense core. The granules always occur in the cytoplasm and are never seen in the intracisternal space. They occur in the main part of the cell and in the cell processes and may be extruded into the extracellular space.

References

1. Anderson, C. E., and Parker, J. (1968). Electron microscopy of the epiphyseal cartilage plate. A critical review of electron microscopy observations on enchondral ossification. *Clin. Orthop.* **58**, 225–241.
2. Bhatmagar, R. S., and Prockop, D. J. (1966). Dissociation of the synthesis of sulphated mucopolysaccharides and the synthesis of collagen in embryonic cartilage. *Biochim. Biophys. Acta* **130**, 383–392.
3. Bondareff, W. (1957). Submicroscopic morphology of connective tissue ground substance with particular regard to fibrillogenesis and aging. *Gerontologia* **1**, 222–233.
4. Cooper, G. W., and Prockop, D. J. (1968). Intracellular accumulation of protocollagen and extrusion of collagen by embryonic cartilage cells. *J. Cell Biol.* **38**, 523–537.
5. Fitton Jackson, S. (1964). Connective tissue cells. *In* "The Cell" (J. Brachet and A. E. Mirsky, eds.), Vol. 6, pp. 387–520. Academic Press, New York.
6. Gersh, I., and Catchpole, H. R. (1949). The organization of ground substance and basement membrane and its significance in tissue injury, disease, and growth. *Amer. J. Anat.* **85**, 457–521.
7. Gersh, I., Isenberg, I., Bondareff, W., and Stephenson, J. L. (1957). Submicroscopic structure of frozen-dried liver specifically stained for electron microscopy. II. Biological. *Anat. Rec.* **128**, 149–169.
8. Grant, M. E., and Prockop, D. J. 1972 The biosynthesis of collagen. *New Engl. J. Med.* **286**, I, 194–199; II, 242–249; III, 291–300.

References

9. Godman, G. C., and Porter, K. R. (1960). Chondrogenesis, studied with the electron microscope. *J. Biophys. Biochem. Cytol.* **8**, 719–760.
10. Goldberg, B., and Green, H. (1964). An analysis of collagen secretion by established mouse fibroblast lines. *J. Cell Biol.* **22**, 227–258.
11. Heller-Steinberg, M. (1951). Ground substance, bone salts and cellular activity in bone formation and destruction. *Amer. J. Anat.* **89**, 347–379.
12. Hirschman, A., and McCabe, D. M. (1969). Staining of intracellular granules in fresh epiphyseal cartilage by cationic dyes. *Calcif. Tissue Res.* **4**, 260–268.
13. Movat, H. Z., and Fernando, N. V. P. (1962). The fine structure of connective tissue. I. The fibroblast. *Exp. Mol. Pathol.* **1**, 509–534.
14. Prockop, D. J. (1969). The intracellular biosynthesis of collagen. *Arch. Internal Med.* **124**, 563–570.
15. Revel, J.-P., and Hay, E. D. (1963). An autoradiographic and electron microscopic study of collagen synthesis in differentiating cartilage. *Z. Zellforsch. Microsk. Anat.* **61**, 110–144.
16. Ross, R. R, and Benditt, E. P. (1961). Wound healing and collagen formation. I. Sequential changes in components of guinea pig skin wounds observed in the electron microscope. *J. Biophys. Biochem. Cytol.* **6**, 677–700.
17. Wassermann, F. (1956). The intercellular components of connective tissue: origin, structure and interrelationship of fibers and ground substance. *Ergeb. Anat. Entwicklungsgesch.* **35**, 240–333.

5

Nucleic Acids in Chondrocytes of Epiphyseal Plate of the Rat Tibia

Isidore Gersh

The epiphyseal plate affords an opportunity to study the growth, functioning, and senescence of a differentiated cell, the chondrocyte. Changes in nucleic acids are reflected intimately by these activities. A special advantage of using epiphyseal plate for such studies is the fact that the cells are arranged in order of age and state of activity. Furthermore, there is, built into the extracellular matrix, a record of periods of accelerated growth.

This report recounts the findings with the electron and light microscopes of the distribution of DNA and of soluble and insoluble RNAs in chondrocytes. Some general comments on methods are presented here. More specific details will be given in each of the two parts of this chapter.

These findings became possible with the development of methods for the preservation and staining of nucleic acids for the electron and light microscopes (Vol. I, Chapter 3). The methods consist in freezing and drying of thin slivers of epiphyseal plate, followed by postfixation *in vacuo* with vapors of cross-linking reagents. Nucleic acids remain insoluble after this procedure and are stained with aqueous gallocyanin–chromalum (GCA). Ultrathin sections are examined with the electron microscope, where the chief structures having appreciable density are the nucleic acids. Thicker sections (about $1/3$ μ) are mounted on glass coverslips and restained with toluidine blue for viewing with the light microscope. Quantitative studies were unlikely to be profitable with the electron mcroscope, but were shown to be practical to a limited extent with the light microscope.

Female rats, 3 weeks old, were bled by thoracotomy. The hind leg was removed, and the tibia was freed of attached muscles. In a moist box, while viewing with a dissecting microscope, the epiphyseal plate was separated by two parallel cuts with a fine razor blade. The plate was placed on its broad surface and sectioned transversely 0.1–0.2 mm thick. These slices were placed on thin aluminum foil and frozen ultrarapidly to prevent the formation of detectable ice crystals. The specimens were dried at about −40°C or lower. Some specimens were infiltrated *in vacuo* with 95% alcohol at about −40°C, and embedded in water-soluble Durcupan. Others were postfixed *in vacuo* with vapors of difluorodinitrobenzene (FFDNB), or with a mixture of vapors of FFDNB and difluorodinitrodiphenylsulfone (FFSulfone). The use of the second compound as a cross-linking reagent has been described earlier in this volume (Chapter 4). In specimens, postfixed in alcohol alone, soluble RNA is removed when the specimens are immersed for staining in aqueous gallocyanin–chromalum; hence, only insoluble nucleic acids were stained. In specimens postfixed with a cross-linking agent, insoluble as well as some soluble RNAs are stained.

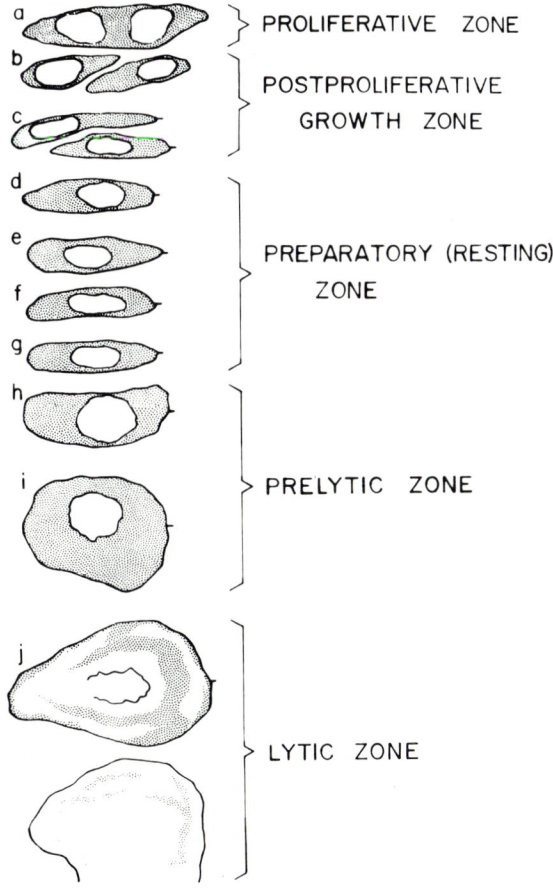

Fig. 5.1. Diagram of chondrocytes in epiphyseal plate of rat tibia and their classification. See p. 88 for discussion.

The material in this chapter is presented in two parts. Part I is concerned with the submicroscopic distribution of nucleic acids, while part II deals with microscopic aspects of the nucleic acids.

As all observations were organized in relation to the growth of the epiphyseal plate, a concept of growth of this structure is presented in the form of a diagram of parts of a cell column in the distal part of the proximal tibial epiphyseal plate (Fig. 5.1), with supporting data given on p. 117. Growth of the epiphyseal plate is considered to take place when the distance between the midpoints of adjacent cells (the half-cell distance) is increased notably, i.e., when there is a net increase in the amount of matrix deposited in the transverse walls between longitudinal cell columns. By this criterion, growth of the epiphyseal plate is notable in the postproliferative zone and is especially prominent in the prelytic and lytic zones. In between, is a relatively quiescent zone, the preparatory (resting) zone. Growth in length of the epiphyseal plate is even less significant in the proliferative zone. These relations may be seen by comparing the separations of the midpoints of the cells, indicated by the short lines which project to the right of the cells (Fig. 5.1). This characterization of the major growth zones of the epiphyseal plate is important for the consideration of secretion of extracellular matrix. The protein components of extracellular matrix are chiefly collagen and protein–polysaccharide complex. One presumes that the nucleic acids of the chondrocytes would be involved in this process of protein synthesis prior to secretion. It is for this reason that the distribution of nucleic acids in chondrocytes is presented in relation to the five zones represented in the diagram. Work with radioactive isotopes supports this interpretation (13). This terminology, as used above, seems to be more pertinent than the customary one based on morphology alone: zone of cell proliferation, zone of cell maturation, zone of hypertrophy, and zone of degeneration or of provisional calcification. In order to shorten the sections on observations, some detailed descriptions are given in the figure legends only.

Part I—Submicroscopic Distribution of Nucleic Acids

Method

Freeze-dried specimens were postfixed in vapors of the cross-linking reagents or were infiltrated with cold 95% alcohol. Both were stained with gallocyanin–chromalum (GCA), dehydrated, and embedded in water-soluble Durcupan. Ultrathin sections were cut with glass or diamond knives, floated briefly on water, mounted rapidly on grids, and examined with an Hitachi HU–11A electron microscope operating at an accelerating voltage of 50 kv. Electron micrographic plates were enlarged $\times 2$ in printing.

GCA is highly selective, and, under the conditions of staining, virtually specific for nucleic acids. Further identification as DNA or RNA was achieved by use of DNase and RNase to hydrolyze the appropriate substrate prior to staining with GCA. The procedure used in, and the results of, enzyme treatment are described in Chapter 3, Vol. I.

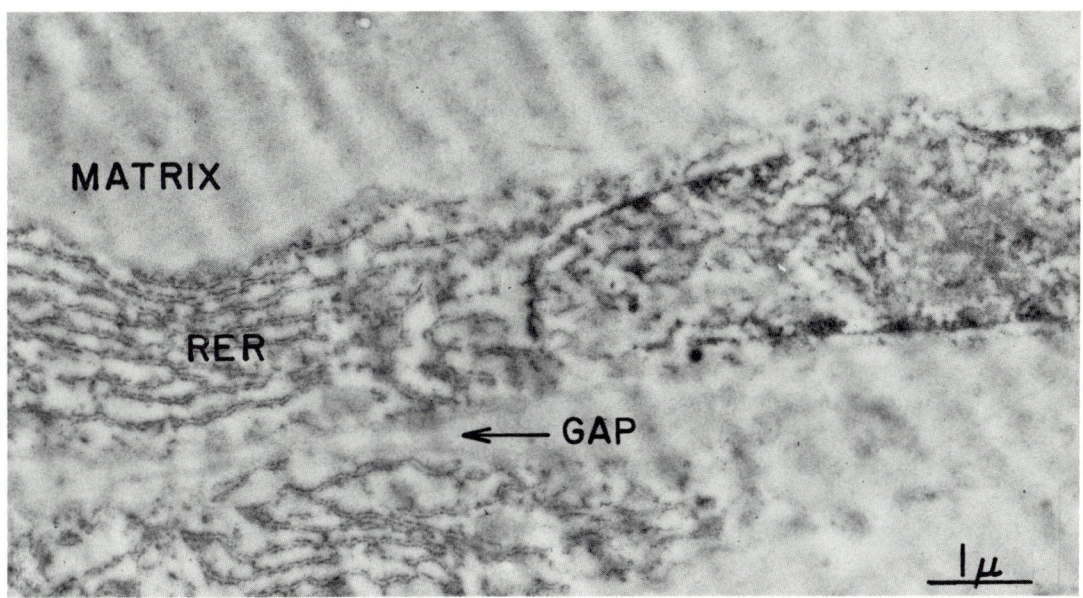

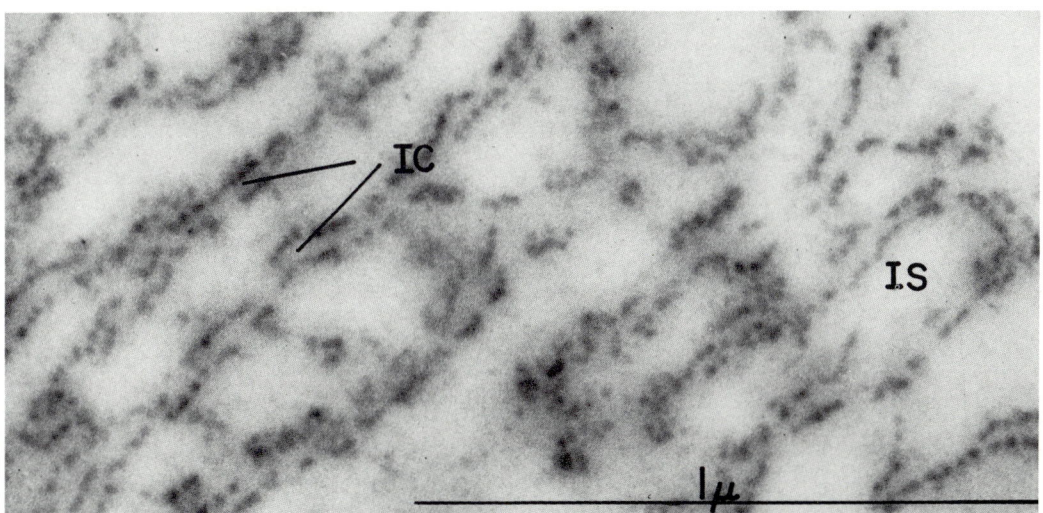

FIGS. 5.2 (TOP) and 5.3 (BOTTOM). Postproliferative cells of proximal epiphyseal cartilage of the tibia of 21-day-old rat, postfixed in 95% alcohol and stained with gallocyanin–chromalum (GCA). The interval between adjacent cells (GAP) is small. Ribosomes are discrete and are confined to the rough endoplasmic reticulum (RER). The intracisternal space (IC) is rather uniform and narrower than the intercisternal space (IS). The nuclear chromatin is aggregated chiefly around the periphery of the nucleus in small dense clumps. Figure 5.2, × 14,000; Fig. 5.3, × 82,000.

OBSERVATIONS

Proliferative Zone. Because sections were thin and orientation was very seldom favorable for definite identification (as in Fig. 5.37), not enough cells of this zone were studied for a meaningful description to be given.

Postproliferative Zone. In specimens postfixed in alcohol, the rough endoplasmic reticulum (RER) is oriented lengthwise with the cell axis (Figs. 5.2 and 5.3). The ribosomes are sharp, dense, and confined to the walls of the cisternae. The granules sometimes form clusters. The intracisternal space is always submicroscopic, while the intercisternal space ranges from the submicroscopic to the microscopic. (The term "microscopic" is used in describing structures which can be seen with the light microscope; when structures cannot be resolved with the light microscope, they are referred to as "submicroscopic.") The nucleus contains numerous irregularly clumped chromatin masses.

In specimens postfixed with cross-linking vapors and stained with GCA, the RER is not distinct, and the ribosomes are not sharp (Figs. 5.4–5.6). Their vagueness is caused by a homogeneous background density between them, so that they have less contrast than after postfixation in alcohol. In the nucleus, are numerous chromatin clumps, many of which are microscopic in dimension. Perinucleolar chromatin is prominent. Along much of the nuclear wall is a thin rim of peripheral chromatin masses, which are mostly submicroscopic in thickness, and are separated by gaps containing nucleoplasm. In all sites, the chromatin consists of figures, lines, and granules which are sections in various planes of the second-

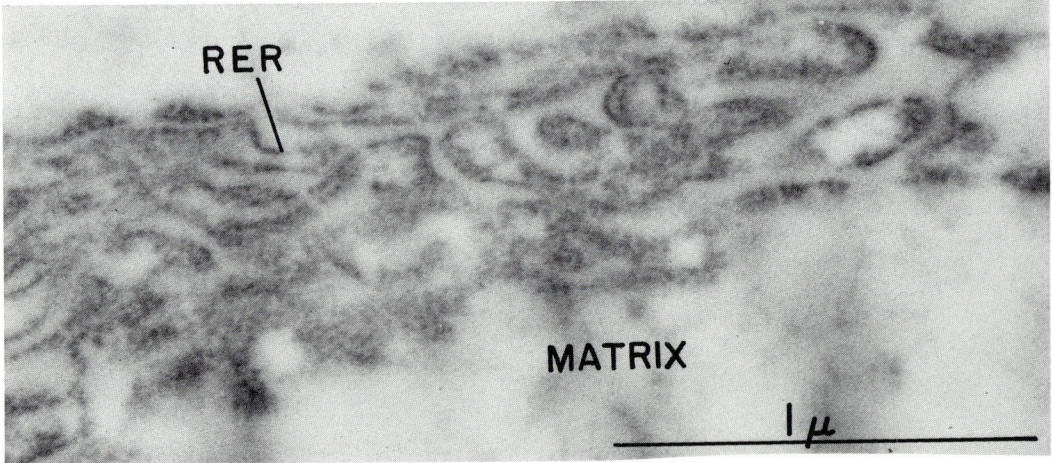

Fig. 5.4. Postproliferative cell stained *in vacuo* with vapors of difluorodinitrobenzene (FFDNB) and difluorodinitrodiphenylsulfone (FFSulfone) followed by aqueous GCA. The RER is clearly marked. At higher magnification, the ribosomes are seen embedded in a diffuse background representing soluble RNA. The consequence is that the ribosomes do not seem to be as clearly demarcated as after alcoholic postfixation followed by GCA, during which procedure soluble RNA is dissolved. × 56,000.

ary coils of DNA with a period and diameter of about 400 Å. The first-order coils of DNA (the double helix) amplified by gallocyanin–chromalum (GCA) are about 45 Å in diameter. The second-order DNA coils are closely apposed in the chromatin clumps. The relations between the first- (the double helix DNA molecule) and the second-order coil may be clarified by referring to Chapter 3, Fig. 3.1, Vol. I.

In the nucleolus, are numerous granules which rarely exceed 120 Å in diameter, embedded in a somewhat dense background. Similar granules are present in the nucleoplasm, where they are grouped in small clumps and rows. The homogeneous background density is appreciable here also.

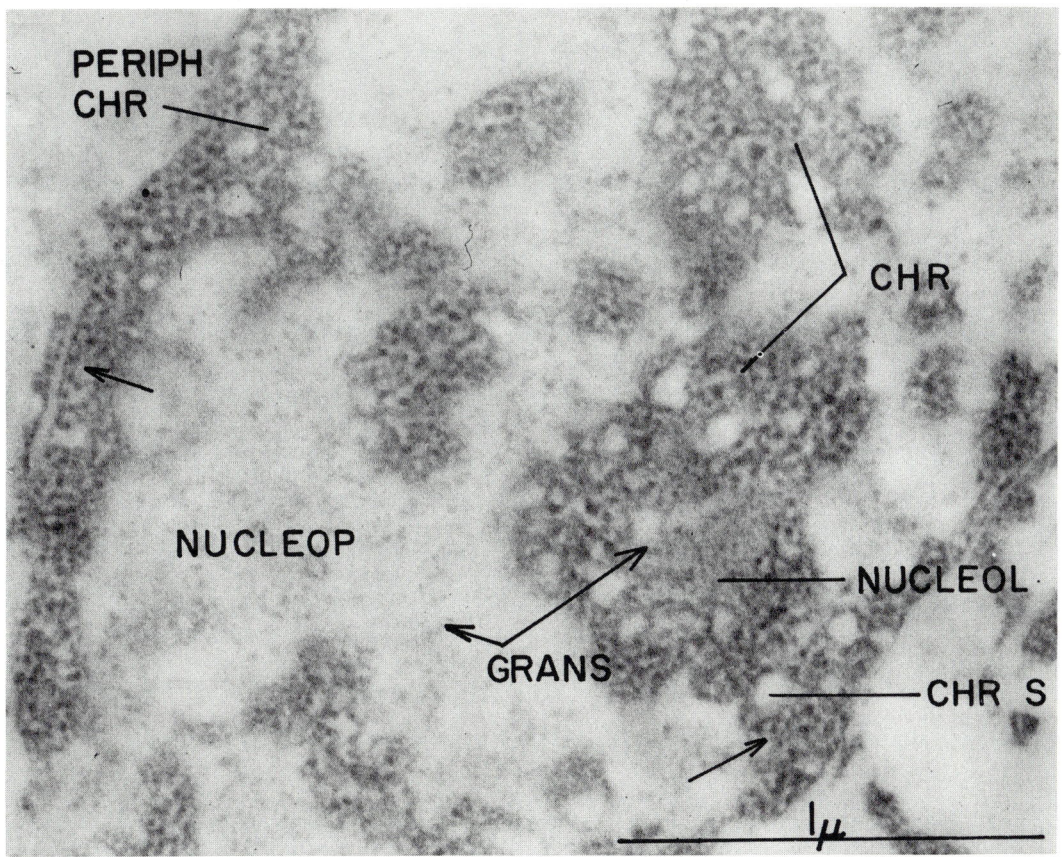

Fig. 5.5. Postproliferative cell. Postfixation and staining *in vacuo* with vapors of FFDNB, followed by aqueous GCA. The second-order DNA coils are aggregated in clumps frequently of microscopic size (PERIPH CHR and CHR). Spaces (CHR S) are present within the clumps. The nucleoplasm (NUCLEOP) contains RNA particles of various sizes (GRANS), all embedded in a moderately dense, homogeneous background. The nucleolus (NUCLEOL) also contains granules (GRANS). × 56,000.

Preparatory (Resting) Zone. After alcoholic postfixation the cells resemble those of the previous stage (Figs. 5.7 and 5.8). After postfixation with vapors of FFDNB, the RER appears somewhat irregularly disposed. The intracisternal space is rather uniform, but gives the impression of expanding as if distended, when, in fact, these are places where the curving RER is sectioned flatwise, rather than transversely (Figs. 5.9–5.13). The chromatin masses are smaller than in the preceding stage. The second-order coils are more prominent. The nucleolus contains numerous granules in a homogeneous background, and some very fine filaments are present. The nucleoplasm also contains numerous small granules in groups and lines, which are embedded in a homogeneous, somewhat dense background.

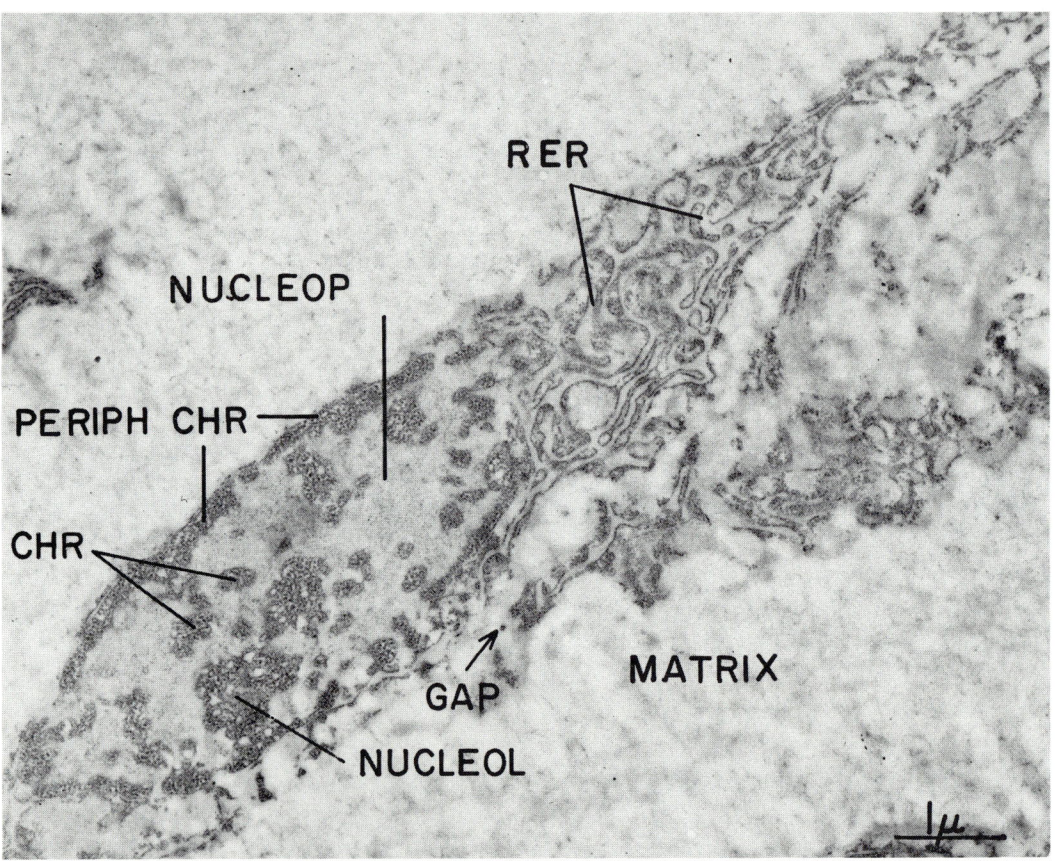

Fig. 5.6. Postproliferative cell. Postfixation and staining with vapors of FFDNB, followed by staining with GCA. Second-order DNA coils are tightly packed as clumps (PERIPH CHR and CHR), which are frequently of microscopic dimensions. The nucleolus (NUCLEOL) and nucleoplasm (NUCLEOP) contain numerous granules. The RER is well developed. Some matrix separates the two cells (GAP). × 14,000.

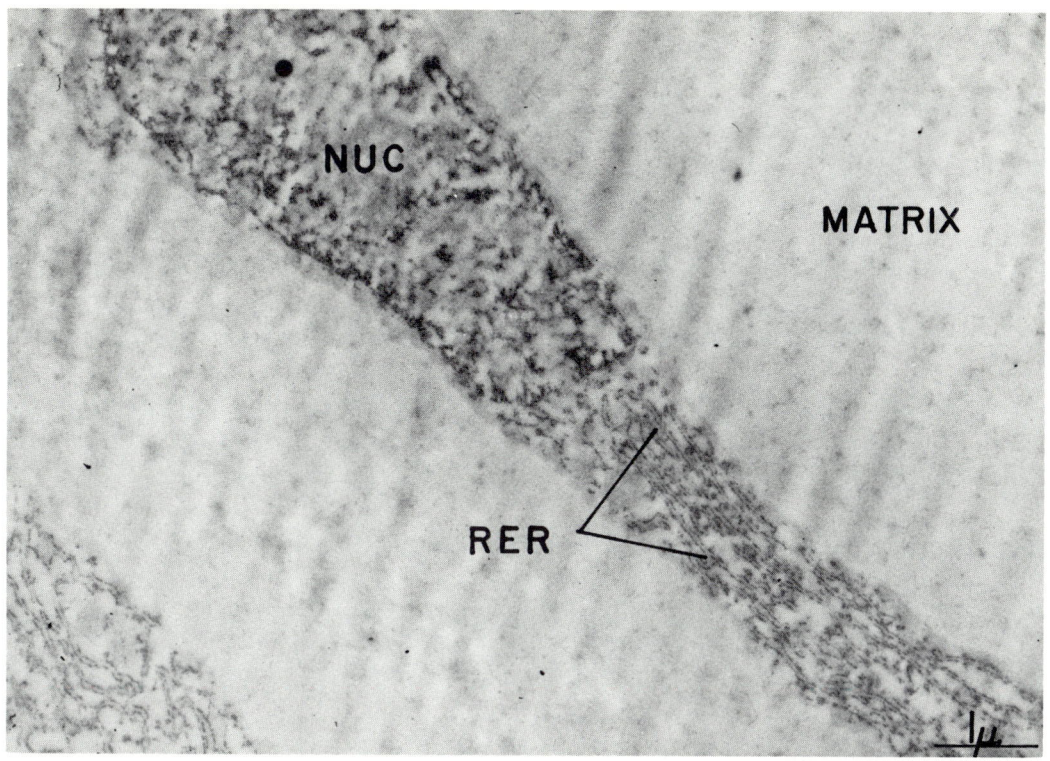

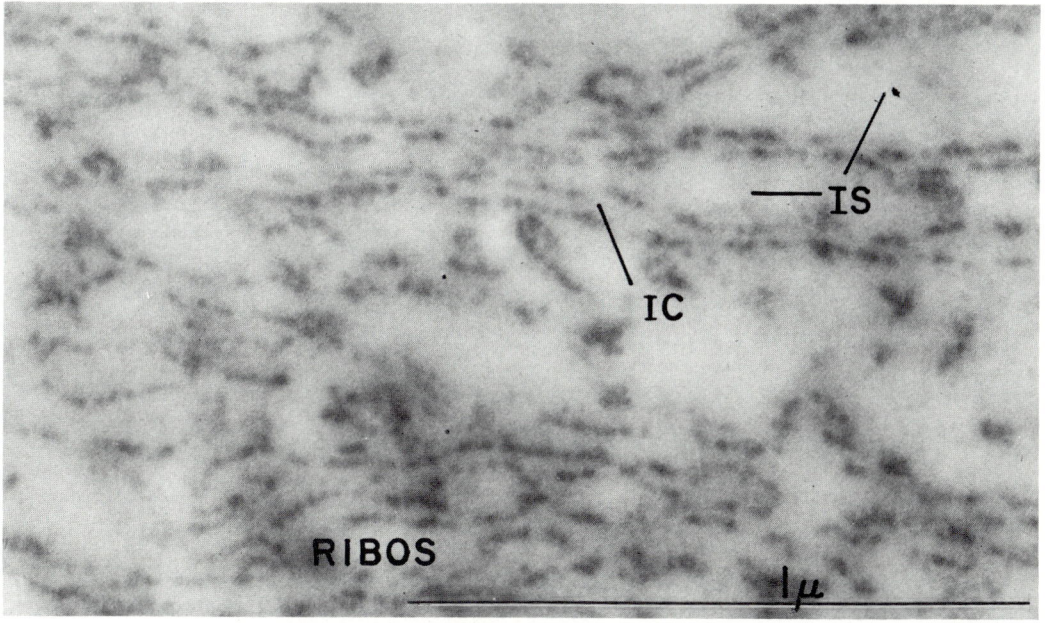

Figs. 5.7 (top) and 5.8 (bottom). Cartilage cell of preparatory zone. Postfixation in 95% alcohol, stained with GCA. The cell does not appear to differ notably from those of the postproliferative zone. The ribosomes (RIBOS), which are confined to the RER, are discrete, though aggregated. The narrow intracisternal spaces (IC) are separated by the generally wider intercisternal spaces (IS). Figure 5.7, × 14,000; Fig. 5.8, × 82,000.

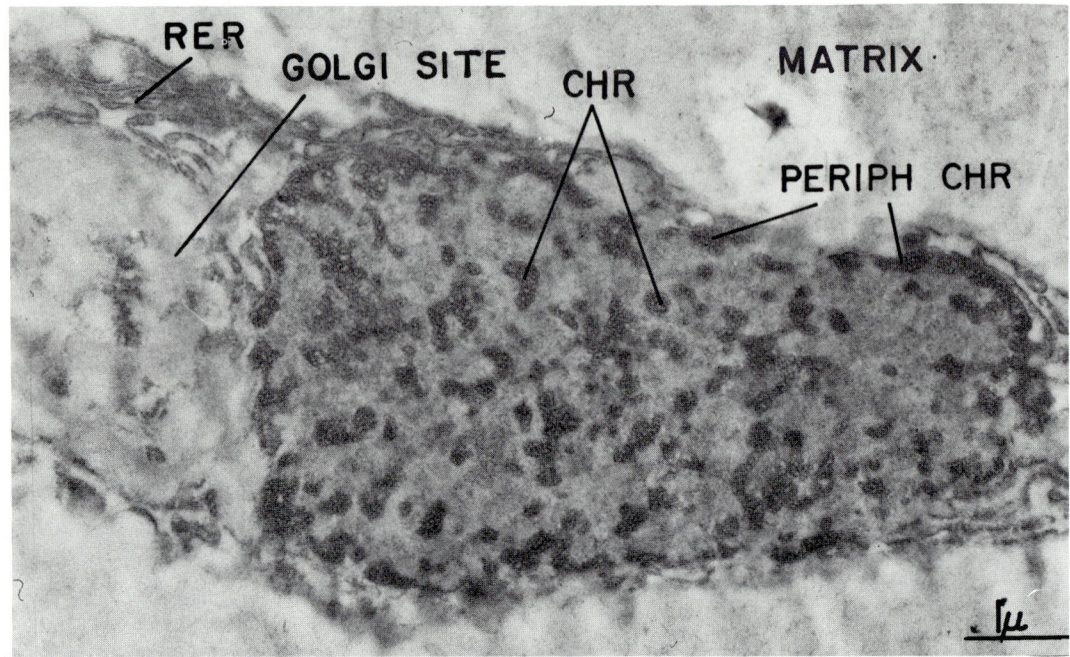

Fig. 5.9. ▲   ▼ Fig. 5.10.

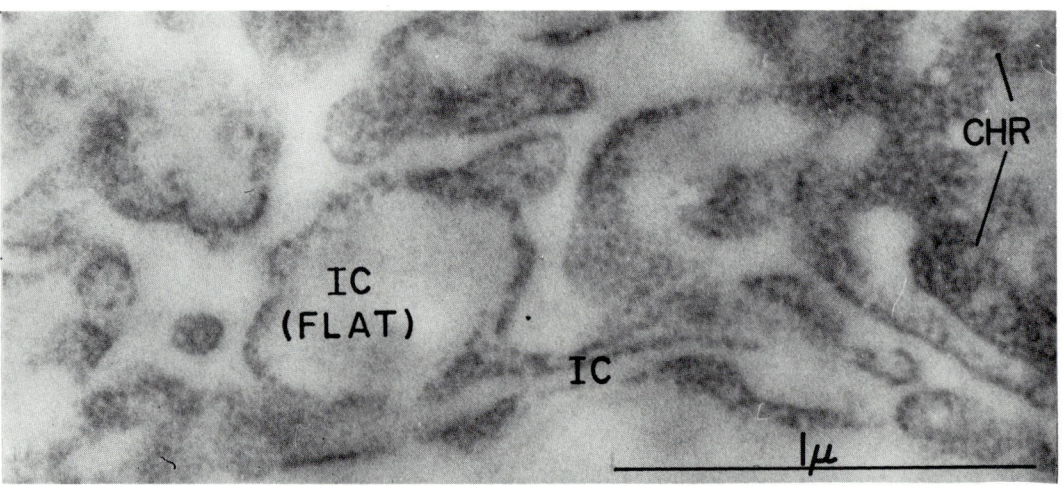

Figs. 5.9–5.11. Cells of preparatory (resting) zone. Postfixation and staining *in vacuo* with vapors of difluorodinitrobenzene, followed by GCA. In the nucleus (NUC), chromatin comprises small clumps of DNA (PERIPH CHR and CHR), separated by nucleoplasm rich in granules containing RNA. In the cytoplasm, the RNA is distributed as ribosomes which are frequently interconnected by (unresolvable) soluble RNA. The RER seems to be markedly distended in some regions, but these are fortuitous sections through the flattened cisternae [IC(FLAT)] which are tilted. The Golgi site is rather free of RNA except when it encloses interdigitating RER. Figures 5.9 and 5.11, × 14,000; Fig. 5.10, × 56,000.

Prelytic Stage. In cells postfixed in alcohol, the cisternae of the RER are widely separated (Figs. 5.14–5.16). The ribosomes vary in density and size, and are no longer more or less regularly disposed on the outer surface of the cisternae. In the nucleus, the chromatin masses are smaller and are now all submicroscopic and the DNA is in the form of granules. In cells postfixed with cross-linking vapors (Figs. 5.17–5.21), the cisternae are more widely separated than in earlier stages, but the intracisternal gap is unchanged. The ribosomes are variable: in some parts, they are clearly delineated because of the absence between them of the homogeneous background density, while, in other parts, they are partly obscured as in earlier stages. Granules of RNA are aggregated along the outer part of the wall of the RER, and these are sometimes not clearly defined because of the increased density of the homogeneous component. In the nucleus, the chromatin is finely divided and the second-order coils of DNA are clear and commonly separated from each other. Nucleoplasmic RNA granules are not numerous, and the background density is largely lacking.

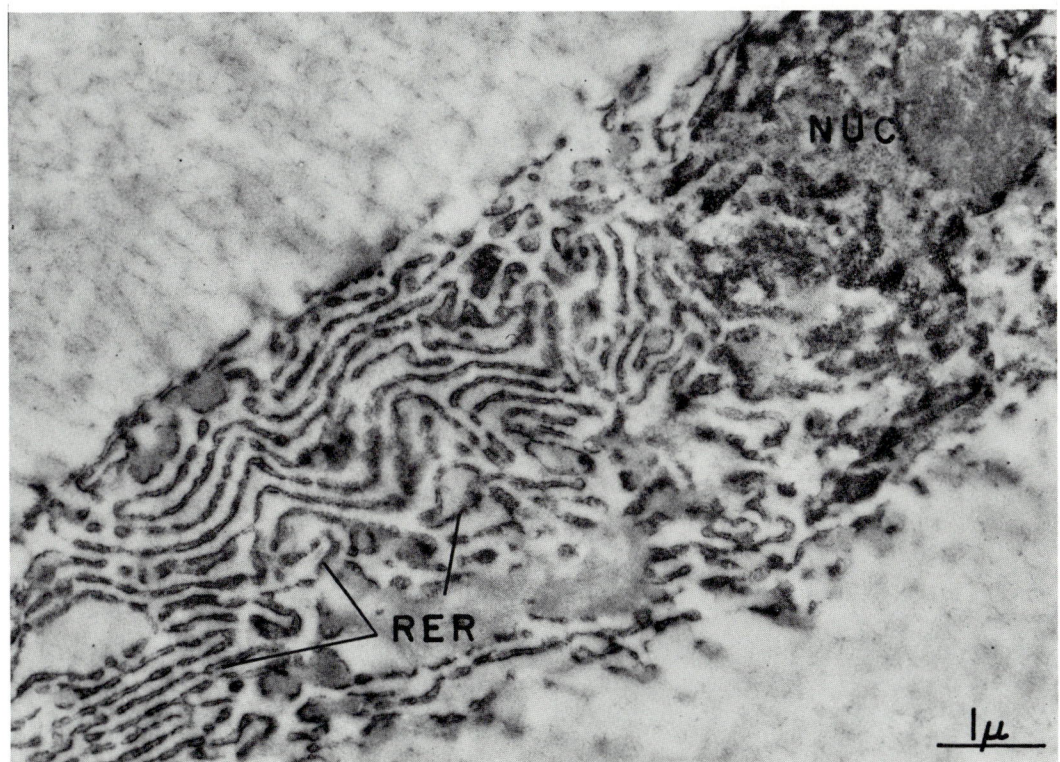

Fig. 5.11.

Lytic Stage. After alcoholic postfixation, clear spaces are the most prominent feature of the protoplasm; what remains of the RER looks like that in the previous stage (Figs. 5.22 and 5.23). After postfixation with cross-linking agents, the RER is widely dispersed and fragmented, though the intracisternal space is unaltered (Figs. 5.24–5.30). In some parts, the ribosomes are clear, while in most, they are tightly clustered in a dense homogeneous background. These clusters are smaller than in the earlier stage. In some parts, the clusters are pale. In a few regions, the RER appears foamy or "bubbly," as if frozen in the process of disintegrating through solution. The nucleus is also marked by very varied pictures, in contrast with the marked uniformity of earlier stages. There may be short seg-

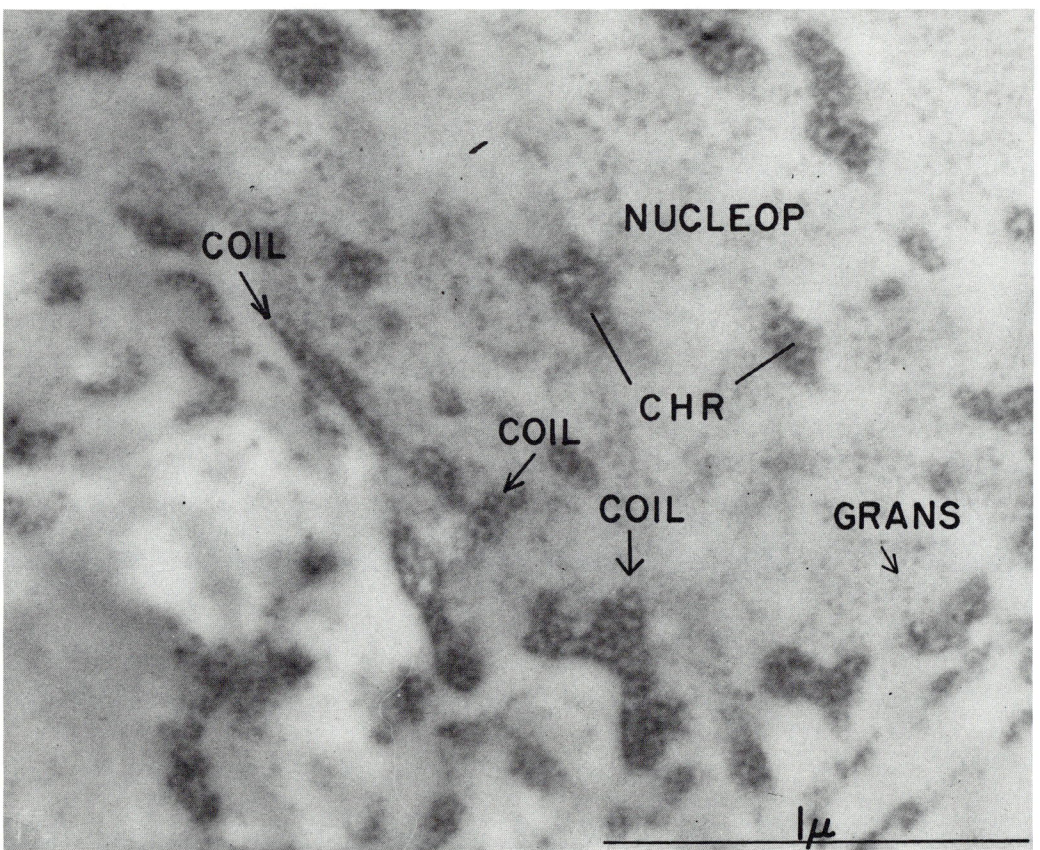

Fig. 5.12.

Figs. 5.12 and 5.13. Preparatory (resting) cell. Postfixation and staining *in vacuo* with vapors of FFDNB followed by staining with GCA. The small chromatin clumps (PERIPH CHR and CHR) are separated by nucleoplasm (NUCLEOP) rich in RNA granules (GRANS), and are arranged as second-order coils (COIL). The background density is less than in earlier stages. Some filaments (FIL) are visible in the nucleolus (NUCLEOL). × 56,000.

ments of second-order DNA coils, or isolated loops, or small or large accumulations of DNA granules of variable density and size. Neither nucleolar nor nucleoplasmic granules can be seen. The background is clear.

Identification of DNA and RNA. The identification of the nucleic acids was based on submicroscopic and microscopic morphological criteria, on cytochemical selectivity of the stain, and on the results of enzyme treatments. As for the morphological criteria, intranuclear coils of a diameter of about 400 Å were regarded as a positive though not complete identification of DNA. Dense granules in the walls of the RER were regarded as containing RNA, as were also granules in the

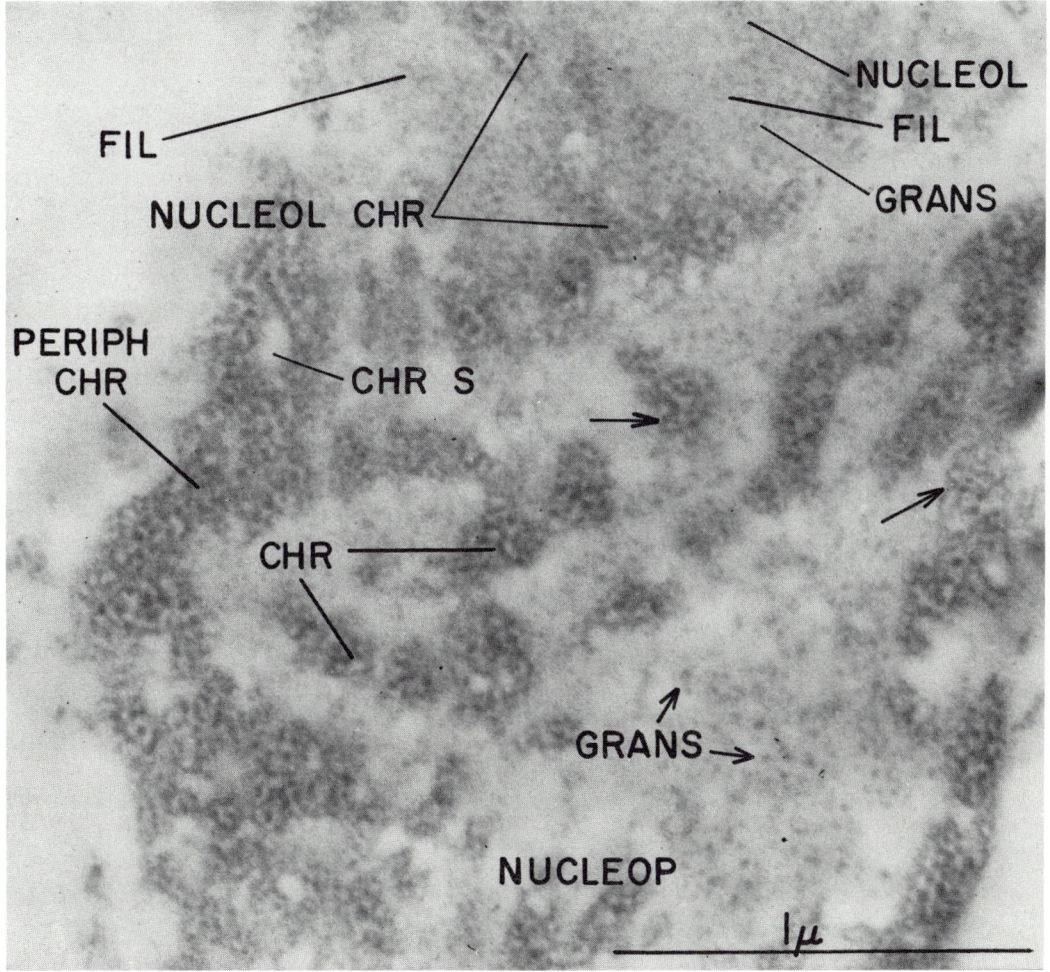

Fig. 5.13.

nucleolus and nucleoplasm. The cytochemical selectivity and virtual specificity of the GCA lake for nucleic acids has been discussed earlier (see Vol. I, Chapter 3). For enzyme treatments, specimens were prepared as described in Chapter 3 (Vol. I). They were freeze-dried and treated *in vacuo* with cross-linking vapors, as for staining with GCA. Excess vapors were withdrawn and the specimens were infiltrated with 95% alcohol at room temperature, where they remained for 1 day. They were then rehydrated during 1 hour in successive changes of alcohol (80, 70, 60, 50, and 40%). They were then treated at room temperature with nine successive changes of pancreatic DNase (Worthington, RNase-free, 1 mg/ml, water) or RNase (Worthington, 1 mg/ml water) during 150 minutes. The specimens were washed rapidly in water, and dehydrated by graded alcohol of increas-

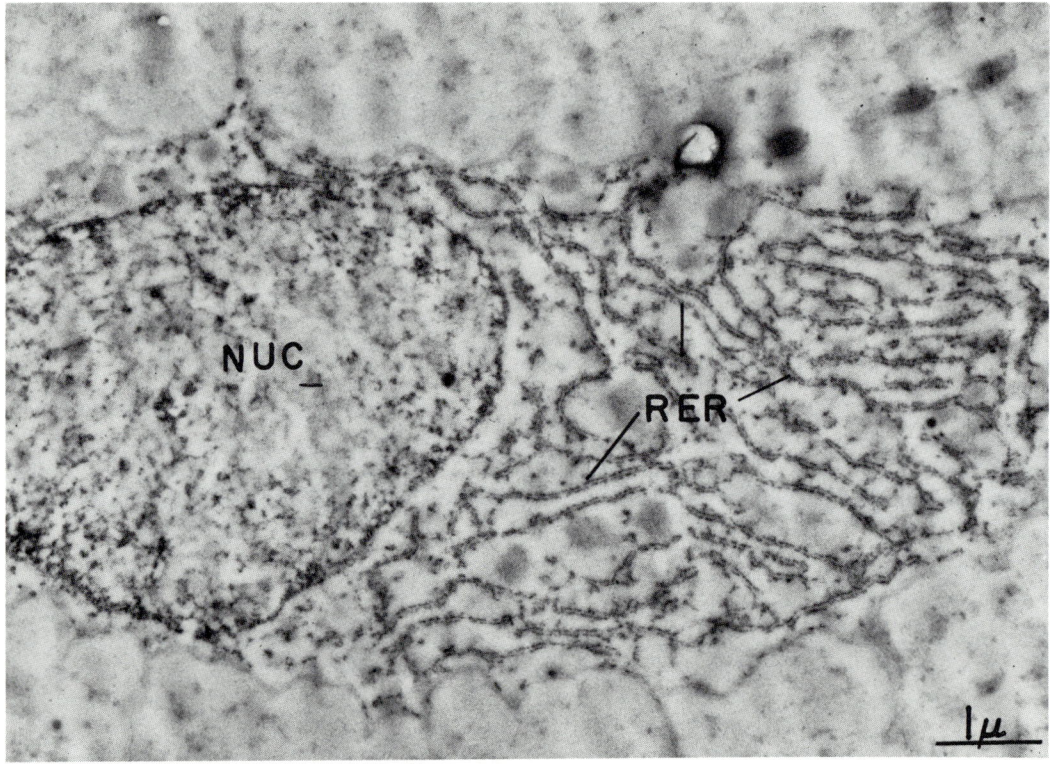

Fig. 5.14.

Figs. 5.14–5.16. Prelytic cell. Postfixation in 95% alcohol, stained with GCA. The RER is more dispersed, with wider intercisternal spaces (IS), but the intracisternal gap (IC) remains about the same as in earlier stages. Ribosomes appear fused in some sites (AGGREG) despite the absence of stained soluble RNA, and are paler in some areas (PALE RIBOS) even in the same field. The DNA in the dispersed nuclear (NUC) chromatin appears granular (DNA GRANS). Nucleoplasmic RNA granules (GRANS) are less common than in earlier stages. Figure 5.14, × 14,000; Fig. 5.15, × 82,000; Fig. 5.16, × 56,000.

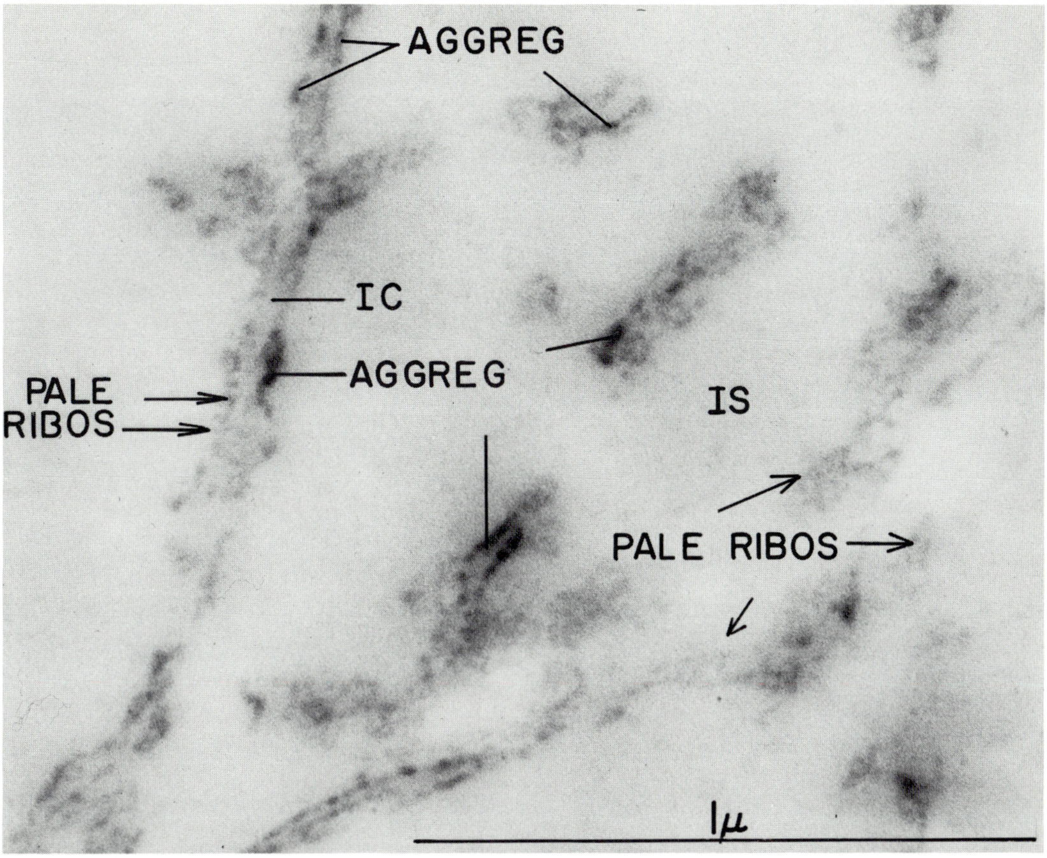

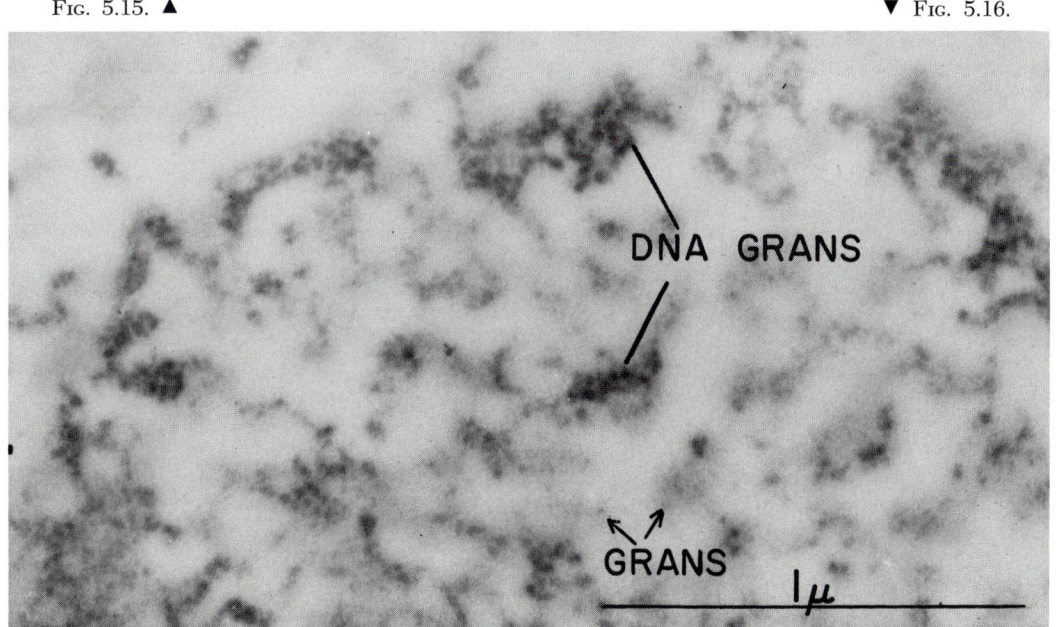

Fig. 5.15. ▲

▼ Fig. 5.16.

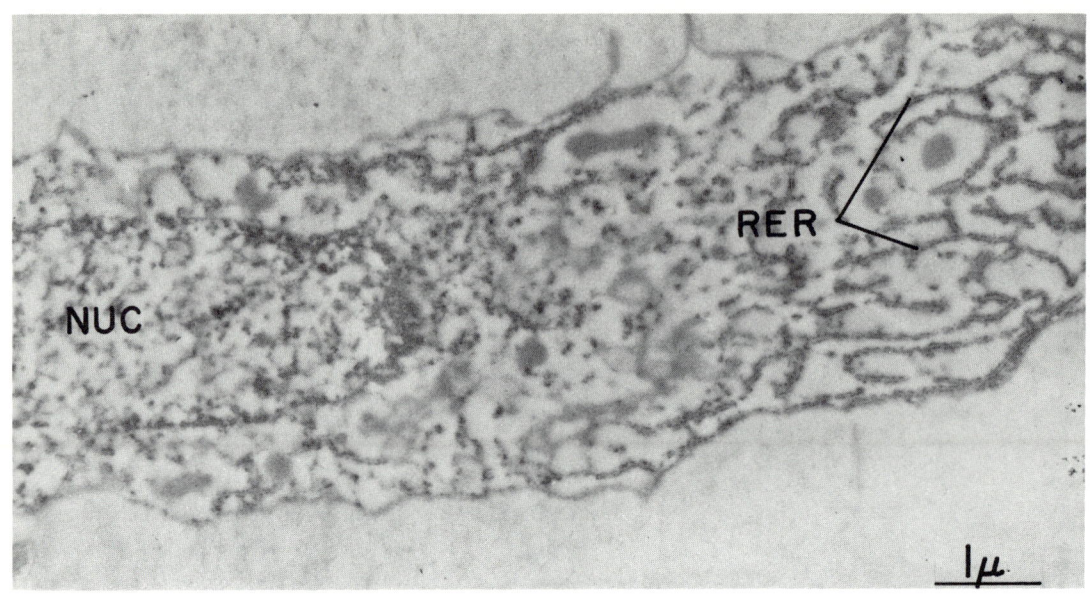

Fig. 5.17. ▲

▼ Fig. 5.18.

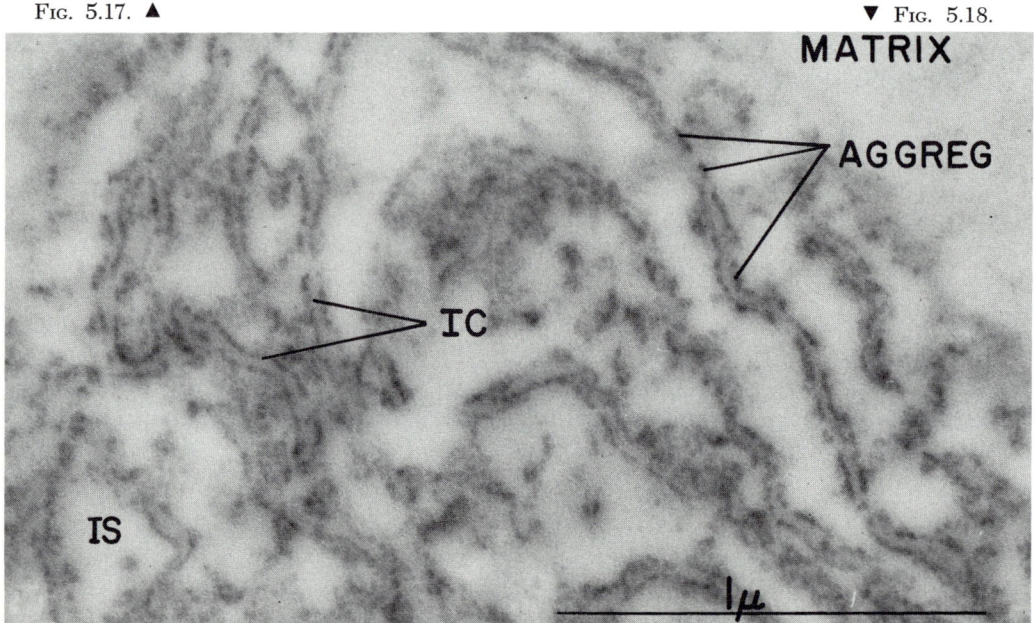

Figs. 5.17–5.21. Prelytic cell. Postfixation and staining *in vacuo* with vapors of FFDNB and FFSulfone followed by staining with GCA. Cell in Figs. 5.17–5.19 is an earlier stage than that in Figs. 5.20 and 5.21. The RER is widely dispersed through enlargement of the intercisternal spaces (IS), but the intracisternal gap (IC) remains about the same as in earlier stages. Ribosomes are aggregated (AGGREG) and are paler in some regions (PALE RIBOS) than in others, even in the same field. In some parts, the ribosomes are discrete, while in others the homogeneous background obscures individual ribosomes. Nuclear DNA is subdivided into bodies too small to be seen with the light microscope, except for the Barr body (BARR), and distributed very largely as isolated second-order DNA coils (COIL). RNA granules in nucleolus and nucleoplasm (GRANS) are relatively few in number and seem to be distributed in chains or clusters. The background density of the nucleolus and nucleoplasm is markedly less than in earlier stages. Figures 5.17 and 5.20, × 14,000; Figs. 5.18 and 5.19, and 5.21, × 56,000.

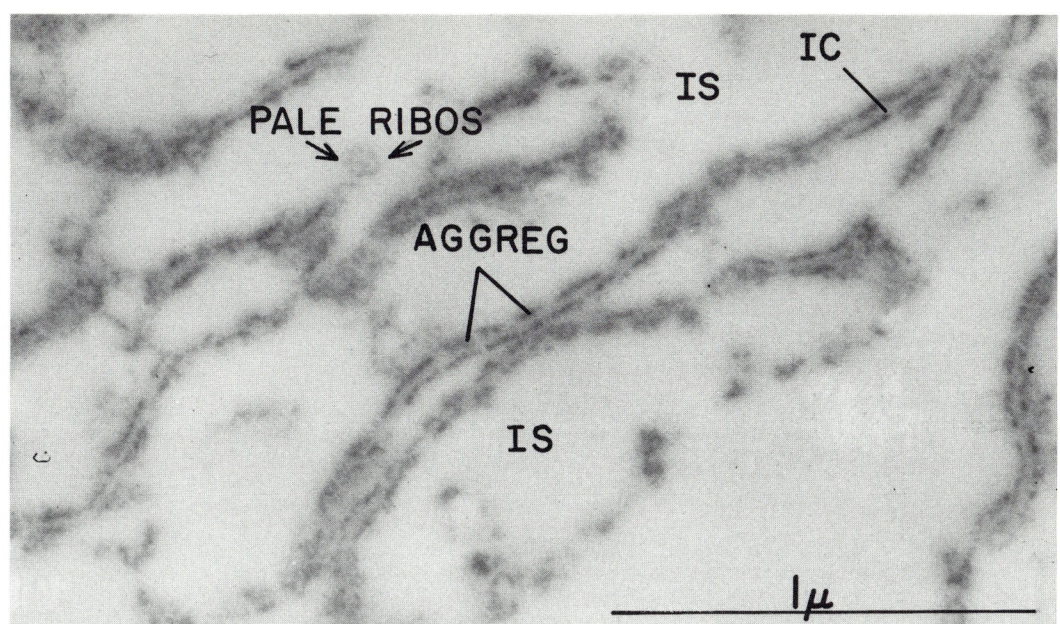

Fig. 5.19.

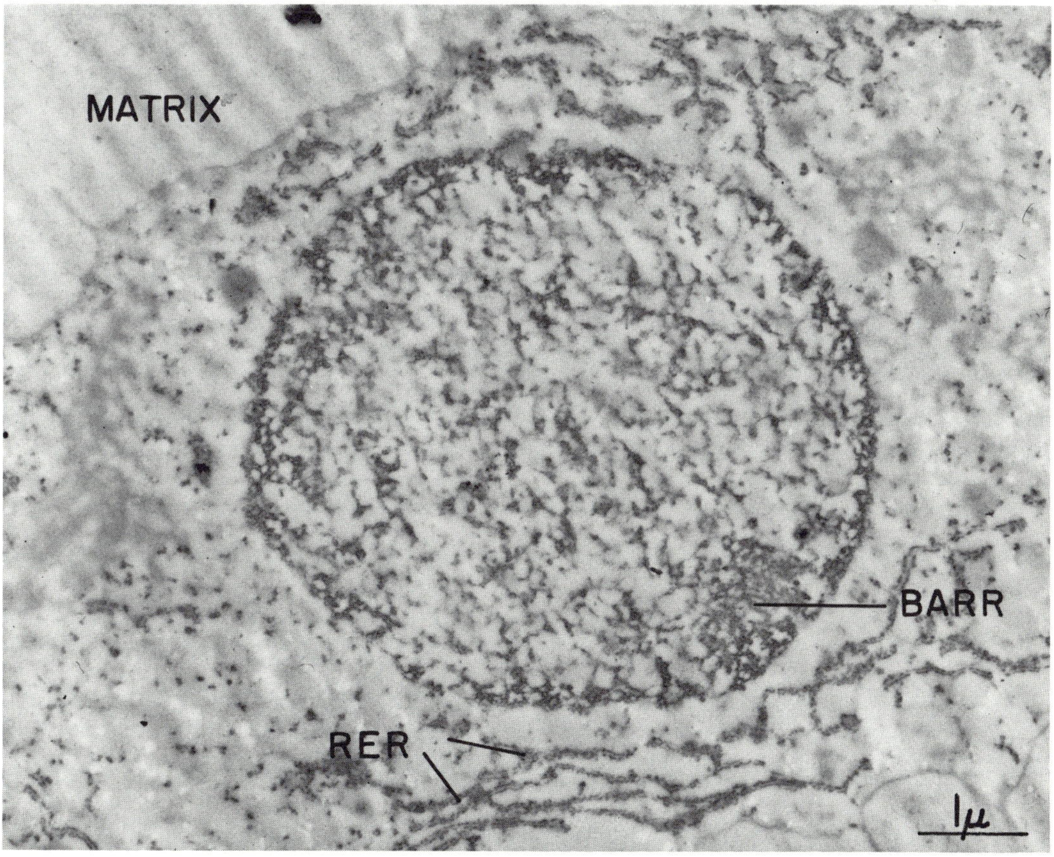

Fig. 5.20.

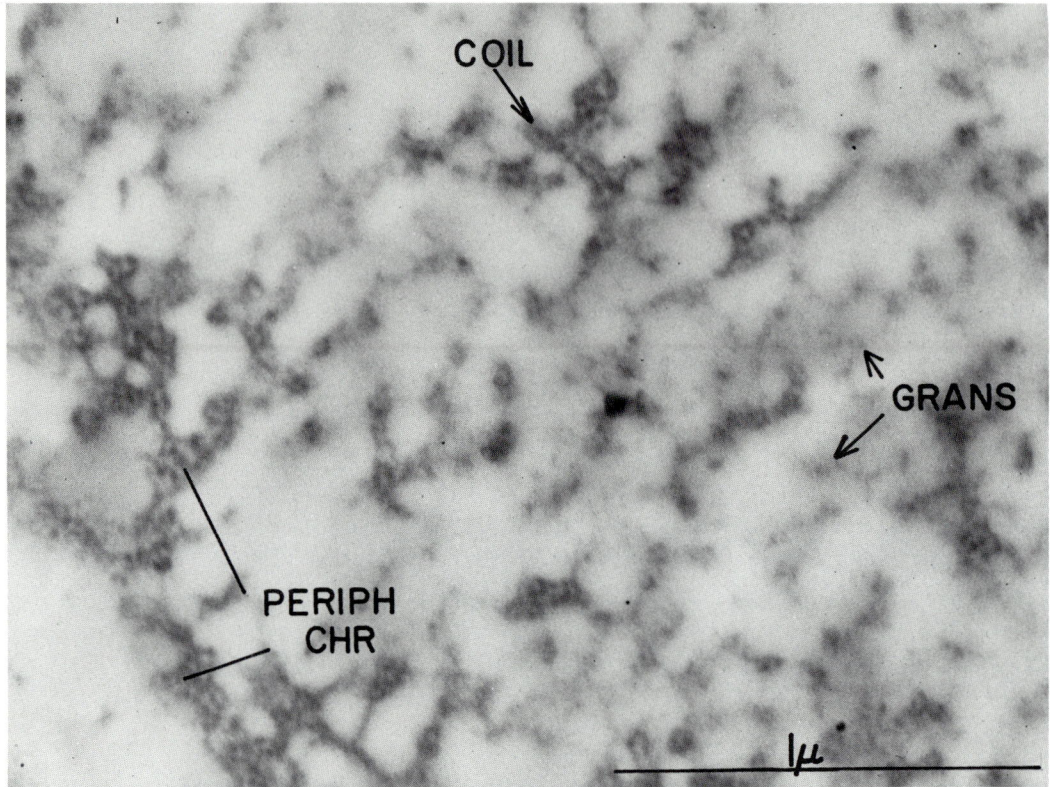

Fig. 5.21. For complete legend see p. 100.

ing concentration (40–95%). After 1 day to harden the specimen, the specimens were rehydrated and stained with gallocyanin–chromalum, embedded in water-soluble Durcupan, sectioned, and observed with the electron microscope in the usual way. After treatment with DNase, chromatin masses and coils were unidentifiable, while nucleoli with their dense granules persisted. Ribosomes were also seen attached to the RER (Figs. 5.31–5.33). By contrast, after treatment with RNase, nucleoplasmic and nucleolar granules were no longer clearly recognizable, and the ribosomes of the RER were extremely pale (Figs. 5.34–5.36). At the same time, the DNA coils in nuclei of young cells were readily recognizable, as were the granules and coil fragments in the nuclear strands of lytic chondrocytes.

Figs. 5.22 and 5.23. Lytic cells. Postfixation in 95% alcohol, stained with GCA. Nuclear DNA is very widely fragmented, and separated by wide spaces. The RER is similarly fragmented. The intracisternal gap (IC), when recognized, is of about the usual dimension. The intercisternal space (IS) is greatly enlarged. Ribosomes are very irregular in size, degree of aggregation (AGGREG), and in density (PALE and DARK RIBOS), even though soluble RNA is not preserved or stained. LYT S, lytic space. Figure 5.22, × 14,000; Fig. 5.23, × 82,000.

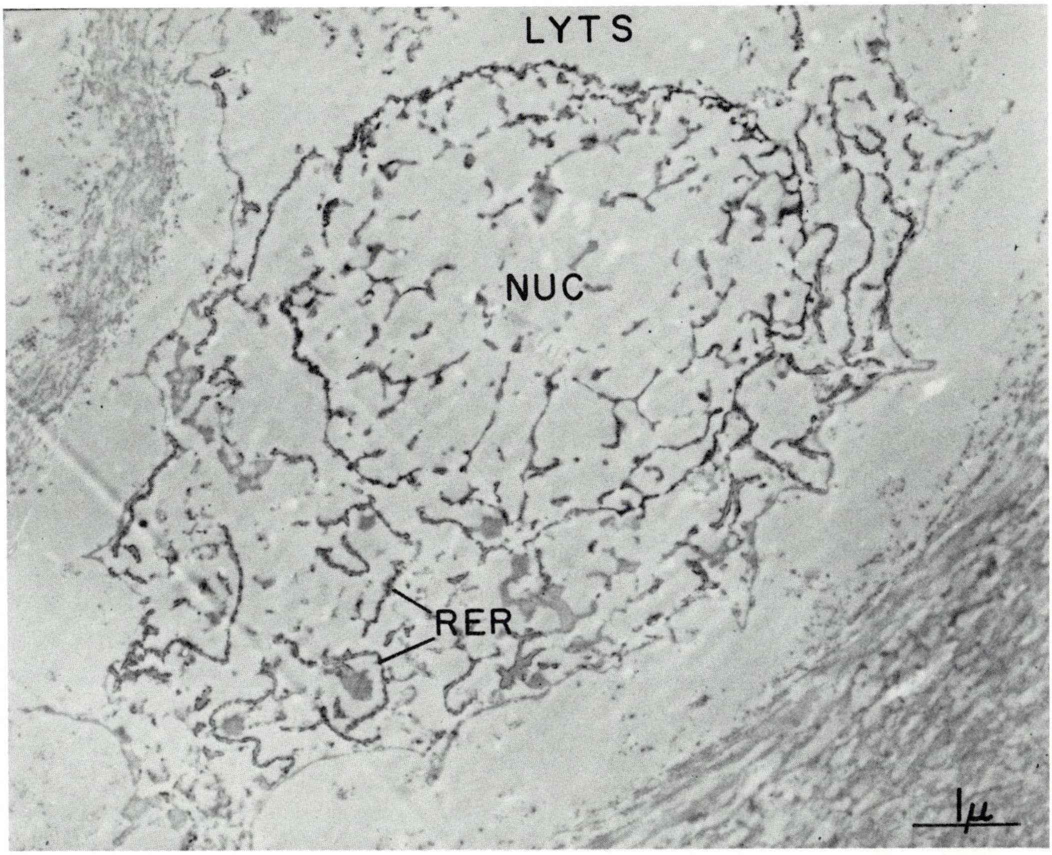

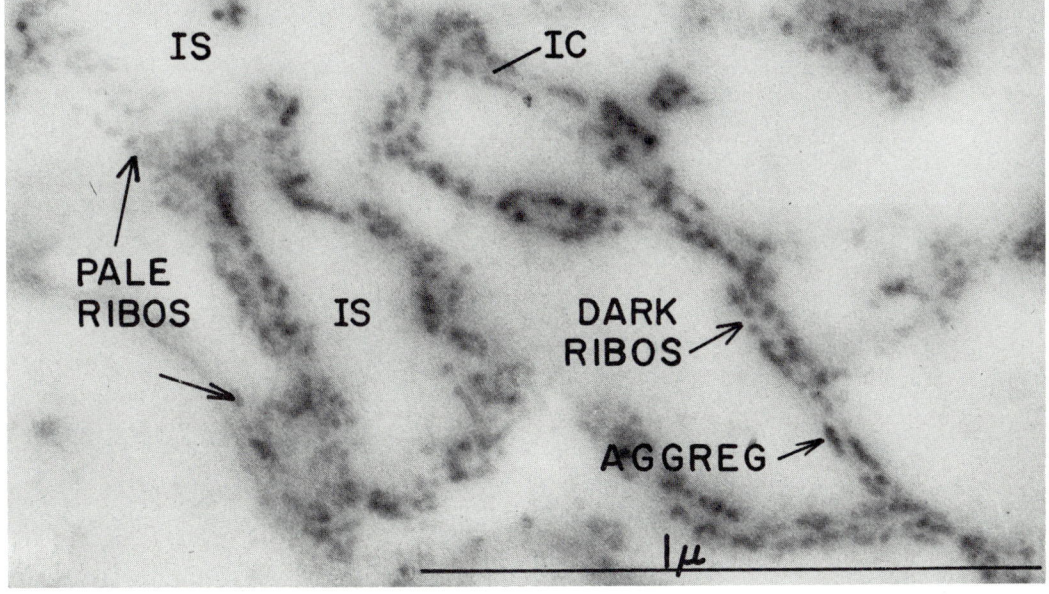

Fig. 5.22. ▲   ▼ Fig. 5.23.

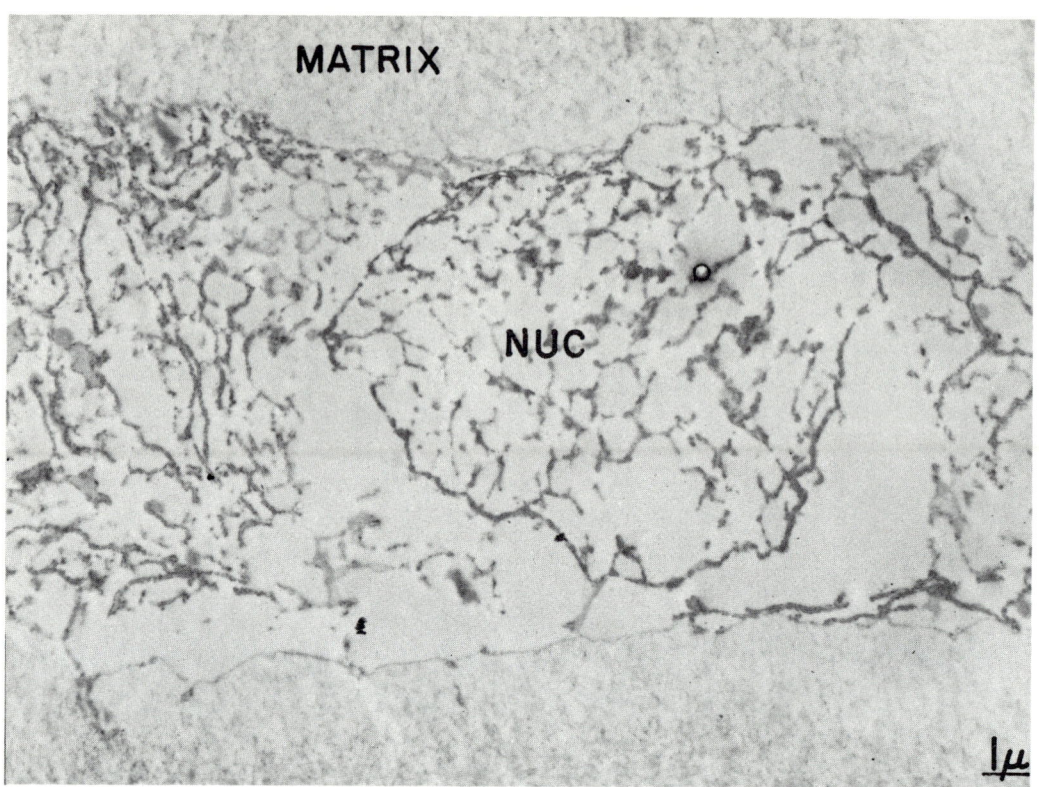

Fig. 5.24. ▲   ▼ Fig. 5.25.

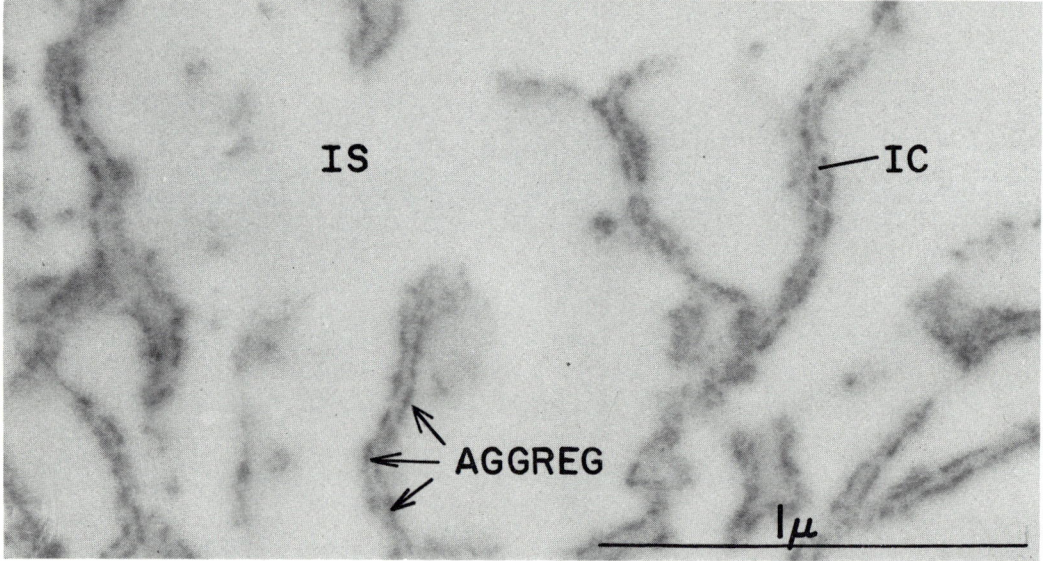

Figs. 5.24–5.26. Lytic cells. Postfixation and staining *in vacuo* with vapors of FFDNB and FFSulfone, followed by staining with GCA. Nucleus and cytoplasm have numerous large, clear spaces. The RER is highly fragmented, but the intracisternal gap (IC) is not markedly different from earlier stages. The intercisternal spaces (IS) are enormous. Ribosomes are irregular in size and density (PALE RIBOS) and are commonly aggregated (AGGREG). In some areas, the ribosomal site appears foamy (BUB). Figure 5.24, × 6000; Figs. 5.25 and 5.26, × 56,000.

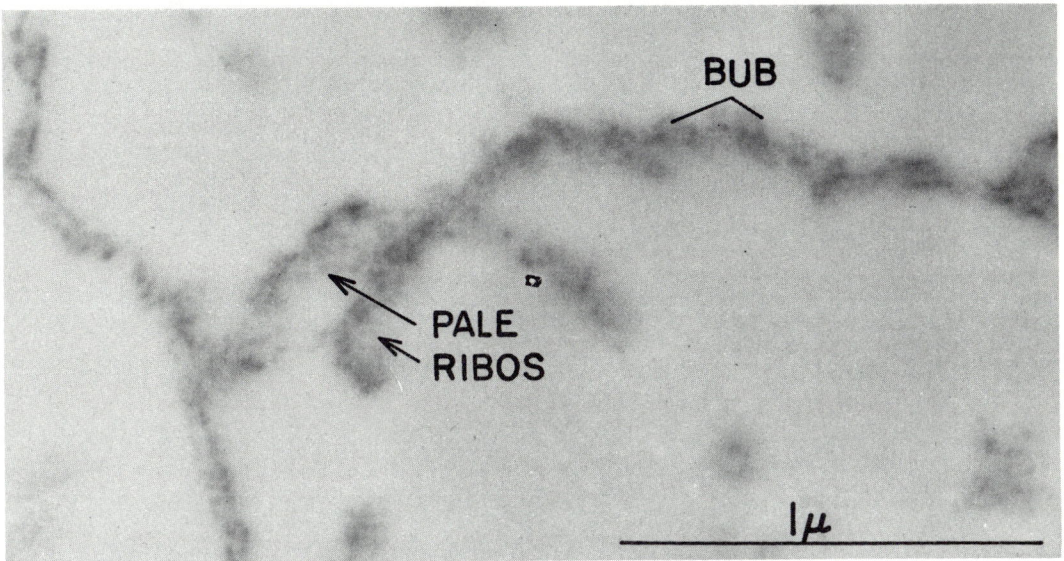

Fig. 5.26.

Part II—Microscopic Distribution of Nucleic Acids

Method

Thin slivers of epiphyseal plate were freeze-dried, postfixed with cross-linking vapors, and embedded in water-soluble Durcupan. Serial sections were cut at about ⅓ μ thick with the Porter–Blum microtome. Up to five sections were cut and picked up on a chip of coverglass. The chips were drained, and placed in numbered depressions of a porcelain spot test plate, where they dried. Each coverslip was covered with a filtered 0.5% aqueous solution of toluidine blue. The whole test plate was covered with a glass plate, and the two were sealed with petroleum jelly. The whole plate was then placed in an oven at 60°C for 20 minutes, after which it was removed and allowed to cool for 10 minutes. The glass plate was removed and each slip was washed with distilled water, drained, and allowed to dry in a correspondingly numbered depression of a second porcelain spot test plate. The glass chips were then mounted in order on glass slides with Permount and studied with the microscope. Photomicrographs were made through orange filters with a Zeiss Photomikroskop on 4 × 5 sheet film Panatomic X at a magnification of 1600 times. They were developed in standard Microdol solution. Negatives were enlarged two times for printing.

A preliminary series of measurements of RER was made. These measurements are of units of RER in a purely morphological sense, and the relation of the num-

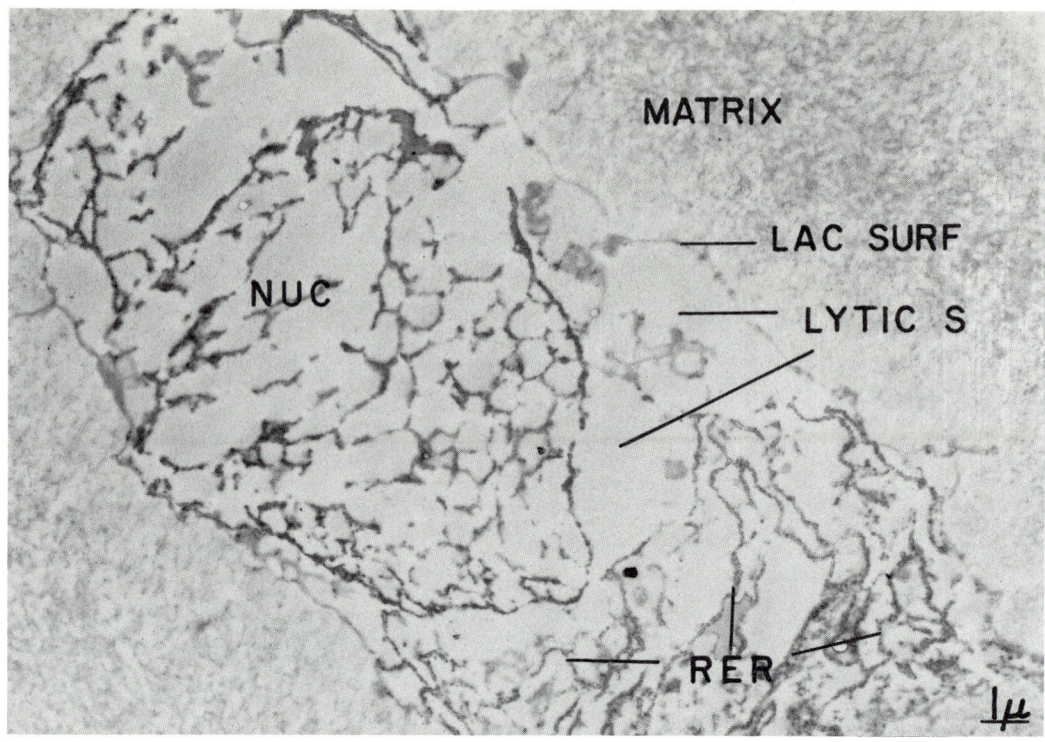

Fig. 5.27. ▲   ▼ Fig. 5.28.

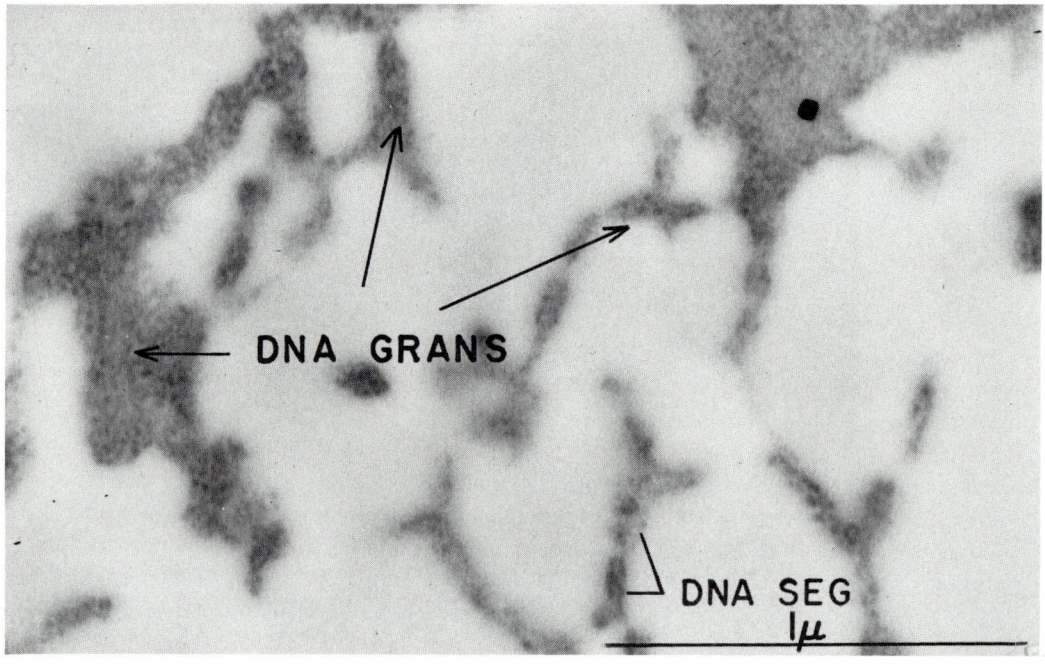

Figs. 5.27–5.30. Lytic cells. Postfixation and staining *in vacuo* with vapors of FFDNB, followed by staining with GCA. Figures 5.27 and 5.28 are earlier stages than Figs. 5.29 and 5.30. Lytic spaces (LYTIC S) appear in both nucleus and cytoplasm. Only in such late lytic stages is there an appreciable space between the cell surface and the matrix (LAC SURF). Nucleus ap-

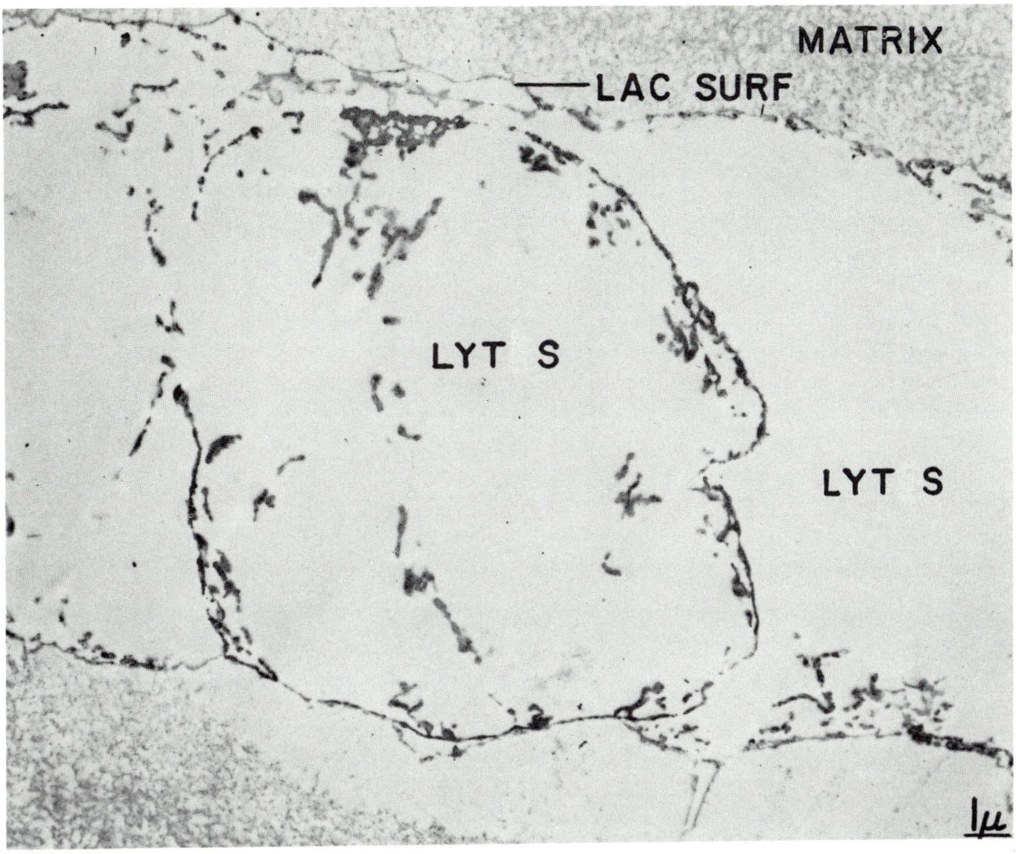

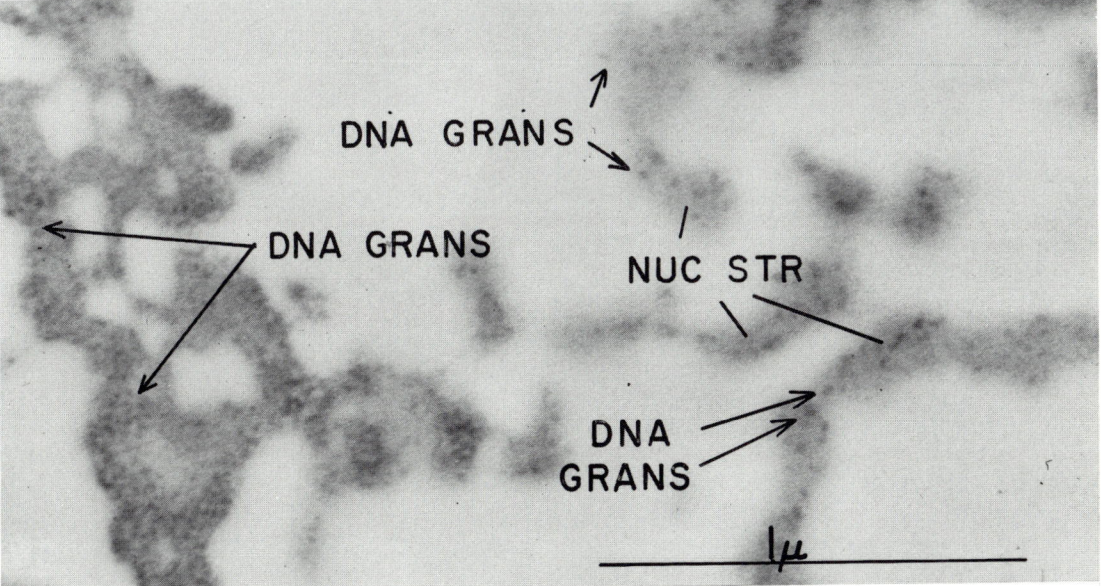

Fig. 5.29. ▲   ▼ Fig. 5.30.

pears in sections to consist of a very wide net of strands (NUC STR) with no intervening background density. The threads consist of overly dark or pale fragments of second-order DNA coils (DNA SEG), and of numerous granules (DNA GRANS). No RNA granules are detectable in the nucleoplasm. Figures 5.27 and 5.29, × 6000; Figs. 5.28 and 5.30, × 56,000.

bers to the number of ribosomes or to soluble RNA molecules, or to the lipid or protein content of the ribosomes is not known. For this purpose, the same negatives were enlarged eight times, and the cytoplasm as well as the RER were outlined on paper. Both were checked carefully visually, and the outlines were corrected when necessary. Comparison with duplicates made at different times showed good agreement. The area of cytoplasm of each cell studied was measured with a planimeter, and the extent of the RER as length was determined with a map measurer. Again duplicates showed very good agreement.

The RER could be stained almost selectively in the cytoplasm of chondrocytes for observation with the light microscope because of the specificity of the stain for nucleic acids, after the particular type of fixation, and because of the particular

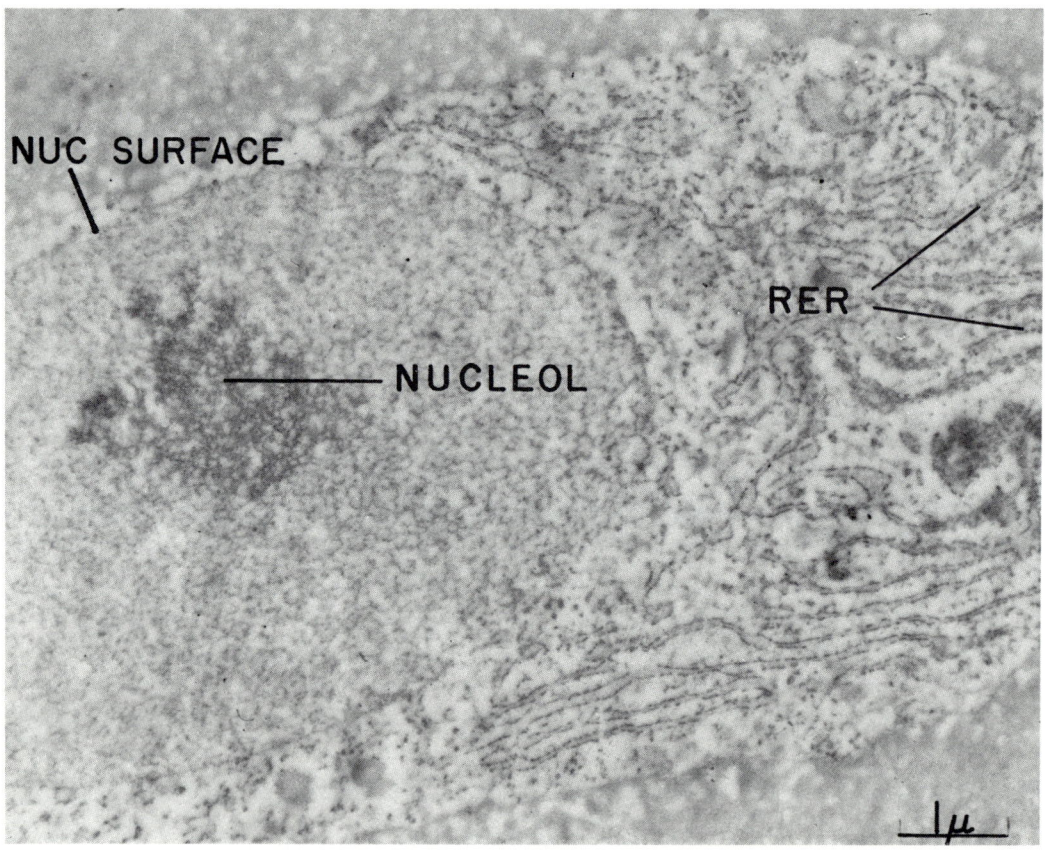

Fig. 5.31.

Figs. 5.31–5.33. Young chondrocyte, treated *in vacuo* with vapors of FFDNB, digested with DNase, and stained with gallocyanin–chromalum. There is no detectable DNA, but the RNA-containing granules of the nucleolus (NUCLEOL) and of the RER are stained. Other abbreviations are given earlier. Figure 5.31, $\times$ 14,000; Figs. 5.32 and 5.33, $\times$ 56,000.

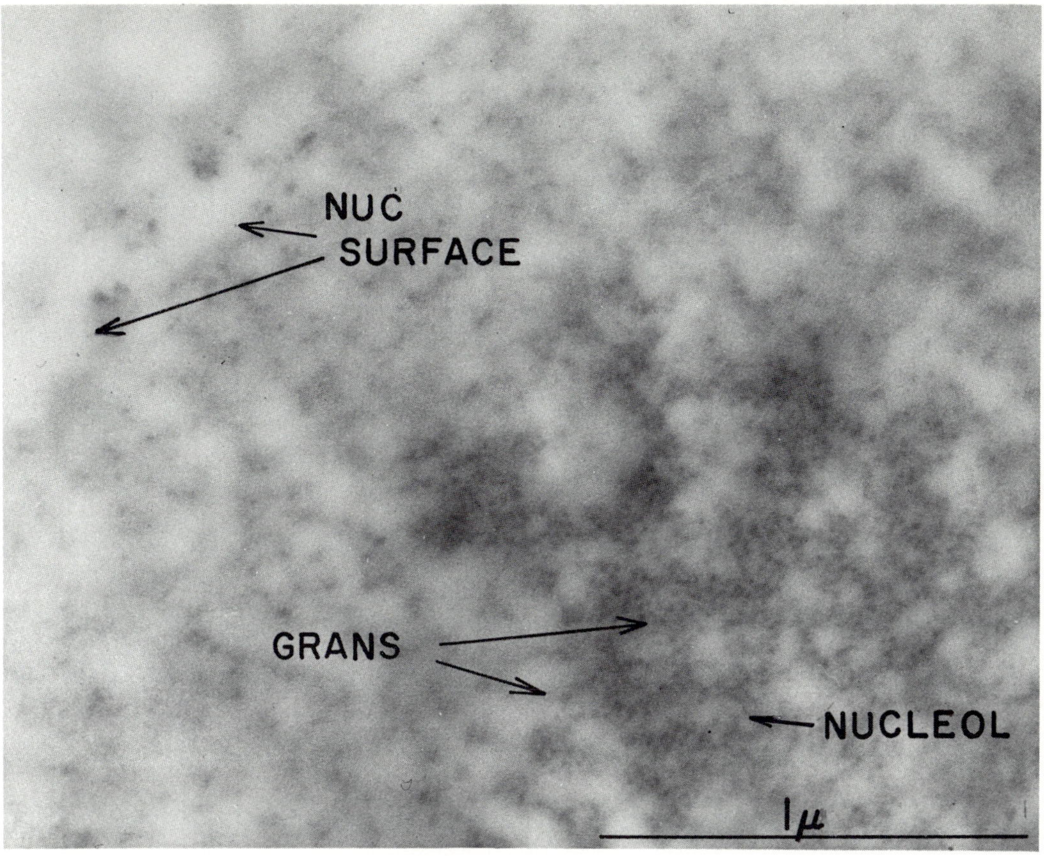

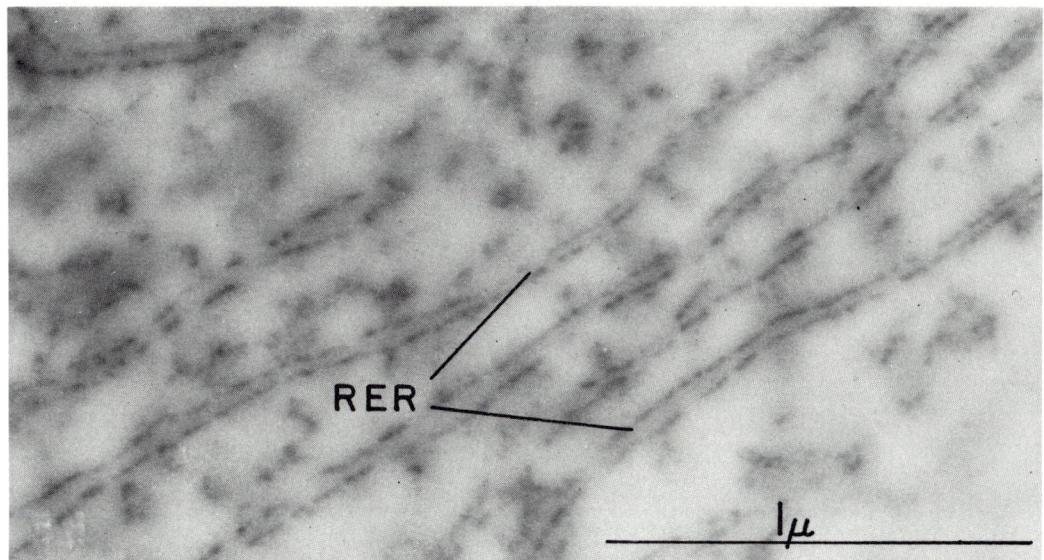

Fig. 5.32.

Fig. 5.33.

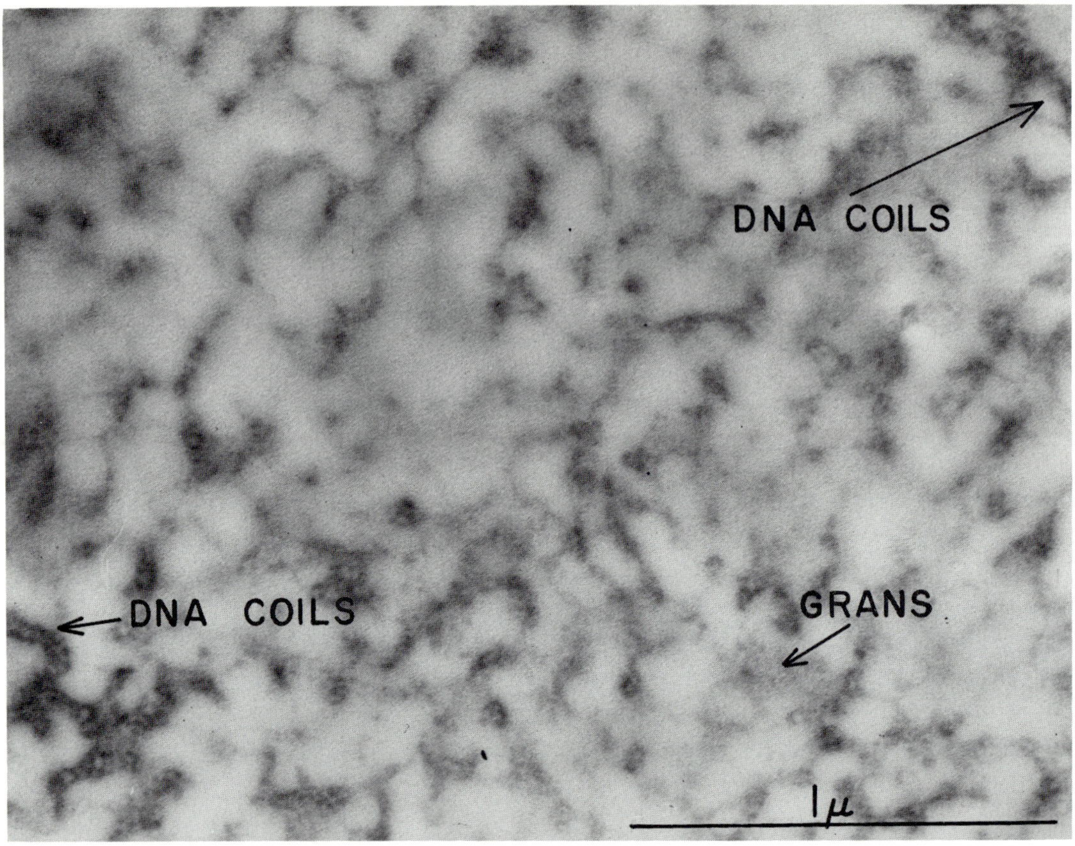

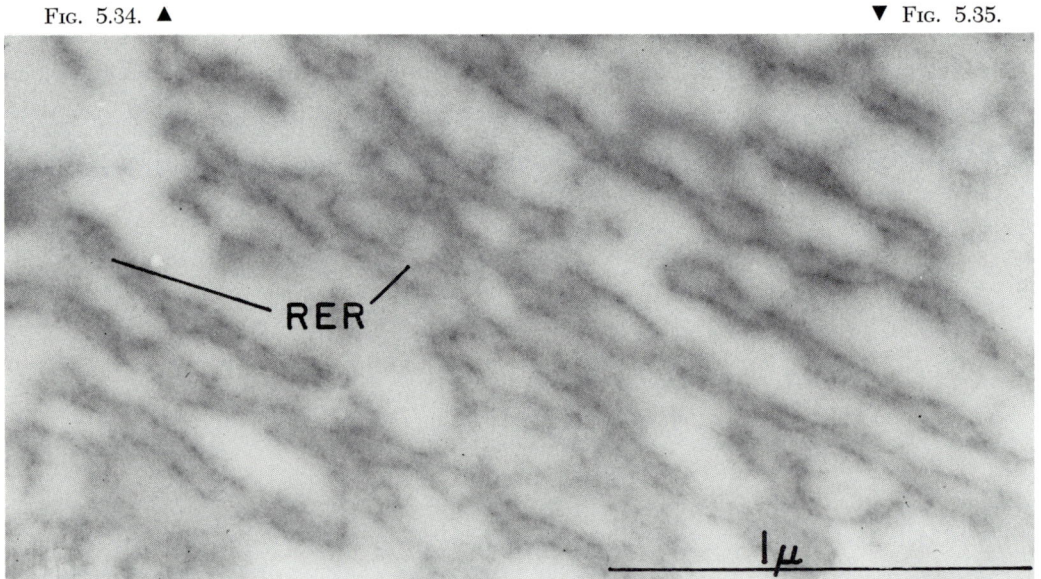

Fig. 5.34. ▲   ▼ Fig. 5.35.

Figs. 5.34 and 5.35. Young chondrocyte, treated *in vacuo* with vapors of FFDNB, digested with RNase, and stained with gallocyanin–chromalum. Although the DNA is dense, and may form various figures including coils, the RNA-containing granules of the nucleoplasm and of the RER are barely discernable. Abbreviations as indicated earlier. × 56,000.

Microscopic Distribution of Nucleic Acids

spatial relations of the cisternae. Study of the RER with the light microscope has been possible hitherto in certain favorable cells *in vitro* (3, 15, 30, 32, 33). It is possible that the RER has been observed also in fixed, sectioned, and stained osteoblasts *(19)*. Section thickness is of critical importance in the studies in chondrocytes. Thicker sections would permit too much overlay of adjacent cisternae when viewed as optical projections. Thinner sections would increase the proportion of cisternae which would become invisible as such because their thickness would be below the limit of resolution of the light microscope when viewed as shadow projections. The most favorable thickness is that in which all segments of the RER could be visible as in shadowgrams except those sectioned strictly transversely or nearly horizontally.

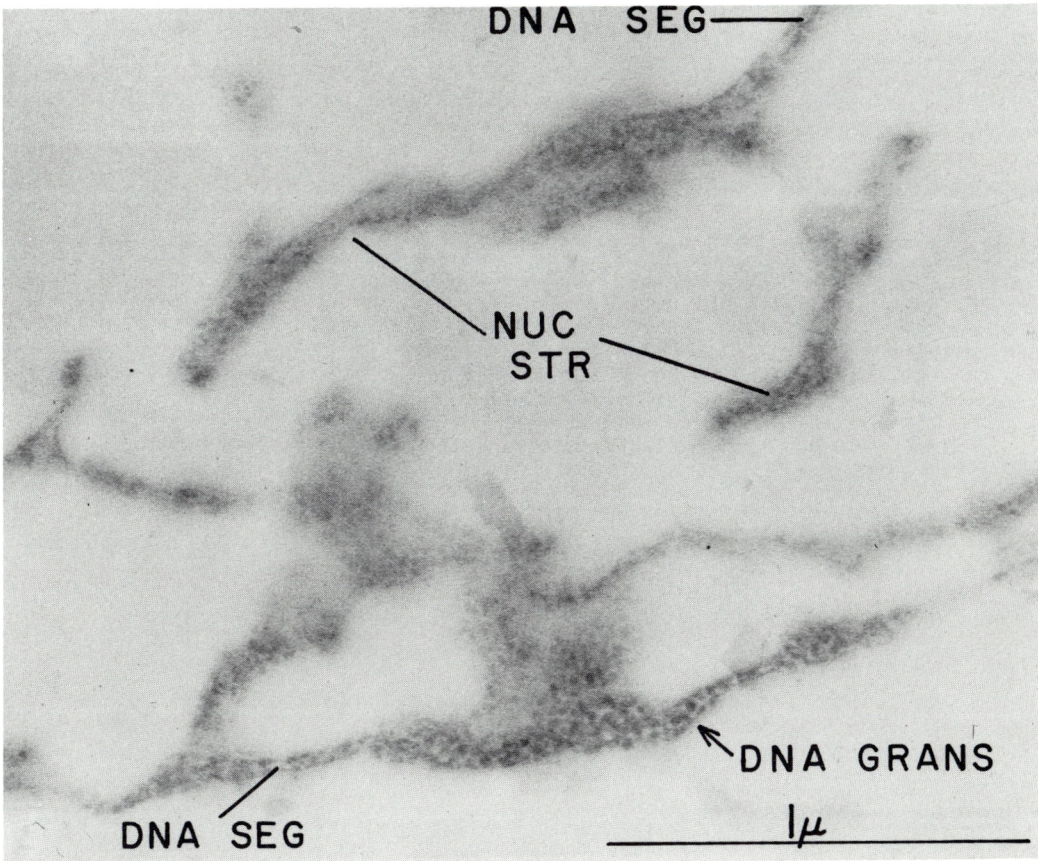

Fig. 5.36. A lytic chondrocyte, treated as described in Figs. 5.34 and 5.35. The nuclear strands (NUC STR) contain dense residual granules (DNA GRANS) and segments (DNA SEG) of DNA coils. × 56,000.

OBSERVATIONS

Descriptive. Chondrocytes prepared for light microscopy, like those prepared for electron microscopy, fill their lacunae. The matrix comprises numerous clear areas close to the limit of resolution, separated by densely stained walls. They are somewhat oriented in the longitudinal columns and the transverse plates of the interterritorial regions. The structure of the matrix, which is in immediate contact with at least part of the surface of immature chondrocytes, is irregular, in contrast with the structure of the matrix enclosing all other chondrocytes. Around the latter, the walls for a distance of one or more compartments are more regularly spherical. These regions correspond to the pericellular regions or cell territories of histological literature (and are so labeled in the photomicrographs), where the ground substance of the matrix is disaggregated *(11)*. The disposition of collagen fibers and of the protein and polysaccharide moieties of the protein–polysaccharide complexes of cartilage matrix, and their relations to its compartmental structure, will be analyzed in Chapter 7 (this volume).

The RER of chondrocytes in the cells of the proliferative zone is frequently not regularly oriented, but in some cells it is oriented (Fig. 5.37). In the cells of the postproliferative growth zone and the (relatively) quiescent preparatory zone, the RER is often markedly oriented, extending lengthwise from the nucleus (in the earlier stage) or the Golgi structures (in the later stage) toward the far edges of the

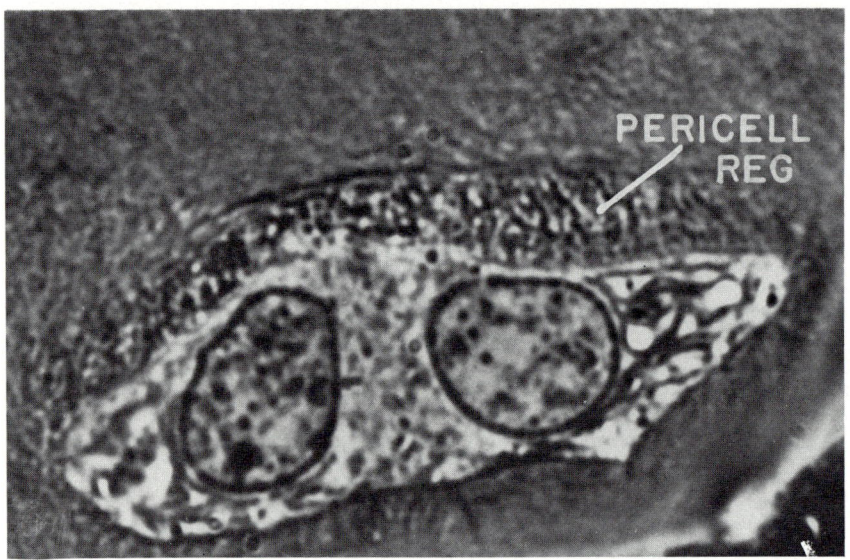

FIG. 5.37. Photomicrograph of a chondrocyte which has divided; the daughter cells have not yet separated. The chromatin bodies are clearly visible. The RER is somewhat oriented. Some compartments in the pericellular region (PERICELL REG) are somewhat larger than most. × 3200.

cells (Figs. 5.38–5.40). In the cells of the prelytic zone, the RER is again irregularly disposed, and the gap between adjacent cisternae is also wider and less regular (Fig. 5.41). In the lytic zone, the RER is fragmented and even more widely dispersed (Fig. 5.42).

The nuclei of cells of the proliferative zone are marked by scattered numerous small chromatin masses, many attached to the nuclear wall. The sex chromatin (Barr body) is very prominent. These chromatin masses remain in the cells of the postproliferative zone (Fig. 5.38) and become smaller in the cells of the preparatory zone. In this stage or in the prelytic and lytic zone, they are no longer resolvable, though the Barr body can still be recognized. The nucleoplasm is structureless in the earlier stages, but in the prelytic stage, this homogeneous appearance is replaced by a netlike structure with paler intervening spaces (Fig. 5.41). This altered nucleoplasmic pattern is emphasized in the lytic stages (Fig. 5.42). Nucleoli are prominent in early stages of maturation, but are no longer visible in later developmental stages of the preparatory (resting) cells or in subsequent stages.

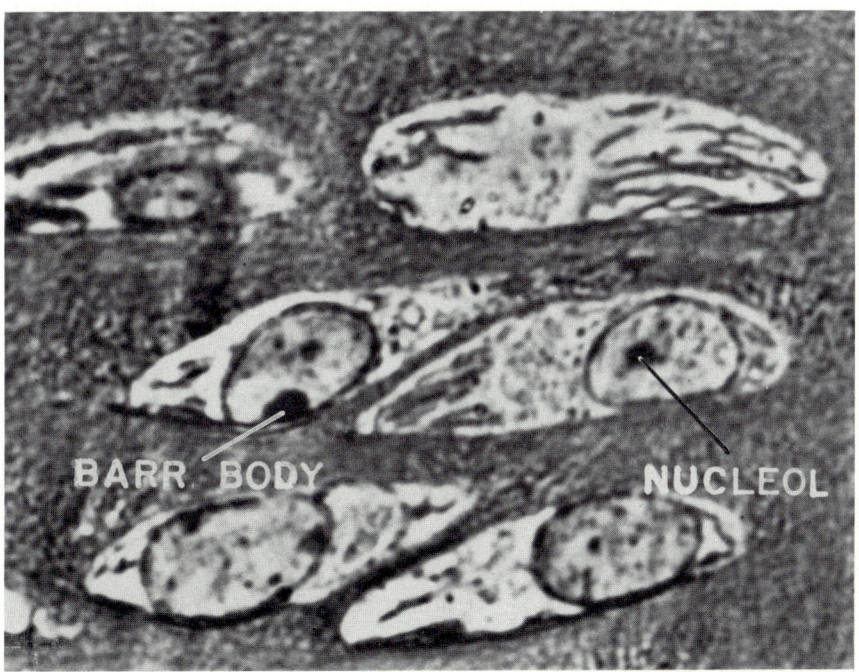

Fig. 5.38. Photomicrograph of several postproliferative cells. The chromatin consists of many small bodies especially aggregated around the nuclear wall. The nucleolus (NUCLEOL) is clear. The RER is irregularly oriented in most cells. A Barr body is clearly visible in one cell. Pericellular compartments are not clearly different from those elsewhere in the matrix. × 3200.

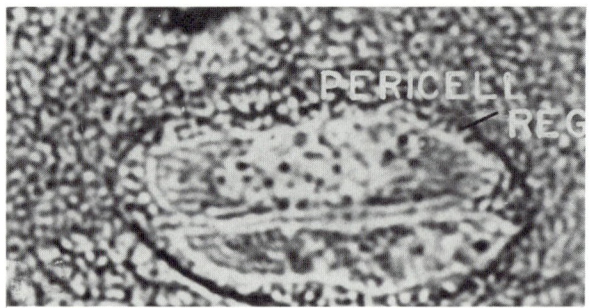

Fig. 5.39. Photomicrograph of a doublet of cartilage cells with some matrix between them. Compartments of the pericellular region (PERICELL REG) of the matrix appear different from those farther from the cells. Nucleus contains many small chromatin bodies. The RER is dense and oriented. × 3200.

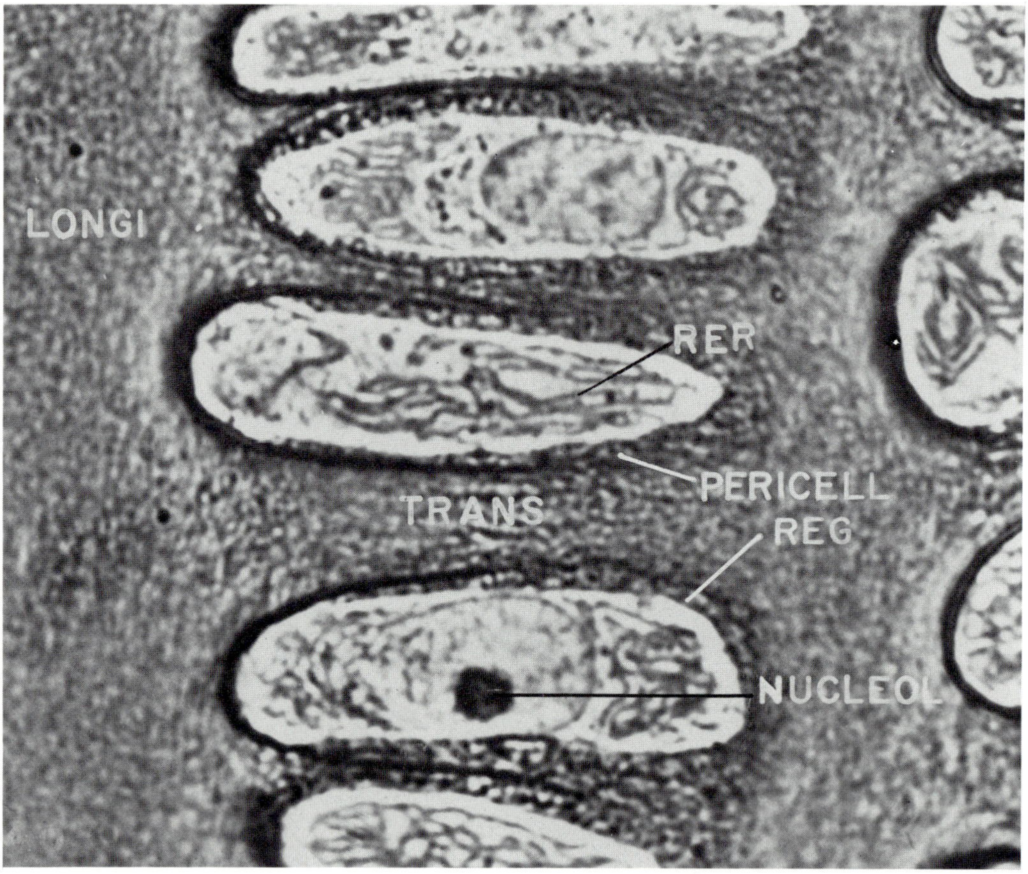

Fig. 5.40. Photomicrograph of part of a stack of chondrocytes in the preparatory (resting) stage. The DNA of all nuclei is finely divided and in some is unresolvable. The nucleolus (NUCLEOL) is very prominent. The RER is also enlarged. The orientation of the compartment walls in the transverse (TRANS) and longitudinal (LONGI) matrix columns differ. Those in pericellular regions (PERICELL REG) are more spherical. × 3200.

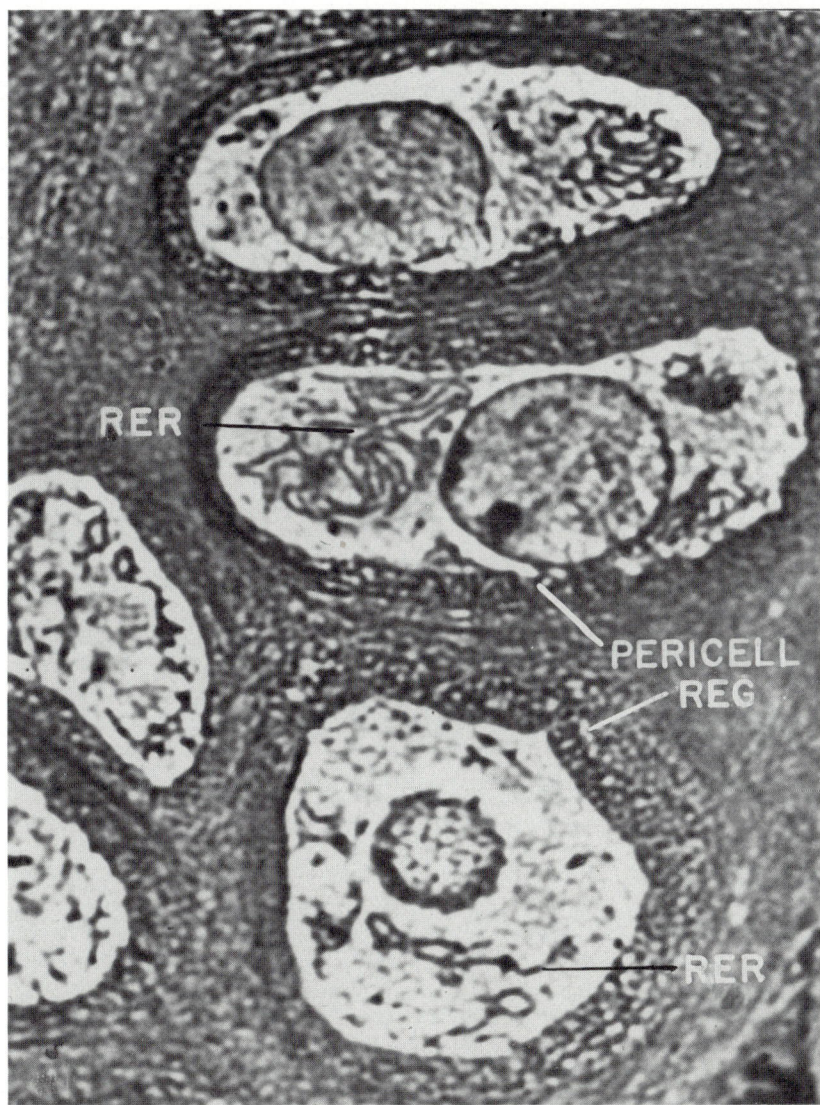

FIG. 5.41. Photomicrograph of part of a stack of chondrocytes showing prelytic and lytic changes. The nuclei contain little or no resolvable chromatin clumps except for the Barr body, and the RER is seemingly reduced in amount. In the lowest cell, the RER is fragmented. The orientation of the compartments in the transverse walls is at about 90° to those in the longitudinal walls. The compartments of the pericellular region (PERICELL REG) of the matrix are enlarged and more spherical. × 3200.

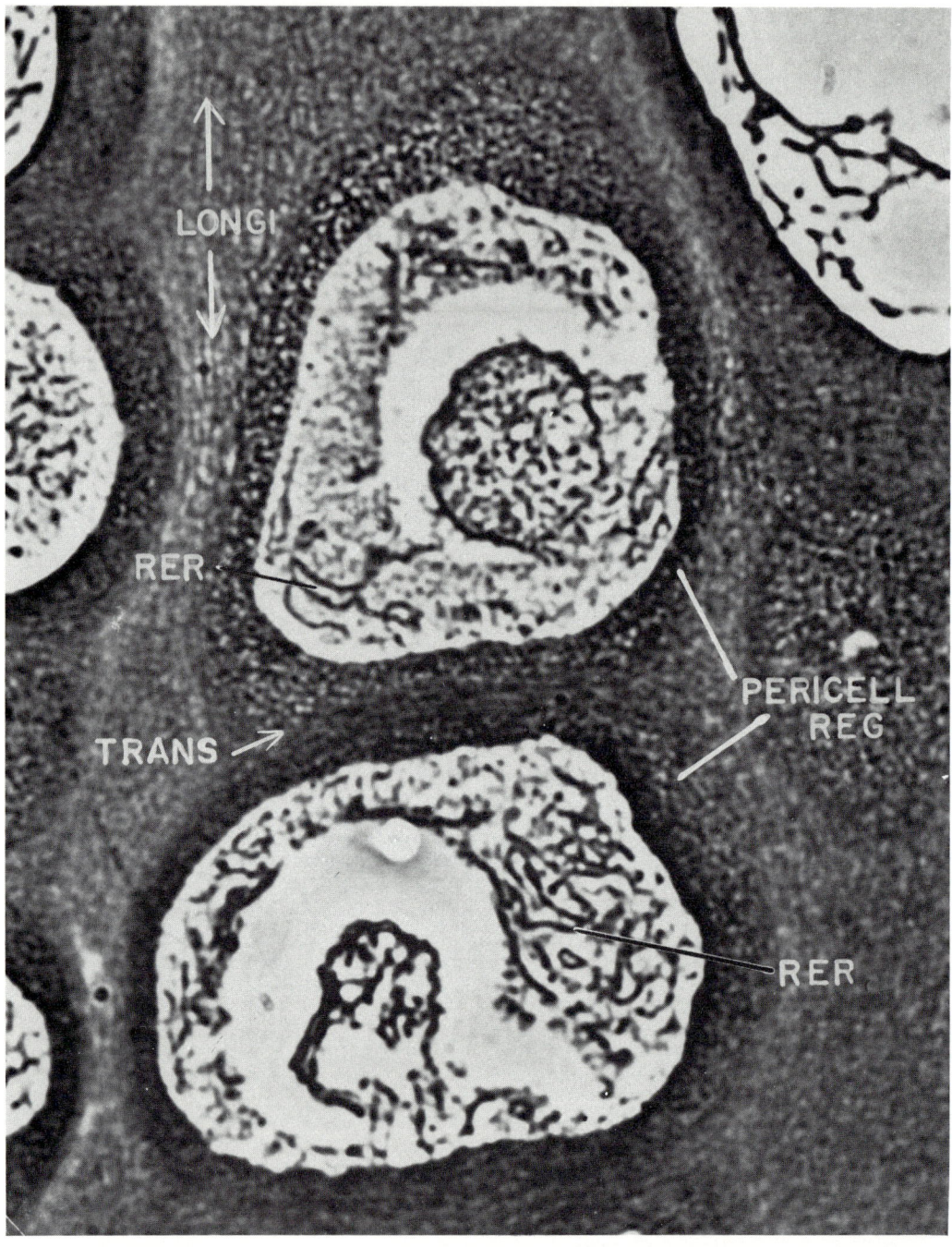

Fig. 5.42. Photomicrograph of cells in the late lytic stage. Nuclear and cytoplasmic changes are more extreme than in earlier stages. Compartments of the pericellular matrix are more spherical than in the longitudinal walls where they are oriented. Abbreviations as listed earlier. × 3200.

TABLE 5.1

RATIO OF LENGTH OF RER TO AREA AND OF RELATIVE AREAS OF CYTOPLASM OF CHONDROCYTES IN DIFFERENT ZONES

Zone	Length/area	Relative area	Number of cells measured
Proliferative zone	0.60		11
Preparatory (resting) zone	0.90	1	17
Prelytic zone	0.76	1.9	6
Lytic zone	0.50	3.3	4

Quantitative. Some measurements were made of the half-cell distance as a measure of growth of the epiphyseal plate. Considering the distance between midpoints of adjacent cells (the half-cell distance) of the proliferative and postproliferative cells as one, then the half-cell distance of the cells in the other zones were, relatively as follows: preparatory zone, 1.8; prelytic zone, 2.5; lytic zone, 8.3.

Measurements were also made of the ratio of the length (in microns) of RER per cell section to the area in square micra of cytoplasm in the same cells (Table 5.1). The measurements were reduced to microns and square microns before the relative values were calculated. The area of the cells in the preparatory (resting) zone was taken as one. The relative area of cells in the proliferative zone is not given. The data show that the relative amount of RER per unit cytoplasm is maximal in the preparatory zone, and then falls in later stages of growth. Because the total amount of cytoplasm in the prelytic and lytic stages is greater, it seems probable that the total amount of RER in these stages is greater than in the earlier preparatory (resting) stage. In stages later than any studied in this research, the total amount of RER would be expected to fall.

Changes in nuclear volume are even greater than those of the cytoplasm. The nuclear volume was estimated by measuring the diameter (when the nuclei appeared spherical) or the long and short axis (when the nucleus was ovoid) with an ocular micrometer. The relative volumes of the nuclei of cells in the various zones of the epiphyseal plate were as follows: postproliferative growth zone, 1; preparative (resting) zone, 4; prelytic zone, 6; lytic zone, 19.

Discussion

It is necessary to comment on some aspects of the terminology used in this report. The characterization of the different stages or zones is based exclusively on the growth in thickness of the epiphyseal plate, that is, on the secretion by the cells of an organized extracellular matrix which consists almost entirely of collagen and protein–polysaccharide complex (P-PC). Thus, cell characterizations are ultimately based on a cellular activity, i.e., secretion. All chondrocytes in epiphyseal plate probably take part in this activity, which is continuous throughout all stages, beginning with the postproliferative stage. The secretory activity is more intense in some stages than in others, but probably never ceases. The term "preparatory (resting) zone" refers to the long column of cells which are not secreting rapidly, even though the cellular activities set the stage for the following periods of rapid growth in thickness of the epiphyseal plate.

In Chapter 3, (Vol. I), I described the rationale of the method of preparation. Though one stain is used to describe DNA and RNA in the light and electron microscope, these are discriminated by their location, by their digestion by DNase and RNase, and by their morphology. It was stated that the form alone of DNA (first- and second-order coils) was so characteristic that it could be used to identify this substance. RNA, on the other hand, occurred as granules ranging from about 200 Å to or below the limits of resolution under the conditions employed. When resolved, the granules are observed in the nucleolus, in the nucleoplasm, and in the RER. When not resolvable, the RNA appears as a homogeneous cloud in the same sites, and is referred to as soluble RNA (sRNA). Soluble RNA in-

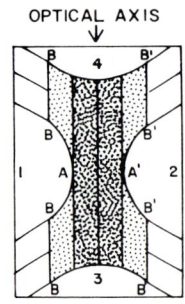

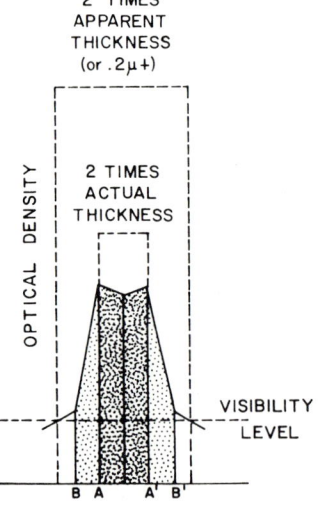

Fig. 5.43. Diagram of a simple arrangement of large matrix compartments to illustrate how their walls, though submicroscopic, may become visible with the light microscope. At the same time, the contents of the compartments are reduced in size. See text for discussion.

cludes perhaps some components of the nucleotide pool, some parts of transfer RNA (tRNA), especially the acylated amino acid fraction (because of the cross-linking effect of the reagent), and perhaps some parts of messenger RNA (mRNA). Filaments were observed in nucleoli and nearby nucleoplasm at only one stage (in the preparatory zone), and they may represent mRNA, or its precursor, or unfolded rRNA, or they may represent the first-order DNA coil. If the latter, this uncoiling of the second-order DNA coil is comparable with a similar process which takes place during chromatolysis (Vol. I, Chapter 12) and during the interphase of the nuclear cycle in eggs of *Drosophila* (Vol. I, Chapter 13). All three examples of extreme uncoiling occur during periods of great chromosomal (nuclear) activity. (See Vol. I, Chapter 12 and 13 for a more detailed discussion of the possible significance of the change in morphology of the DNA.) Filaments were not observed at other stages or other sites, perhaps because they are obscured by the homogeneous and granular background, or because they were not numerous enough at other stages to be visible in the ultrathin sections used in electron microscopy.

There are two notable periods of growth in thickness of the epiphyseal plate: (1) a relatively minor one at the postproliferative zone and (2) a major one during the prelytic and lytic zone. During the first growth period, the secretory products (tropocollagen and P-PC) must be released preferentially between the cell doublets, and also, to a lesser degree, laterally. Growth in thickness of the epiphyseal plate at this period is thus coincidental with an increase in thickness of the transverse wall between the doublet cells and a corresponding increase in length of the longitudinal wall between the cell columns. During the second growth period, the greater increase in length of the epiphyseal plate is associated with secretion apparently primarily into the longitudinal walls of the cartilage matrix. However, in reality, to the extent that lacunae are enlarged and reorganized, the matrix of the transverse wall must also be reshaped and replaced by newly deposited matrix. In contrast with the first growth period, in the second there is no marked orientation since the secretion product must be deposited around the entire periphery of the cell. However, lytic activity must be oriented if the lacunae are to persist as separate units without lateral extension into, and consequent weakening or breakage of, the longitudinal walls of cartilage. In still later stages, which were not studied extensively here, it is assumed that erosion of the transverse walls reflects the continued orientation of the lytic process. Only in the latest stages, when longitudinal walls of the matrix are also fragmented, is the lytic process no longer oriented. Numerous lytic enzymes have been suggested: cathepsins, chondromucoproteinase, β-glucuronidase, phosphatases, and perhaps others. Whether the lytic enzyme(s) is secreted by the chondrocyte, or is the result of controlled or uncontrolled release of lysosomes is unknown *(1, 2, 6, 7, 9, 10, 12, 14, 17, 20–22, 27, 29, 31, 36, 37, 41, 42)*. Although the occurrence of these enzymes in hypertrophic regions of epiphyseal plate is clear from chiefly biochemical studies, only a few deal with the possible occurrence of lysosomes in and around cartilage cells *(16, 25, 37, 39, 40)*.

The matrix of epiphyseal cartilage is characterized by its compartmentalization,

TABLE 5.2

BRIEF SUMMARY OF OBSERVATIONS ON MICROSCOPIC AND SUBMICROSCOPIC DISPOSITION OF NUCLEIC ACIDS IN CHONDROCYTES OF THE DIFFERENT GROWTH ZONES OF PROXIMAL TIBIAL EPIPHYSEAL PLATE OF RAT

Zone	Relative growth of epiphyseal plate	RER						Nucleolus and nucleoplasm		
		Relative amount per unit area of cytoplasm	Degree of dispersion	Degree of orientation	Soluble RNA	Ribosomes	DNA State of aggregation	RNA granules	Soluble RNA
Proliferative	1	0.60	Compact	Poor			Microscopic		
Post-proliferative	1–1.8		Compact	Present	Present	Usually uniform, uniformly distributed	Microscopic, smaller	Numerous	Present
Preparatory	1.8	0.90	Compact	Marked	Present	Uniform, uniformly distributed	Submicroscopic, some second-order coils, separate	Numerous	Present
Prelytic	2.5	0.76	Widely dispersed	Irregular	Present in some areas, absent in others	Irregularly distributed. Varied in size and density	Many separate second-order coils	Scarce	Almost absent
Lytic	8.3	0.50 and less	Widely dispersed	More irregular	Present in some areas, absent in others	Very irregularly distributed. Varied in size and density	Short segments, fragmented coils, granules	Absent	Absent

which appears in all sections viewed with the light or electron microscope. The compartments vary in shape, but are, in general, elongated and oriented at right angles to each other according to whether they are located in the transverse or longitudinal matrix walls. Those which are in contact with or are in the near vicinity of cartilage cells are more spherical and uniform than the majority of compartments elsewhere, and constitute the pericellular regions or cell territories of histologists. The pericellular compartments in their entirety correspond with the more intensely stained regions visible after staining with the PAS method. It is possible that in these regions the polysaccharide is more depolymerized (in later terminology, disaggregated) than the remainder of the matrix, and hence less viscous or more fluid *(11)*. If this disaggregation does take place, the pericellular region reflects the continual readjustments in matrix which must take place if growth in length is to take place. New matrix is presumed to be secreted as old matrix is lysed. It is also possible, of course, that a different P-PC, with different mucopolysaccharides may be present in the pericellular region *(38)*. Or both possibilities may coexist. It is interesting, and consistent with the hypothesis, that the compartments of the matrix are largely nonspherical when they enclose chondrocytes of the proliferative zones (which have not yet been rearranged into columns).

Superficially, the visualization of the large matrix compartments presents no problem—they are simply large enough (0.2–0.3 μ) to be resolved with the light microscope. This explanation, however, is difficult to accept, because the walls of the large compartments (which are the only parts stained) are submicroscopic (say about 400 Å). A simple illustration which resolves this apparent paradox is presented in Fig. 5.43. In the upper part of this figure, are parts of four compartments (1–4) where they meet in a close packing pattern. A density curve as viewed from above (compartment 4 to 3) would show a peak which traverses twice the wall thickness (AA') and then slopes gradually toward a low point, as appears in the lower part of Fig. 5.43. The wall is intensely stained and will become visible when a certain density level is achieved and sustained through a distance which reaches or exceeds the minimum dimension visible with the light microscope. This is achieved because other parts of the walls (e.g., BB and B'B') contribute sufficient density to increase the apparent thickness. To the same degree as these parts are included in the apparent thickness of the walls, the contents of the compartments are reduced. The thickness of the section is critical in these considerations, and if the section were too thick or too thin, the morphology of the large matrix compartments, as seen with the light microscope, would be markedly affected. Parenthetically, there seems to be no structural basis for a capsule, a structure thought by some earlier histologists to be between the surface of chondrocytes and their pericellular or territorial region. Refraction of light by the curved surface of the lacunar wall, as suggested by Schaffer *(35)*, seems to be an adequate explanation.

The most prominent growth region of the epiphyseal plate (Table 5.2) comprises the prelytic and lytic zones. As growth here is represented by secreted matrix, and as the most prominent proteins secreted in the matrix are collagen and

P-PC, cellular correlations or accompaniments should be sought in the disposition and amounts of nucleic acids. Examination of the brief summary in Table 5.2, points to the following cellular changes in respect to nucleic acids. In the preparatory zone, preceding the growth period, the second-order intact coils of DNA are largely separated from each other. They had reached this state gradually, having separated from aggregations or clumps which were microscopic in dimension in early stages. In the same period (preparatory zone), nucleoli are prominent and both nucleoli and nucleoplasm contain numerous minute RNA granules, as well as unresolvable or soluble RNA. This is the only stage when fine filaments were observed in nucleoli and adjacent nucleoplasm. This is the same period, also, when the relative amount of RER is at a maximum. Ribosomes and soluble RNA are regularly distributed and the former are rather uniform.

In contrast with the lively appearance of the cells preceding the maximal growth period, is the deteriorated appearance of the cells during these periods. At the very end of the preparatory stage and in later stages, nucleoli are no longer visible, suggesting that the RNA granules have been dispersed in the nucleoplasm and probably in the cytoplasm as ribosomes. In the prelytic stages also the nuclear volume is increased sharply. This corresponds with increases in nuclear volume which take place reversibly during activity, a topic which is discussed in Vol. I, Chapter 10. In later stages, during lysis, the nuclear volume is increased very greatly. Unlike the earlier, relatively moderate enlargement, this gross swelling is probably to be regarded as part of the picture of irreversible cell lysis, rather than of increased activity. This is presumed to be so since the DNA is now in the form of granules and fragments of short segments and broken coils. At the same time, RNA granules and soluble RNA in the nucleoplasm are reduced and eventually absent. In addition, the RER is relatively scarcer than in earlier stages, more widely dispersed, and marked by irregular dense aggregations and other variations of ribosomes as well as a reduction in sRNA. Eventually, of course, the cells disintegrate.

The interpretation of these changes is undoubtedly extremely complex. In the absence of refined quantitative and kinetic information, the interpretation must remain general. It is suggested that during the preparatory stage, the chondrocytes are being programmed for a rapid synthesis and secretion of the protein constituents of cartilage matrix. This correlates first with the fine separation of DNA coils so that they can act efficiently as a template for RNA transcription and so that the transcribed RNAs have fewer obstacles to their transport to the rest of the cell. The separation of the second-order DNA coils from the more compact arrangement is a general phenomenon associated with increased cellular activity (protein synthesis). Further examples are cited in studies on salivary gland chromosomes of *Drosophila* (Vol. I, Chapter 4), on stimulation of certain cells (Vol. I, Chapter 10), and on recovery of spinal ganglion cells during chromatolysis (Vol. I, Chapter 12). Then it correlates with the enlarged nucleoli bulging with RNA granules which spill out into the nucleoplasm en route to the cytoplasm. These granules are nearly all smaller than, and perhaps contribute to the formation of, ribosomes of the RER. At the same time, filaments stained for nucleic acid ap-

Discussion

pear in and around nucleoli and both nucleoli and nucleoplasm contain large quantities of sRNA (as the homogeneous component). Presumably DNase breaks and eventually hydrolyzes the DNA coils. The RNA, which bears the transcribed information, passes into the cytoplasm where it is translated at the RER through the mediation of rRNA and sRNA into the major protein products of cellular activity, which are then secreted to form the matrix. As this reaches completion, changes appear in the ribosomes (as if they were denatured) and the nucleotide pool and tRNA pool appear exhausted. Thus it appears that the program for an extreme of cellular protein synthesis and secretion is at the same time a program for cell death. That the enlarged prelytic and lytic cells probably synthesize more protein-polysaccharide complex than other cells are shown very recently (12) by ascertaining the incorporation of radioactive sulfur (as sulfate) at different levels of the epiphyseal plate.

The hypothesis presented above requires that the activity of very markedly hypertrophied and even "lysed" chondrocytes be continuous, at least with respect to synthesis and secretion of extracellular lytic enzymes. This possibility is supported by the findings that both DNA and RNA are synthesized by all cells of the columns of the epiphyseal plate, even by the disembodied residues of disintegrated chondrocytes (18, 23, 24). Apparently the fragmented DNA and associated DNase provide the requisite broken ends or nicks required for effective separation of the double helix which precedes the action of polymerizing enzymes in the synthesis of (DNA and) RNA and protein enzymes.

In the discussion of growth of the epiphyseal plate, I suggested that an essential part of the process was the continual remodelling of the matrix through selective lysis. The synthesis of lytic enzyme(s) must also involve the sequence of changes in nucleic acids, but their degree of involvement in this functon may be relatively slight.

It is pertinent to mention briefly some morphological aspects of the picture of epiphyseal cartilage which emerges from this report. These differ from those of all recent reports on the submicroscopic aspects of epiphyseal cartilage, with the exception of Durning's, which is, like mine, based on fixation by freezing-drying (8). Meachim (26) does find meshwork patterns in the matrix of articular cartilage, but regards them as fixation artifacts. In doing so, he repeats the errors made since 1890, which at various times, for similar reasons, denied the existence in the living condition of mitochondria, fibrils of various sorts, Golgi apparatus, etc.

1. The cell surface of chondrocytes is closely apposed to the pericellular matrix, and there is no appreciable space between them. A pericellular space can be produced at will in living or surviving cartilage by changing the tonicity of the surrounding fluid. [See Schaffer (35) for early references.] The generally amorphous deposits which may be visible in the pericellular space (which is formed during fixation) could represent protein and/or polysaccharide components of matrix and/or protoplasm dissolved in an early stage of fixation and reprecipitated as the process of fixation continued to completion. Such changes are prevented by freezing and drying.

2. The intracisternal space is uniformly narrow; rather, it is the intercisternal space which is progressively enlarged during maturation of chondrocytes. By contrast, after the use of fluid fixatives, the intracisternal space may be markedly dilated and may contain dense material or granules. In Chapter 6 (this volume), evidence will be presented which shows that presumptive secretory granules are visible, after freeze-drying, exclusively in the intercisternal space or in the Golgi structures and never in the intracisternal space.

3. Large matrix compartments of microscopic dimensions are characteristic of epiphyseal plate after freezing and drying. By contrast, after fixation with fluid fixatives the matrix is described as a delicate fibrillar net (with interspersed amorphous deposits and/or granules) traversing a considerable amount of empty space. So far as I know, only light microscopists have described the large matrix compartments—Bütschli *(4,5)*, Nowikoff *(28)*, and Ruppricht *(34)*. I shall present evidence in this volume (Chapter 7) of the disposition in the walls of the large matrix compartments and their contents, of collagen fibrils and the protein and polysaccharide components of the protein-polysaccharide complex(es) (P-PC).

Summary

Methods for the fixation and staining of DNA and of soluble and insoluble RNAs for study at the level of the electron microscope are applied to chondrocytes in the tibial epiphyseal plate of the growing rat. The DNA double helix of young chondrocytes is organized as second-order coils (with a repeating unit and diameter of about 400 Å), which are aggregated to form chromatin clumps large enough to be seen with the light microscope. In later stages, the coil aggregates are separated into smaller submicroscopic aggregates, and later still, into individual coils, which finally break up into segments and granules in lytic cells. The nucleolar and nucleoplasmic RNA (both insoluble and soluble) are plentiful in early stages of cell development and become progressively reduced to the near vanishing point in prelytic cells. Cytoplasmic RNA (insoluble and soluble) is confined to the RER, and dramatic changes take place during maturation and degeneration in the arrangement, aggregation, and stainability of the ribosomes. Through it all, the intracisternal space persists relatively unaltered, almost as long as the RER is recognizable, while the intercisternal spaces expand greatly. These findings are interpreted in the light of current concepts of transcription and translation of the DNA code for synthesis and secretion of proteins (of cartilage matrix).

References

1. Ali, S. Y. (1964). The degradation of cartilage matrix by an intracellular protease. *Biochem. J.* **93**, 611–618.
2. Ali, S. Y., and Evans, L. (1969). Studies on the cathepsins in elastic cartilage. *Biochem. J.* **112**, 427–433.

3. Buckley, I. K. (1965). Phase contrast observations on the endoplasmic reticulum of living cells in culture. *Protoplasma* **59**, 569–588.
4. Bütschli, O. (1898). "Untersuchungen über Strukturen insbesondere über Strukturen nichtzelliger Erzeugnisse des Organismus und über ihre Beziehungen zu Strukturen, welche ausserhalb des Organismus entstehen." W. Engelmann, Leipzig.
5. Bütschli, O. (1898). "Atlas zu den Untersuchungen über Strukturen." W. Engelmann, Leipzig.
6. Dingle, J. T., Fell, H. B., and Coombs, R. R. A. (1967). The breakdown of embryonic cartilage and bone cultivated in the presence of complement-sufficient antiserum. 2. Biochemical changes and the role of the lysosomal system. *Int. Arch. Allergy Appl. Immunol.* **31**, 283–303.
7. Dingle, J. T., Barrett, A. J., and Weston, P. D. (1971). Cathepsin D. Characteristics of immunoinhibition and the confirmation of a role in cartilage breakdown. *Biochem. J.* **123**, 1–13.
8. Durning, W. C. (1958). Submicroscopic structure of frozen-dried epiphyseal plate and adjacent spongiosa of the rat. *J. Ultrastruct. Res.* **2**, 245–260.
9. Dziewiatkowski, D. D., Tourtellotte, C. D., and Campo, R. D. (1968). Degradation of protein-polysaccharide (chondromucoprotein) by an enzyme extracted from cartilage. *In* "The Chemical Physiology of Mucopolysaccharides" (G. Quintarelli, ed.), pp. 63–76. Little, Brown, Boston, Massachusetts.
10. Fell, H. B., and Dingle, J. T. (1963). Studies on the mode of action of excess vitamin A. 6. Lysosomal protease and the degradation of cartilage matrix. *Biochem. J.* **87**, 403–408.
11. Gersh, I., and Catchpole, H. R. (1960). The nature of ground substance of connective tissue. *Perspect. Biol. Med.* **3**, 282–319.
12. Granda, J. L., and Posner, A. S. (1971). Distribution of four hydrolases in the epiphyseal plate. *Clin. Orthop.* **74**, 269–272.
13. Greer, R. B., Janicke, G. H., and Mankin, H. J. (1968). Protein polysaccharide synthesis at three levels of the normal growth plate. *Calcif. Tissue Res.* **2**, 157–164.
14. Havivi, E. (1971). Lysosomal enzymes in cartilage and new bone in rachitic chicks. *Isr. J. Med. Sci.* **7**, 532–533.
15. Ito, S. (1962). Light and electron microscopic study of membranous cytoplasmic organelles. *In* "The Interpretation of Ultrastructure" (R. J. C. Harris, ed.), pp. 129–148. Academic Press, New York.
16. Iwano, K. (1970). A cytochemical study of nonspecific alkaline phosphatase activity in epiphyseal cartilage. *J. Jap. Orthop. Ass.* **44**, 525–535.
17. Jibril, A. O. (1967). Phosphates and phosphatases in preosseous cartilage. *Biochem. Biophys. Acta* **141**, 605–613.
18. Kember, N. F. (1960). Cell division in endochondral ossification. A study of cell proliferation in rat bones by the method of tritiated thymidine autoradiography. *J. Bone Joint Surg.* **42B**, 824–839.
19. Knese, K.-H., and Knoop, A.-M. (1961). Chondrogenese und Osteogenese. Elektronenmikroskopische und Lichtmikroskopische Untersuchungen. *Z. Zellforsch. Mikrosk. Anat.* **55**, 413–468.
20. Kuhlman, R. E. (1965). Phosphatases in epiphyseal cartilage. *J. Bone Joint Surg.* **47A**, 545–550.
21. Kuhlman, R. E., and McNamee, M. J. (1970). The biochemical importance of the hypertrophic cartilage cell area to enchondral bone formation. *J. Bone Joint Surg.* **52A**, 1025–1032.
22. Lucy, J. A., Dingle, J. T., and Fell, H. B. (1961). Studies on the mode of action of excess of vitamin A. 2. A possible role of intracellular proteases in the degradation of cartilage matrix. *Biochem. J.* **79**, 500–508.
23. Mankin, H. J., Revak, C., and Lippiello, L. (1968). Ribonucleic acid synthesis in the epiphseal plate of the rat: an autoradiographic study. *Bull. Hosp. Joint. Dis.* **29**, 111–118.
24. Mazhuga, P. M., Zhitnikov, A. Y., and Kharchuk, L. N. (1970). Differentiation and reproduction of cells in chondrogenesis. *Anat. Anz.* **126**, 172–181.

25. Matsuzawa, T., and Anderson, H. C. (1971). Phosphatases of epiphyseal cartilage studied by electron microscopic cytochemical methods. *J. Histochem. Cytochem.* **19**, 801–808.
26. Meachim, G. (1972). Meshwork patterns in the ground substance of articular cartilage and nucleus pulposus. *J. Anat.* **111**, 219–227.
27. Morrison, R. I. G. (1970). The breakdown of proteoglycans by lysosomal enzymes and its specific inhibition by an antiserum to cathepsin D. In "Chemistry and Molecular Biology of the Intercellular Matrix" (E. A. Balazs, ed.), Vol. 3, pp. 1683–1706. Academic Press, New York.
28. Nowikoff, M. (1908). Beobachtungen über die Vermehrung der Knorpelzellen, nebst einigen Bemerkungen über die Struktur der "hyalinen" Knorpelgrundsubstanz. *Z. f. Wiss. Zool.* **90**, 205–257.
29. Poole, A. R. (1970). The degradation of cartilage matrix by a lysosomal preparation, isolated from a malignant tumour, and its inhibition by an antiserum to this preparation. *Histochem. J.* **2**, 431–439.
30. Porter, K. R. (1953). Observations on a submicroscopic basophilic component of cytoplasm. *J. Exp. Med.* **97**, 727–750.
31. Quintarelli, G., Sajdera, S., and Dziewiatkowski, D. D. (1968). Modifications of connective tissue matrices by an enzyme extracted from cartilage. *Histochemie* **15**, 1–20.
32. Rose, G. G. (1961). The Golgi complex and endoplasmic reticulum in tissue-cultured human melanoma cells with phase contrast microscopy. *Cancer Res.* **21**, 706–711.
33. Rose, G. G., and Pomerat, C. M. (1960). Phase contrast observations of the endoplasmic reticulum in living tissue cultures. *J. Biophys. Biochem. Cytol.* **8**, 423–430.
34. Ruppricht, W. (1910). Über Fibrillen und Kittensubstanz des Hyalinknorpels. *Arch. Mikrosk. Anat.* **75**, 748–771.
35. Schaffer, J. (1930). Das Knorpelgewebe. In "Handbuch der mikroskopischen Anatomie des Menschen" (W. H. W. v. Möllendorff, ed.), Vol. 2, Part 2, pp. 210–390. Springer, Berlin.
36. Sledge, C. B. (1967). The role of lysosomal enzymes in skeletogenesis. *J. Bone Joint Surg.* **49A**, 794 (Abstr.).
37. Sledge, C. B. (1968). Biochemical events in the epiphyseal plate and their physiologic control. *Clin. Orthop.* **61**, 37–47.
38. Stockwell, R. (1968). Discussion of J. E. Scott, Patterns of specificity in the interaction of organic cations with acid mucopolysaccharides. In "The Chemical Physiology of Mucopolysaccharides" G. Quintarelli, ed.), pp. 229–230. Little, Brown, Boston, Massachusetts.
39. Tanaka, S. (1965). Electron histochemical demonstration on the localization of activities of alkaline and acid phosphatases in the cartilage of mice. *Arch. Jap. Chir.* **34**, 587–590.
40. Thyberg, J. (1972). Ultrastructural localization of aryl sulfatase activity in epiphyseal plate. *J. Ultrastruct. Res.* **38**, 332–342.
41. Thomas, L. (1964). The effects of papain, vitamin A, and cortisone on cartilage matrix in vivo. *Biophys. J. Suppl.* **4**, 207–213.
42. Woessner, J. F., Jr. (1967). Acid cathepsins of cartilage. In "Cartilage Degradation and Repair" (C. A. L. Bassett, ed.), pp. 99–106. Proceedings of a Workshop sponsored by The Committee on the Skeletal System, Division of the Medical Sciences, National Research Council, Washington, D.C.

6

Possible Precursor Granules in Chondrocytes and Osteoblasts

Zelma Molnar, Isidore Gersh, and Giovanni L. Rossi

In Chapter 4, certain cytoplasmic granules were described in fibroblasts, and we suggested that these granules were possibly precursors of collagen, protropocollagen. In this chapter, identical granules will be described in chondrocytes and osteoblasts, and further evidence will be presented to support this suggestion. As in the earlier report, particular attention will be placed on the relations of these granules to the RER and the surface of the cell. In addition, the relation to the Golgi apparatus will be discussed. This chapter is presented in two parts. Part I deals with the granules in chondrocytes, and Part II deals with identical granules in osteoblasts. Part I is further divided into two parts dealing with submicroscopic aspects and with microscopic (quantitative) aspects, respectively.

Part I—Possible Precursor Granules in Chondrocytes

SUBMICROSCOPIC STUDIES

Methods. The specimens are from the proximal tibial epiphyseal plate of young female rats as described in Chapter 5 (this volume) and Chapter 7 (Vol. I), where the details of preparation are also given. Thin slivers of epiphyseal plate were frozen ultrarapidly and dried at $-40°C$ or lower and were treated in one of the following ways:

1. Most of the specimens were treated *in vacuo* with vapors of the cross-linking reagent, mercury (II) hexafluoroacetylacetonate (Hg hfac). After excess reagent was removed, specimens were infiltrated slowly *in vacuo* with chilled 95% alcohol and then with water-soluble Durcupan as described in Chapter 2 (Vol. I).

2. Some freeze-dried specimens were infiltrated slowly *in vacuo* with chilled 95% alcohol, and then with water-soluble Durcupan without staining as described in Chapter 2 (Vol. I).

3. Other freeze-dried preparations were postfixed by infiltrating slowly *in vacuo* with a concentrated, chilled solution of hydrated ferric chloride in 95% alcohol while the tissue was at $-35°C$. The vacuum was broken, and the specimens were removed rapidly from the small squares of aluminum foil on which they had been frozen. They were washed with several changes of 95% alcohol, and infiltrated with water-soluble Durcupan as described in Chapter 8 (this volume).

4. Some specimens were freeze-dried, and infiltrated slowly *in vacuo* with 95% alcohol. After 2 days at atmospheric pressure and room temperature, specimens were stained for 5 hours with a 5% solution of platinum tetrabromide. After washing in several changes of 95% alcohol, specimens were infiltrated with ab-

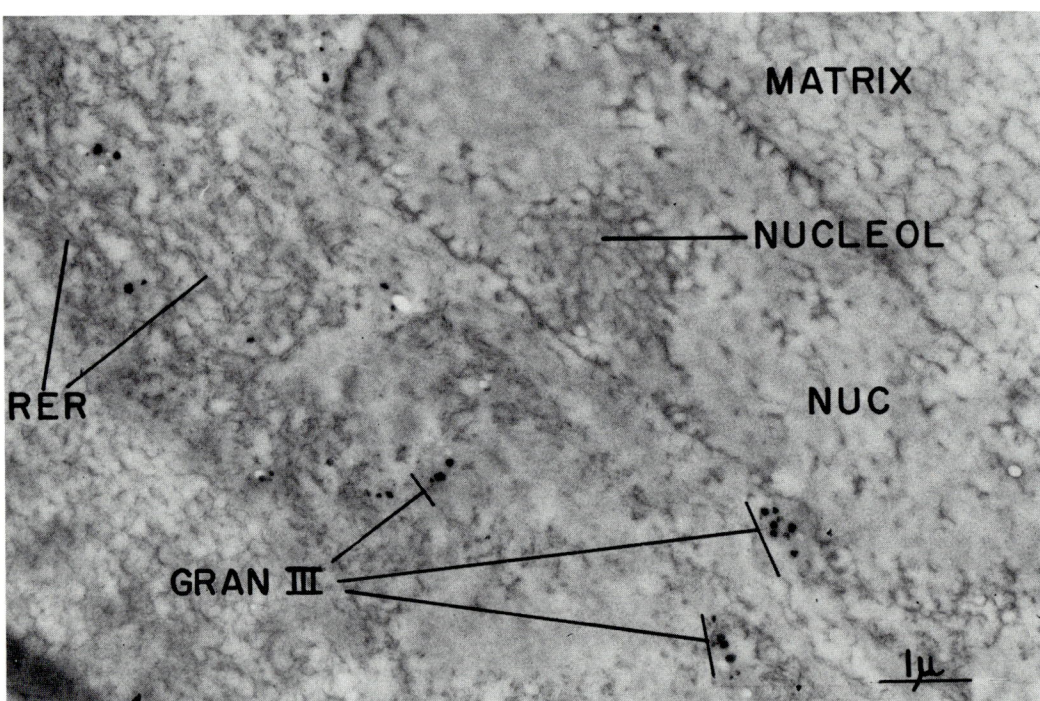

Fig. 6.1. Proximal epiphyseal cartilage of the tibia of 21-day-old rat, postfixed and stained *in vacuo* with vapors of mercury (II) hexafluoroacetylacetonate (Hg hfac). The third-order granules (GRAN III) are in the cytoplasm. The rough endoplasmic reticulum (RER) is denser than the remainder of the cytoplasm. In the nucleus (NUC), the chromatin consists of small clumps of dense material especially prominent near the nuclear surface. The nucleolus (NUCLEOL) is nearly as dense. The extracellular matrix comprises numerous compartments whose walls are stained. × 12,400.

solute alcohol, and then infiltrated with water-insoluble Durcupan, as described in Chapter 2 (Vol. I).

5. Finally, some specimens were postfixed with vapors of Hg hfac and FFDNB and embedded in water-soluble Durcupan as described earlier (Chapter 2, Vol. I). Ultrathin sections were stained with uranyl acetate. Sections were cut with diamond knives, floated on water, mounted on grids, and examined with an Hitachi HU-11A electron microscope operating at an accelerating voltage of 50 kv. Electron micrographs were enlarged two times in printing.

Observations. These are presented in a framework developed in Chapter 5 (this volume), based on chondrocytes in relation to the secretion of extracellular matrix. The cell development and secretory stages or zones are (1) proliferative

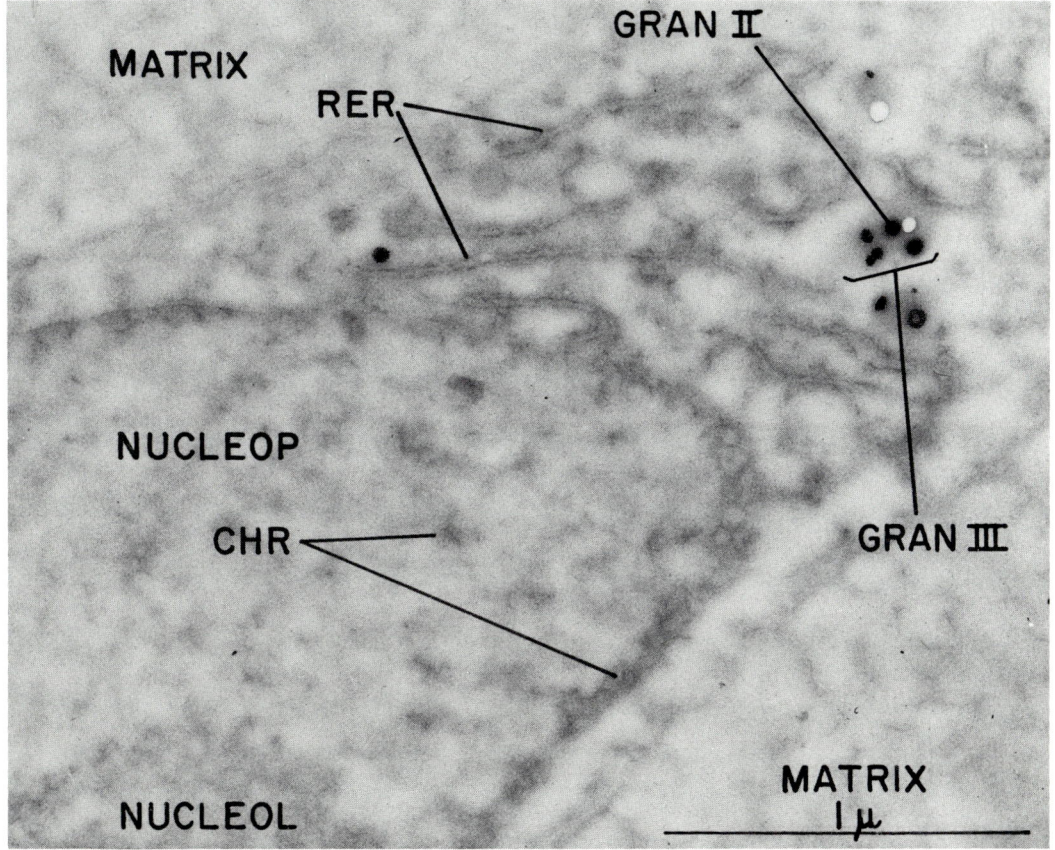

FIG. 6.2. Portion of another chondrocyte at a similar stage, postfixed as above. The holes represent sites occupied by second-order granules (GRAN II) before sectioning. Third-order granules (GRAN III) occur between cisternae of the RER. Nuclear chromatin (CHR) is present chiefly at the periphery of the nucleus, but also occurs as small clumps in the nucleoplasm (NUCLEOP). The nucleolus (NUCLEOL) is of nearly the same contrast. × 48,000.

stage, when the cells are relatively inactive or only mitotically active; (2) postproliferative stage, when the cells are separating from each other, in a column, through the secretion between them of extracellular matrix; (3) preparatory (resting) stage, when the cells are aligned, but cartilage matrix is not being secreted or is being secreted in minimal amounts; (4) prelytic stage, when the cells are hypertrophied and matrix is being secreted and remodelled at a high rate; and (5) the lytic stage, when the cells are progressively lysed and growth of cartilage through deposition of matrix is maximal. Evidence was presented that the epiphyseal plate progresses continuously through two periods of growth: a relatively minor one in the postproliferative period and a major one in the prelytic and lytic periods.

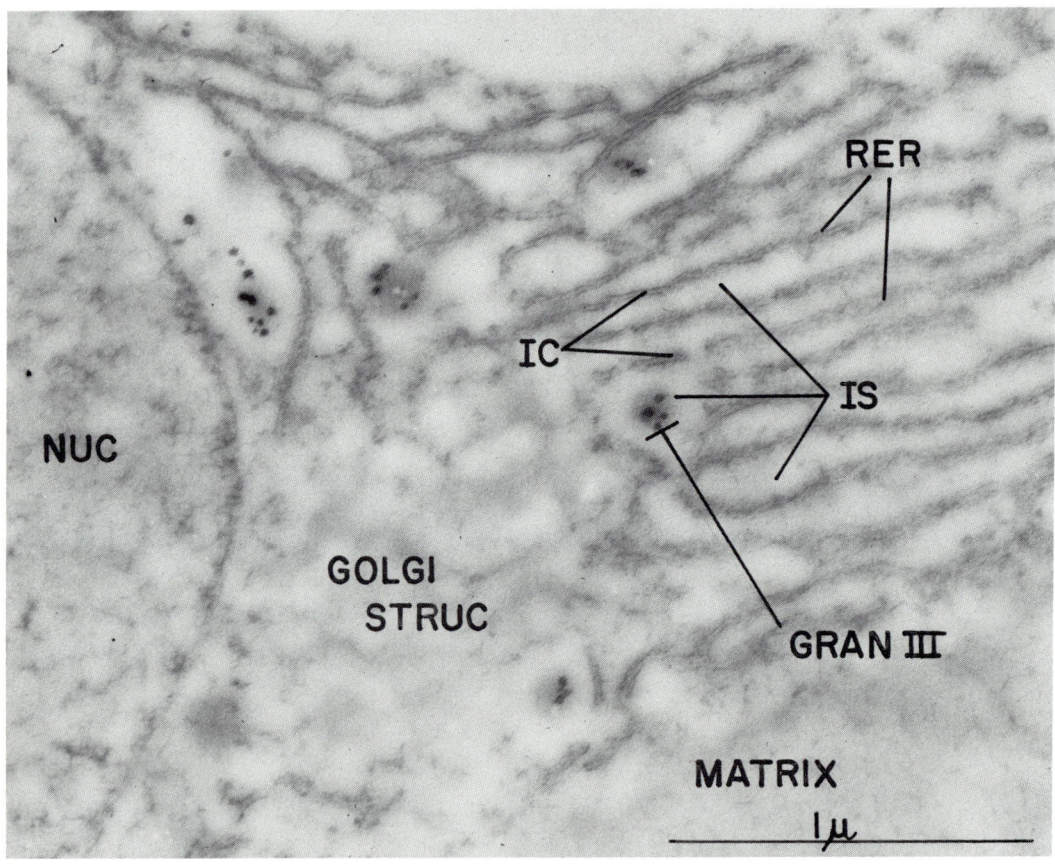

Fig. 6.3.

Figs. 6.3–6.5. Portions of cells in early preparatory stage, postfixed and stained as above. The chromatin of the nucleus (NUC) is more dispersed than in earlier stages. Third-order granules (GRAN III) are present outside the RER, in the intercisternal space (IS), not in the consistently narrow intracisternal gap (IC). They are especially numerous near the periphery of the Golgi apparatus (GOLGI STRUC) as in Figs. 6.3 and 6.4, but also occur in the intercisternal space (Fig. 6.5). × 48,000.

MATRIX
RER
IC
IS
GOLGI STRUC
GRAN III
1μ

Fig. 6.4. ▲

▼ Fig. 6.5.

MATRIX
IS
RER
GRAN III
IC
1μ

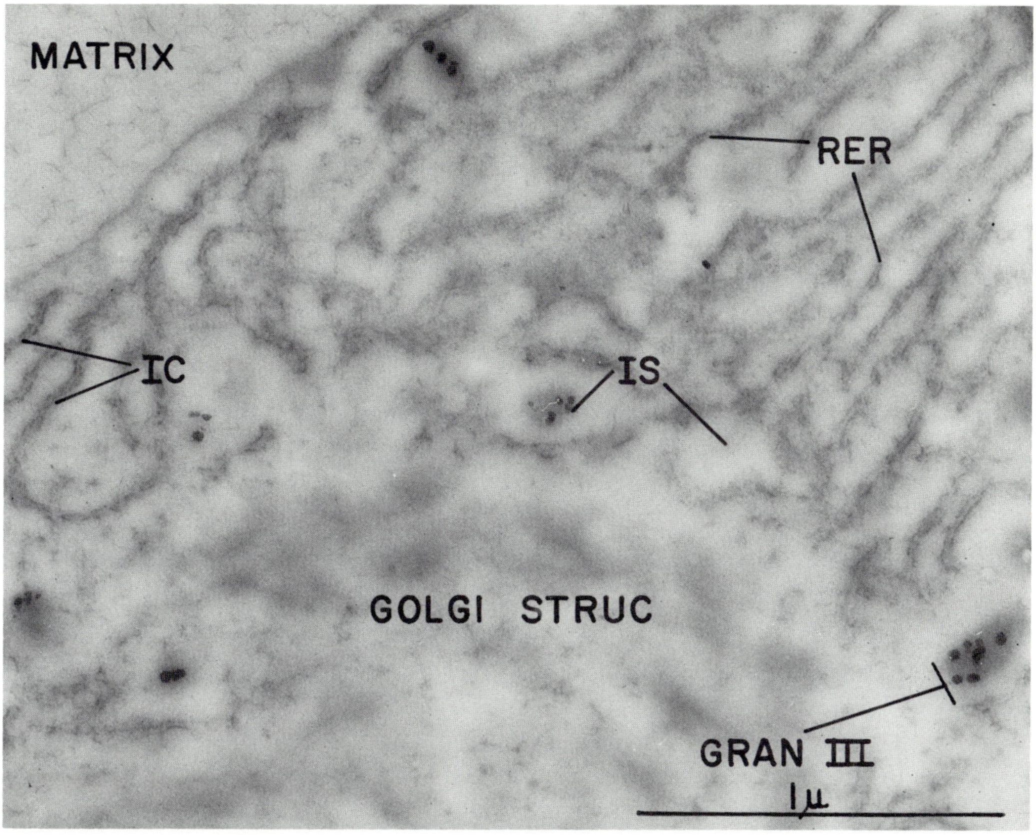

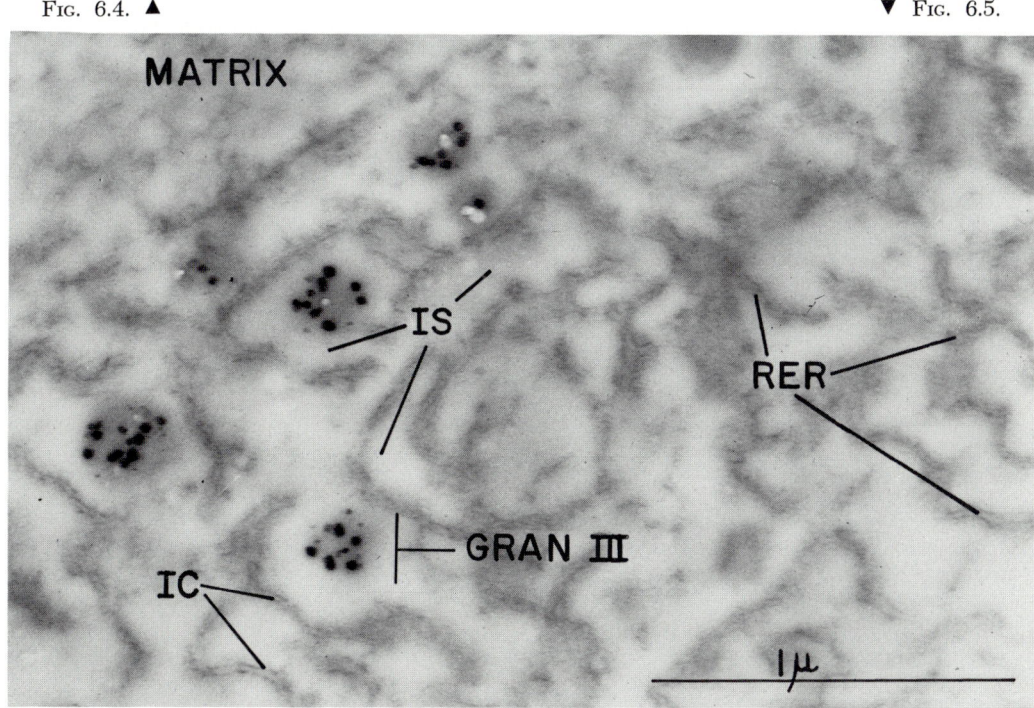

132   6. Possible Protropocollagen Precursors

After treatment with vapors of Hg hfac alone, cells frequently contain granules which are indistinguishable from those of fibroblasts. These are observed in postproliferative cells (Figs. 6.1 and 6.2). The third-order granules are about ½ μ or more in diameter. They consist of a homogeneous material of low contrast in which are embedded the second-order granules, which are of extremely high contrast. They occur also in cells of the preparatory zone (Figs. 6.3–6.5) where they seem to be especially numerous near the border between the Golgi apparatus and the adjacent RER (Figs. 6.3 and 6.4), but they may lie in other parts of the cytoplasm (Fig. 6.5). The outline of the second-order granule is usually not smooth, but irregular (Fig. 6.6). The cytoplasmic granules occur also in prelytic and advanced lytic cells (Fig. 6.7). The cytoplasmic granules are always in the cyto-

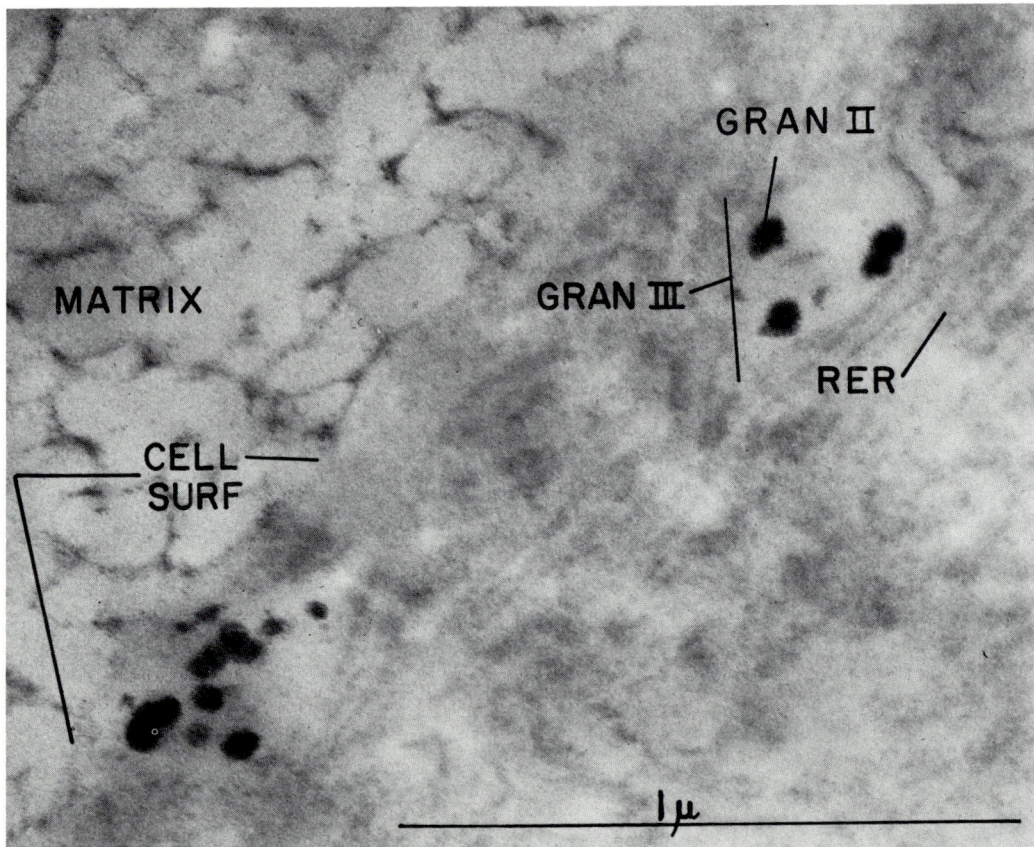

FIG. 6.6. Proximal epiphyseal cartilage of the tibia of a 21-day-old rat, postfixed and stained with an alcoholic solution of ferric chloride. The third-order granules (GRAN III) include second-order granules (GRAN II). One third-order granule is very close to the cell surface (CELL SURF). The RER is not of high contrast, while the walls of the large matrix compartments are dense (MATRIX). × 82,000.

Possible Precursor Granules in Chondrocytes

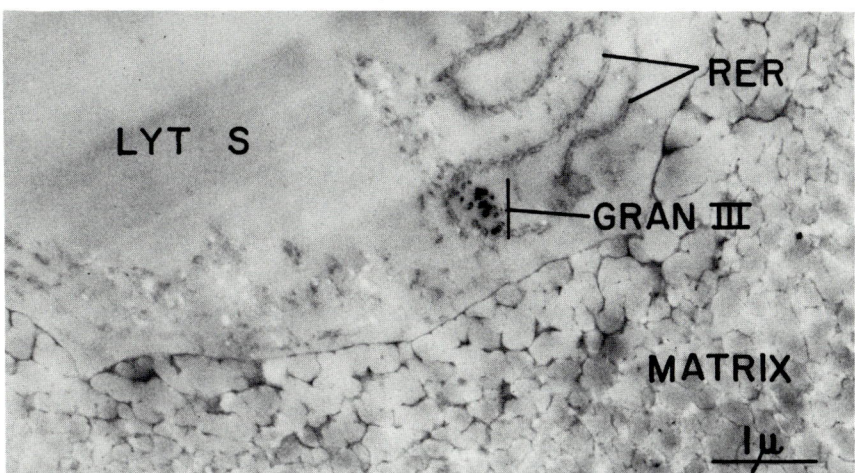

Fig. 6.7. Same preparation as above. Lytic cell with a large space (LYT S) and residual RER contains a third-order granule (GRAN III). The walls of the large matrix compartments between the lytic cells are of high contrast (MATRIX). × 14,200.

plasm outside the cisternal space of the RER. Sometimes they may be seen close to the surface of the cell (Fig. 6.6). The cytoplasmic granules have never been seen in the extracellular matrix.

Although an alcoholic solution of platinum tetrabromide dissolves the cytoplasmic granules and is not useful for their study, the postfixative is useful for illustrating two aspects of an underlying pattern of cartilage cells. This process of postfixation emphasizes the pseudovacuolar nature of protoplasm. In the nucleus, there are many sharply outlined compartments whose walls are thickened in places to accommodate chromatin (Fig. 6.8). In the cytoplasm, numerous partitions extend between the walls of the RER, and subdivide the cisternae into smaller compartments (Figs. 6.8–6.10). The protoplasm extending between adjacent cisternae is subdivided also by numerous thin walls. A second feature, which this method of postfixation emphasizes, is the sharply terminating surface and the numerous delicate processes which extend from the surface, with no discernable gap between them and the surrounding matrix.

Microscopic (Quantitative) Studies

Method. Thin slivers of epiphyseal plate were prepared as for nucleic acids for microscopic study (this volume, Chapter 5). They were frozen and dried, treated *in vacuo* with vapors of the cross-linking reagent (FFDNB), and embedded in water-soluble Durcupan. Sections were cut serially about $\frac{1}{3}$ μ thick with the Porter–Blum microtome. Up to five sections at a time were picked up on chips of a coverslip and dried. For study with the light microscope, the mounted sections

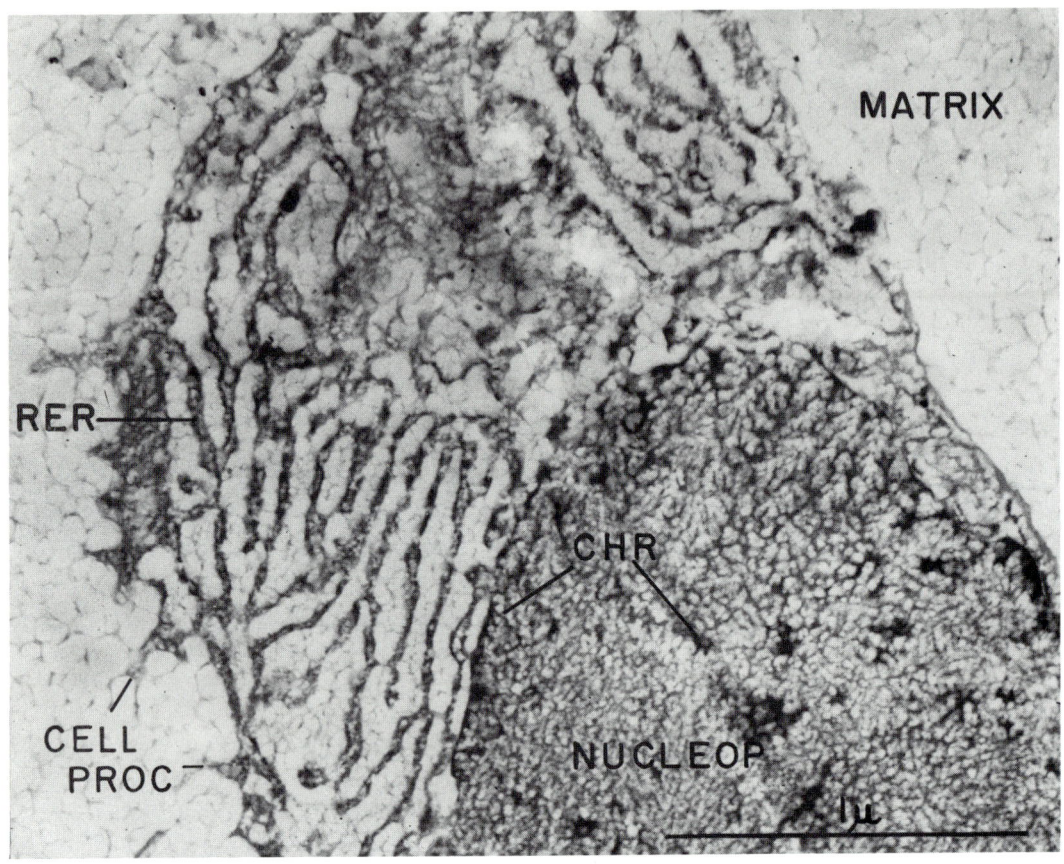

Fig. 6.8.

Figs. 6.8–6.10. Early postproliferative (Fig. 6.10A) and preparatory stage (Figs 6.8 and 6.9) chondrocytes postfixed and stained with alcoholic platinum tetrabromide. The nucleus is filled with dense and less dense parts which constitute the spongelike background structure of the nucleoplasm (NUCLEOP). The very dense chromatin (CHR) is finely dispersed and contributes to the thickness of the walls of the compartments. The RER is subdivided into smaller compartments. Between the cisternae of the RER are more delicate compartments, some of which extend to or through the surface. The extracellular matrix is in contact with the cell surface and consists of large matrix compartments which are of microscopic size. Occasionally large vacuoles (VAC) occur in the cytoplasm, but some of them may be the consequence of a peculiar plane of sectioning through a flattened intercisternal space. Cell processes (CELL PROC) are clearly stained. × 48,000. Figure 6.10B shows a third-order granule (GRAN III) of a cartilage cell of epiphyseal plate postfixed *in vacuo* by successive exposure to vapors of Hg hfac and FFDNB, with section staining with uranyl acetate. The matrix of the third-order granule has a vacuolated appearance. The hydrophilic parts are stained, and outline the hydrophobic components of the protein molecules, which are clear. The same structure in the matrix of the third-order granule was observed also in fibroblasts. × 78,000. G2 print.

Possible Precursor Granules in Chondrocytes

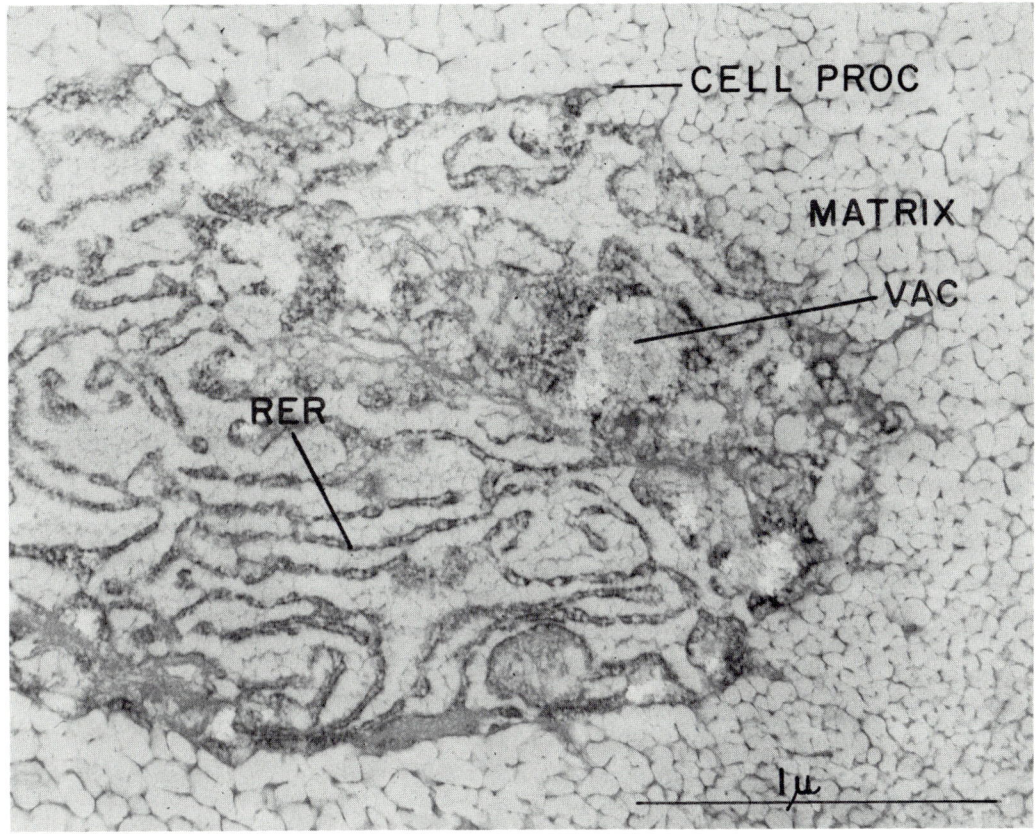

Fig. 6.9.

were stained with toluidine blue, washed, drained, dried, and mounted in order on slides with Permount. Photomicrographs were made through orange filters with a Zeiss Photomikroskop on Panatomic X film at a magnification of 400 times. They were developed in Microdol solution, diluted 1:3. Photomicrographs were made of 13 to 44 sections through the same cells. The films were enlarged four times in printing and every cytoplasmic granule visible on the prints was checked with the microscope by direct viewing, to exclude mitochondria, dirt, or some other artifact. In this way, it was possible to estimate the number of cytoplasmic granules per section through a cell. The cytoplasmic area of a section through a cell was estimated by planimetric measurements of enlargements of the photomicrographs, and the number of granules per unit of cytoplasmic area in a section could be calculated. Both groups of figures are presented for a series of cells in Table 6.1.

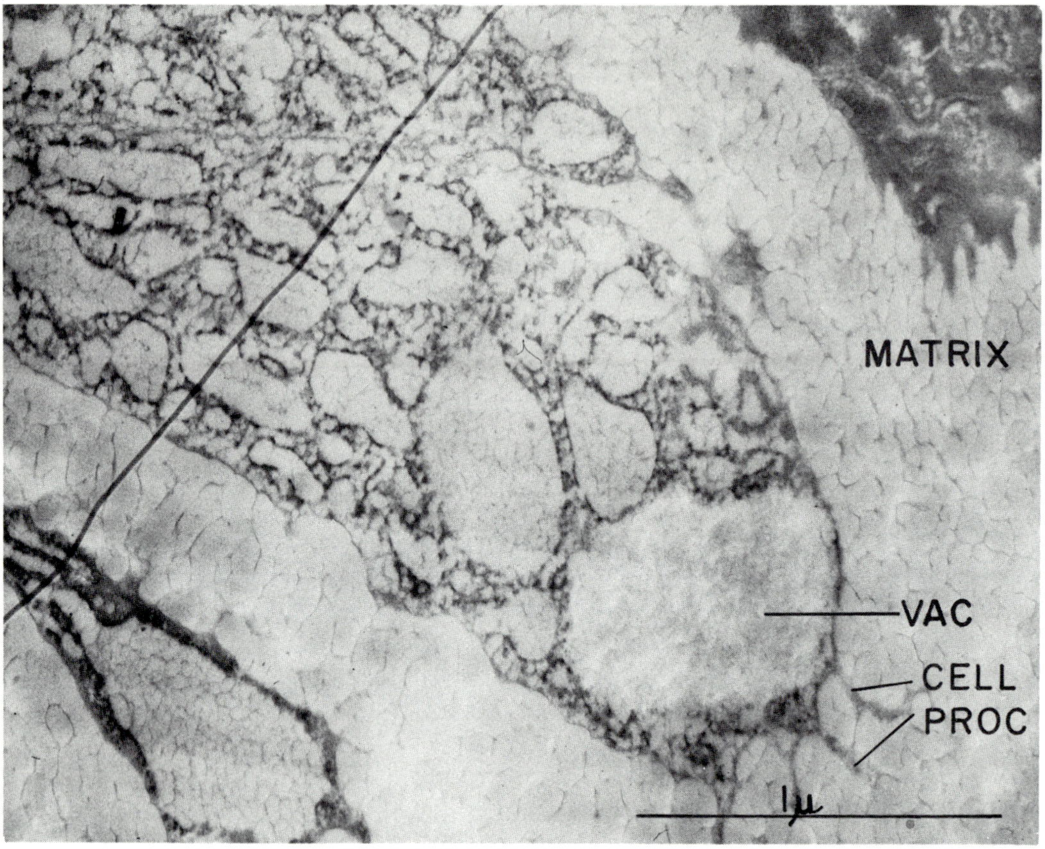

136 6. Possible Protropocollagen Precursors

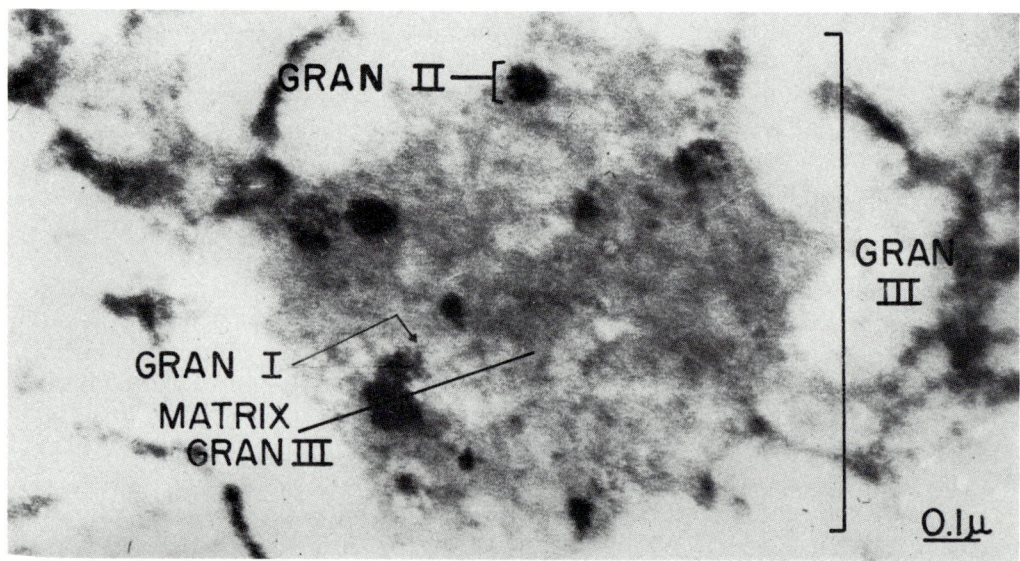

Fig. 6.10A. ▲   See p. 134 for complete legend.   ▼ Fig. 6.10B.

TABLE 6.1

NUMBER OF GRANULES PER SECTION OF A CELL AND PER UNIT OF CYTOPLASMIC AREA IN THREE KINDS OF CHONDROCYTES[a]

	Specific cell (by letter)											
	Postmitotic cells	Preparatory cells							Prelytic cells			
	A–D	E	F	G	H	I	J	K	L	M	N	O	P
Average number of granules per section through a cell	3.8	4.5	2.4	3.0	3.1	2.8	3.6	3.8	2.2	2.3	3.7	0.6	1.2
Number of granules per section through a cell per unit of cytoplasmic area	3.5	1.4	1.6	1.4	1.8	1.7	1.8	1.4	0.7	1.0	0.6	0.2	0.3

[a] A total of 1234 granules was counted.

Observations. When considered on the basis of density of granules per section or per unit area of cytoplasm in a section, it is clear that there are two periods when the number of cytoplasmic granules drops abruptly, and these two periods correspond with the times when the epiphyseal plate is growing rapidly through secretion of matrix, as deduced from different data in Chapter 5 (this volume). These periods occur at times between postmitotic and preparatory stages, and between the latter and the prelytic stage as seen from the data in Table 6.1.

It is also clear that most of the granules occupy a restricted part of the cytoplasm. In one cell, for example, which contained 68 granules, at least 42% of them were located in or around the Golgi apparatus. This confirms the impression obtained from examination of the electron micrographs that a rather large proportion of the granules seemed to be in, or near the periphery of the Golgi apparatus.

Part II—Bone Cells

Methods

Mice, 7 days old, were decapitated, and the calvarium was removed in a moist box. Thin slivers of parietal bone near the midline sulcus were cut coronally and placed on squares of thin aluminum foil, and the whole was frozen ultrarapidly *(12)*. The specimens were dried at a low temperature and postfixed for 5 hours in vapors of FFDNB at 50°C or of FFSulfone at 90°C. Excess reagent was removed by evacuation overnight at a temperature 10°C higher. The vacuum tubes were brought to room temperature and the specimens were infiltrated *in vacuo* slowly with 95% alcohol. After removal from the foil, specimens remained in alcohol for

1 day. After that, they were trimmed and the smaller pieces were infiltrated with methacrylate or with Epon. After polymerization, the blocks were sectioned undecalcified with diamond or glass knives. Sections were viewed in a Philips EM100A electron microscope with a rated resolution of 20 Å, operating at 100 kv. Electron micrographs were made on Kodak spectroscopic film type 649GH. Bone spicules of the long bones of the hind limb of young adult mice were pre-

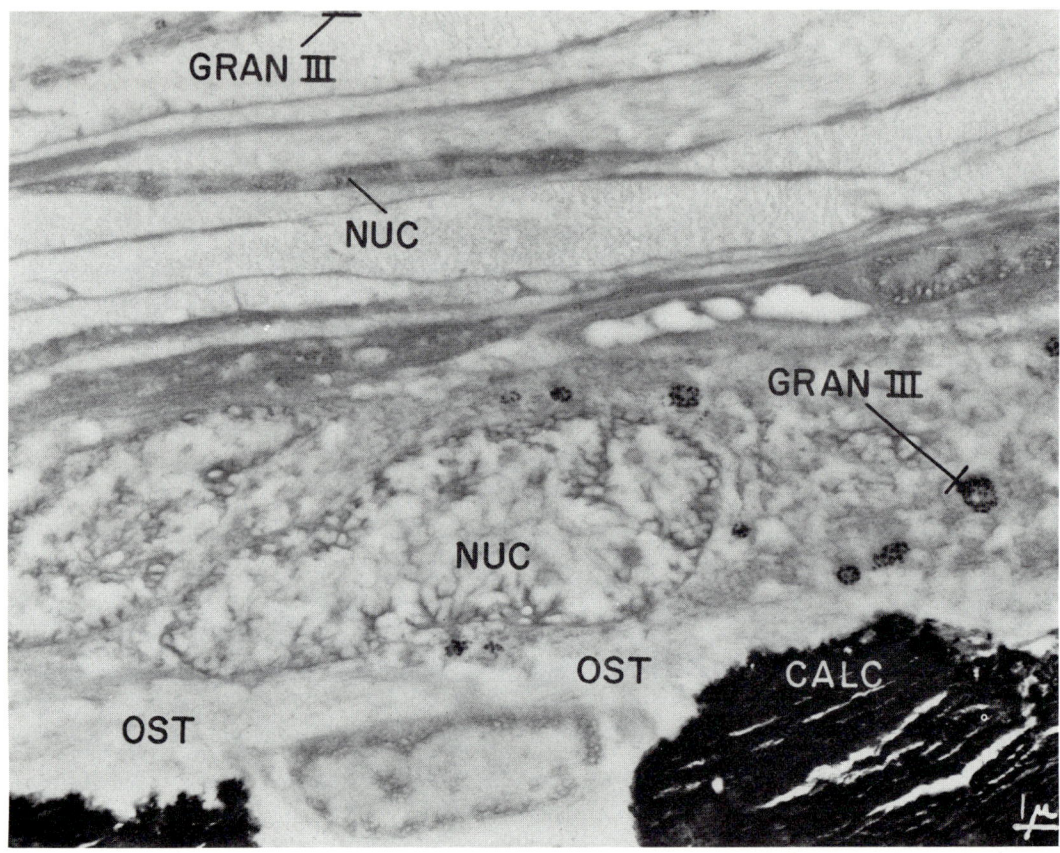

FIG. 6.11.

FIGS. 6.11–6.14. Outer surface of parietal bone of 7-day-old rat, postfixed and stained *in vacuo* with vapors of FFDNB. Calcified matrix (CALC) appears black because of its intrinsic mass owing to the apatite crystals. Cytoplasm of some periosteal cells and osteoblasts contains third-order granules marked by the high contrast second-order granules (GRAN II) (Fig. 6.11). These appear at a higher magnification in Figs. 6.12 and 6.13, and at the highest magnification (Fig. 6.14) are resolvable as clusters of first-order granules (GRAN I). The granules have no clear relation to the RER which is stained weakly. The nuclear chromatin (CHR) is dense. The various orders of cytoplasmic granules are indistinguishable from those in fibroblasts and cartilage cells. Osteoid (OST) appears faintly stained in Figs. 6.11–6.13. Asterisk marks an artifact. NUC, nucleus. Figure 6.11, × 6000; Fig. 6.12, × 21,000; Fig. 6.13, × 42,000; Fig. 6.14, × 84,000.

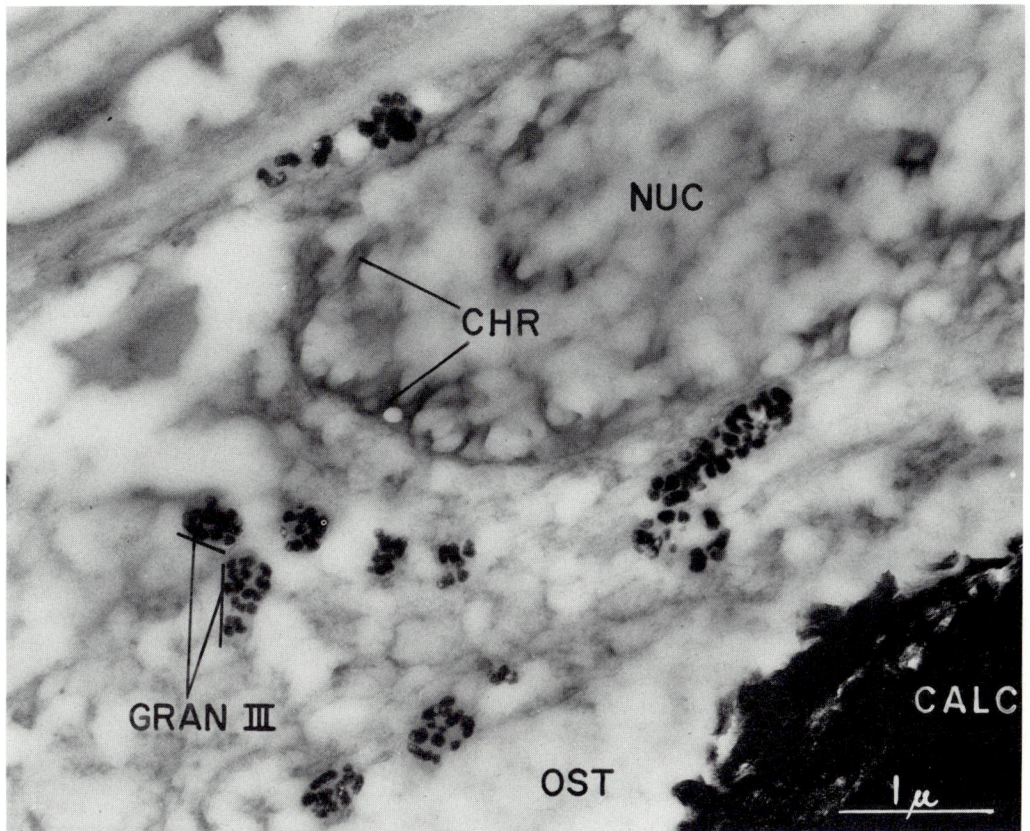

Fig. 6.12.

pared for the study of osteocytes in the same way. For the study of osteoclasts, mice 6 days old were injected intraperitoneally four times during 1 day with a total of 10 units of Parathormone (Eli Lilly), and their parietal bones were prepared like those of normals 30 hours after the last injection.

Some final specimens of parietal bones of normal mice were embedded directly in paraffin. Other pieces were prepared by freezing and drying and treated with vapors of cross-linking reagents *in vacuo*. Some of these were embedded in paraffin, and others were prepared in the usual way for electron microscopy. Sections of both kinds of material in paraffin were cut, mounted on slides, and stained with the PAS method and compared with granules visible in electron micrographs of sections of osteoblasts treated only with the cross-linking reagents.

Observations. Third-order granules are present in the cytoplasm of periosteal cells and osteoblasts (Figs. 6.11–6.14). At higher magnification the second-order granules are resolved into clusters of first-order granules (Fig. 6.14). The third-order granules occur in osteocytes of spicules from the diaphysis of young adult mice, but only rarely, and in osteoblasts nearly totally enclosed in bone matrix

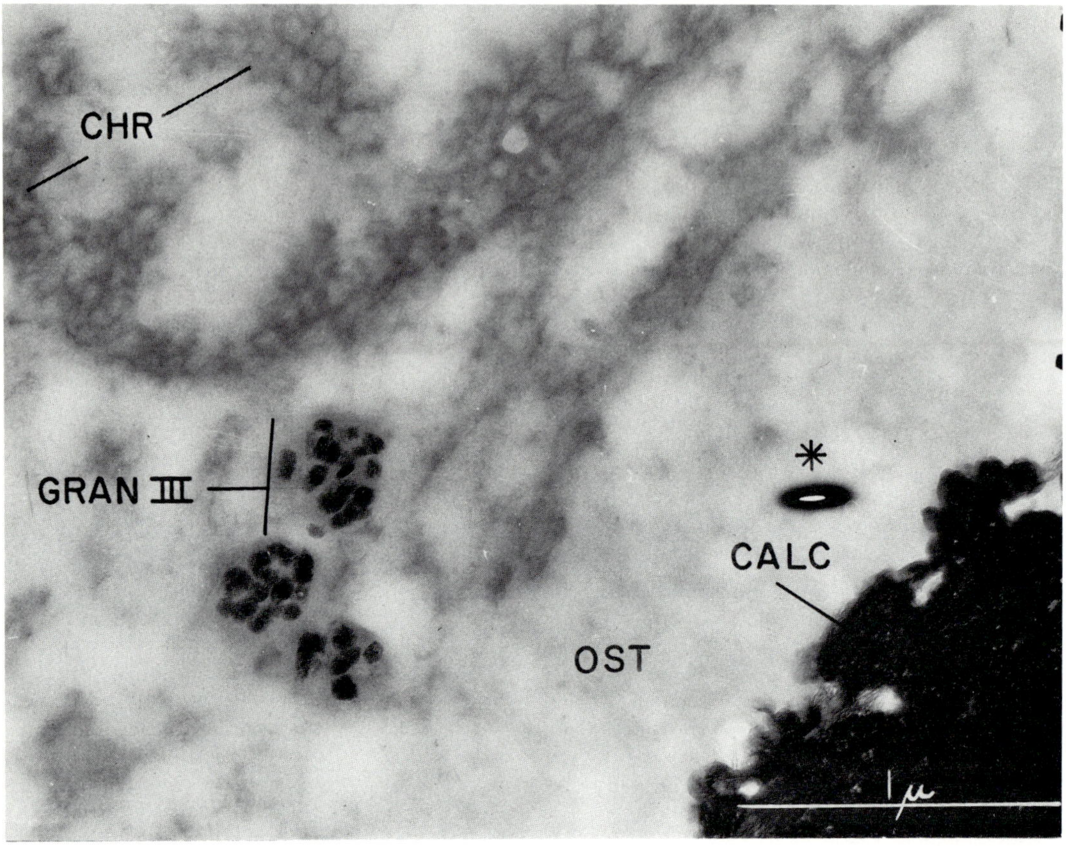

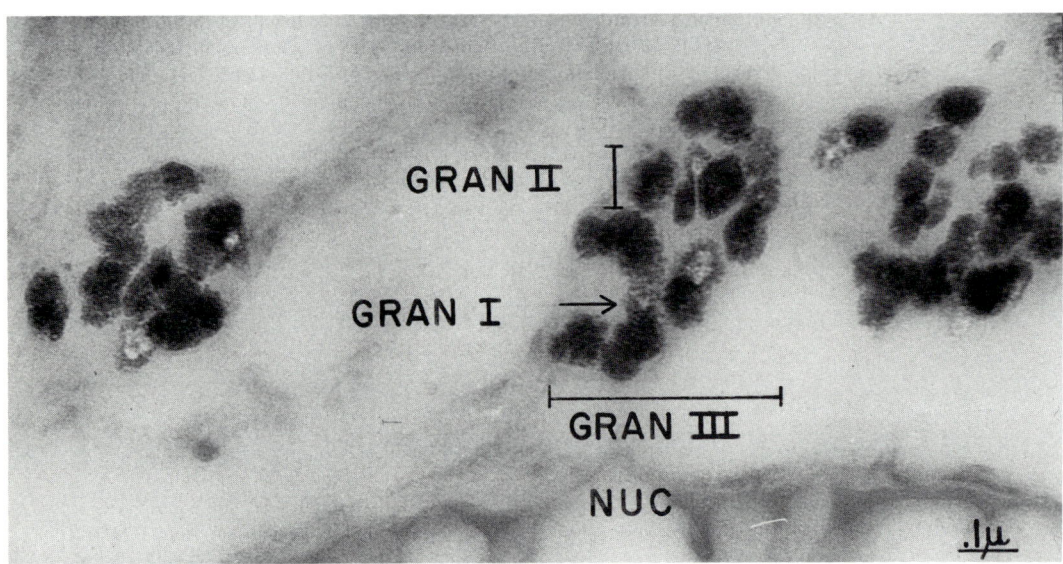

Fig. 6.13. ▲   See p. 138 for complete legend.   ▼ Fig. 6.14.

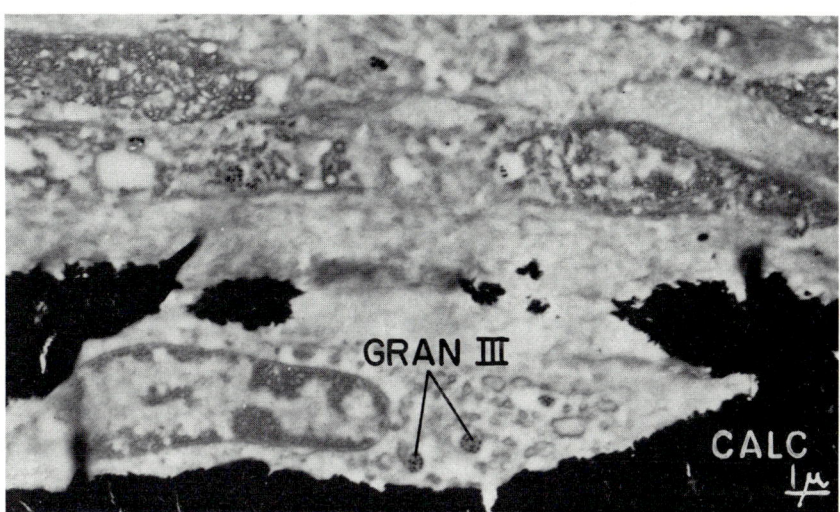

Fig. 6.15. Outer surface of parietal bone of 7-day-old rat, postfixed and stained *in vacuo* with vapors of FFDNB. An osteoblast is almost enclosed by bone matrix and presumably is almost an osteocyte. Two third-order granules are visible in the cytoplasm (GRAN III), identified as clusters of second-order granules. Such granules are very rarely observed in osteocytes. × 6000.

(Fig. 6.15). Osteoclasts, observed in parietal bones of mice treated with Parathormone, also contain cytoplasmic granules (Fig. 6.16).

In the sections prepared from paraffin blocks, PAS-positive granules were observed in specimens prepared by freeze-drying alone and in those postfixed with the cross-linking reagents. However, these granules, which were identical with those described by Heller-Steinberg *(10)*, are different in number and distribution from the third-order cytoplasmic granules visible in the electron microscope.

Discussion

The general comments in this section are based on observations from this and two preceding chapters. Certain features are shared by fibroblasts, chondrocytes, and osteoblasts. At the same time, we recognize that some features are more obvious in one or another cell type. For example, cartilage of epiphyseal plate excels in providing an orderly progression of cell maturation and activity which is more predictable and objective than can be provided by bone or tail tendon. Again, fibroblasts are superb for analyzing the relations of the cell surface to the origin of collagen fibrils. An attempt is made in this section to integrate all our findings and to relate them to published reports by others.

Certainly the major finding is that certain identical granules (the third-order

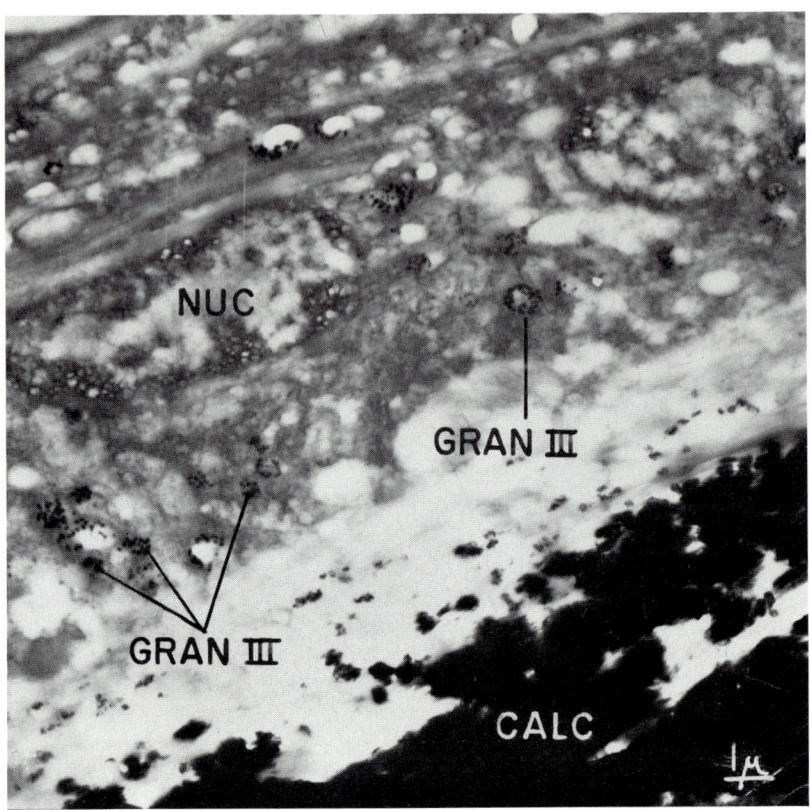

Fig. 6.16. Inner surface of parietal bone of 7-day-old rat, postfixed and stained *in vacuo* with vapors of FFDNB. An osteoclast is present adjacent to eroded and fragmented calcified bone matrix (CALC). The cytoplasm contains several third-order granules (GRAN III), which may reflect the common origin or close relationships of osteoblasts and osteoclasts. NUC, nucleus. × 6000.

granules) are present in the cytoplasm of all three (major) collagen-producing cells, and that they have not been observed in any other cell types studied with the same or other methods. The presence of the same kind of granules in osteocytes (rarely) and in osteoclasts is hardly an exception, since the granules may have originated only a short time before, when the cell may have been secreting collagen as an osteoblast (2, 9). It is, in a sense, as if the program for synthesizing the proteins involved and producing the third-order granule were running its course to completion at the same time as a new program had gotten underway for different activities.

These identical granules are postfixed primarily through the cross-linking action of reagents in a vapor state in the absence of liquid water. These include difluorodinitrobenzene (FFDNB), difluorodinitrodiphenylsulfone (FFSulfone), and mercury(II) hexafluoroacetylacetonate (Hg hfac). The granules are also preserved by 95% alcohol, and by an alcoholic solution of ferric chloride. Granules postfixed and stained with ferric chloride are illustrated in Chapter 8, Figs. 8.2 and 8.6 (this volume). The increased contrast, which these granules acquire after treatment with the vapor reagents, is caused by increased mass resulting from the combination and incorporation of the reacted reagents. It is highly probable that any aqueous or alcoholic solution acting as a fixative of fresh tissue would cause solution of these granules. Bondareff (3) was unable to preserve these granules in freeze-dried specimens, although he tested about two hundred fixative preparations, all in solution. The specific granules of the fibrogenic cells do not correspond with PAS-positive granules in osteoblasts and fibroblasts, or with the granules described by Fitton Jackson (5, 6). The PAS-positive granules are lacking in cartilage cells which are fixed by freezing and drying. By actual experiment, the PAS-positive granules were found to differ in distribution from the third-order granules in osteoblasts. For these reasons, we regard the specific granules as precursors of tropocollagen, possibly protropocollagen. These granules have two components: (1) first-order granules of about 80 Å in diameter which have high contrast and are aggregated into clusters about 1000 Å in diameter, which comprise the second-order granules, and (2) a less dense matrix, which is of very much lower contrast and in which the latter are embedded, the totality constituting the third-order granule which is microscopically visible. The relations of the third-order granules to those which were stained by toluidine blue and other stains when applied to fresh, unfixed cartilage cells is not known (11).

While the matrix of the third-order granule appears quite homogeneous with most methods of postfixation, it appears highly structured after successive exposure of frozen-dried specimens to vapors of Hg hfac and FFDNB (Fig. 6.10A). The units within the matrix may represent the inert, hydrophobic protein [like that described in other cells (Vol. I, Chapter 2) and discussed in this volume (Chapter 11)], which may be synthesized by the ribosomal system and shed into the intercisternal space.

The first- and second-order granules are of such high contrast that it is possible to print the plates so that only they appear on the prints, all other stained portions of the cytoplasm appearing blank. This must mean that the mass of the first-order granules has increased enormously and that the granules must be rich in reactive amino and other groups. Were these reactive groups free to react in the living cell, they could tie up other cellular proteins by combining with them. It would seem probable that in the living cell, the active sites of the third-order granules (which are in the first-order units) are somehow prevented from causing such disasters. This apparent insulation of the active sites of the first-order granules could be achieved in several ways:

1. The matrix of the third-order granule could be rich in phospholipid or phospholipoprotein, with the aliphatic fatty acids organized to form lipid membrane. This possibility seems unlikely from the weak or negative reaction for lipids obtained in some preliminary studies, but is not excluded.

2. The matrix protein may be rich in hydrophobic amino acids, which could

effectively act as a barrier. Such amino acids in proteins would react poorly with cross-linking reagents. In fact, the contrast of the matrix of the third-order granules remains low after staining with the cross-linking reagents or even after the use of uranyl acetate as a stain for sections. It must be admitted, however, that such evidence is negative and may be circumstantial.

3. The matrix protein may be so organized as to induce a high degree of orientation of its water molecules, which would, in effect, limit severely or nearly completely the free activity of such water. This kind of mechanism could effectively insulate the active sites of the first- and second-order granules, but there is no evidence bearing on this possibility. It is clear that the removal of water by sublimation during freeze-drying and the use of the reagent in vapor form enables the latter to approach the active sites on the first-order granules and combine with them.

The third-order granules have been seen in many parts of the cytoplasm of fibrogenic cells—in perinuclear areas, in regions rich in RER or in ribosome aggregates, in or at the periphery of the Golgi apparatus, and in cell processes. In cartilage cells, where the Golgi zone is readily recognized, even when the preparations are not specifically stained, the possible precursor granules are most numerous at the junction of the Golgi and RER zones. When the possible precursor granules occur in the RER region, the relation may be one of accidental contiguity and of confusing plane of section causing difficulties of interpretation due to overlay or underlay. When present in the area rich in RER in fibroblasts and chondrocytes, the possible protropocollagen granules are found in the intercisternal space of the RER, never within the intracisternal space. The intracisternal space of the RER in fibroblasts and chondrocytes is always narrow and relatively uniform in freeze-dried specimens, as has been observed in nearly all other cells studied, without the wide dilations described in the literature. The gap between the adjacent leaflets of the RER is always too narrow to accommodate the third-order granules.

The third-order granules are confined to the cytoplasm—they have not been observed extracellularly in the dense connective tissue of tail tendon or in cartilage or bone. Cell surfaces of fibroblasts and chondrocytes appear in sections to be sharply delimited from the extracellular components, with no deposits or accumulations of any homogeneous or fibrillar material. This is contrary to many reports in the literature which have led their authors to suggest that collagen is somehow formed extracellularly, with the surface acting as a kind of template. This is also contrary to the reported descriptions of fibrils below the surface of cells or in vacuoles, which have led their authors to suggest that fibrils are formed alone or in conjunction with some ground substance component at these sites and then extruded extracellularly. It is possible that these observations recorded in the literature may be artifacts caused by fixation inadequate for the problems studied. They do not occur in our preparations, perhaps because we have avoided these particular artifacts by using different methods of preparations. The voluminous literature on the origin of collagen fibers has been adequately reviewed in recent years, and need not be summarized again here *(1, 5, 8, 13–16)*.

As the first-order granules are of appreciable size and contrast, a determined effort was made to find them in the cisternae of the RER, especially where the ribosomes are located on/in the Golgi region in fibroblasts and chondrocytes or extracellularly, but the results were always negative. We sought unsuccessfully also to find filaments of high contrast in these sites. Perhaps the search might have been successful if the sections were very much thinner (perhaps 20 Å thick or so), and the resolution (and magnification) higher. However, it is also possible that the negative findings are real and not caused by inherent technical limits. We are inclined toward the second possibility as being more plausible.

It is necessary to review certain findings bearing on the growth of the epiphyseal plate. It was stated in Chapter 5 (this volume) that there are two prominent growth periods, a brief one in the postmitotic period, and a more extensive one in the prelytic and lytic stages. As these growth periods are the consequence of the secretion of extracellular matrix, of which the most prominent protein constituents are collagen and the protein component of protein–polysaccharide complex(es), cellular changes in nucleic acids were expected and, indeed, found. The changes were most prominent during the preparatory stage, just preceding the period of greatest growth of the epiphyseal plate. During this preparatory period, the second-order coils of DNA were largely separated from each other, so that neither microscopic nor submicroscopic clumps of nuclear chromatin could be detected. At the same time, the prominent nucleoli of cells in early stages of the preparatory zone disappear in later stages and the numerous RNA granules in the nucleoplasm of the earlier stages are nearly absent. Also, the RER is maximally extended in the cytoplasm. Degenerative changes in the cells take place only during the prelytic and lytic periods, when marked irreversible alterations in nuclear and cytoplasmic nucleic acids are visible, i.e., breaking of second-order coils of DNA into small segments and granules and irregular variations in aggregations of ribosomes (see this volume, Chapter 5). The end of the preparatory stage is accompanied by an abrupt, drastic reduction in the cytoplasm of the possible protropocollagen precursor granules (see Table 6.1). Some possible precursor granules are visible in the lytic cells, but their number is reduced to the vanishing point.

All of the phenomena mentioned up to this point in the discussion have been integrated into a speculative statement of the main features of the secretion of the major extracellular protein components by connective tissue cells. It is suggested that the possible protropocollagen granules (or first-order granules) are synthesized in the cytoplasm in the presence of ribosomes, many of which are attached to the RER, in conjunction with messenger RNA and transfer RNA. All the RNAs arise in the nucleus in a great surge of activity, which is correlated with the separation of the DNA molecules, previously aggregated as chromatin, into individually discrete molecules of DNA which can be recognized as second-order coils with a repeating unit and diameter of approximately 400 Å. The relatively inert protein characteristic of the matrix of the third-order granule is synthesized almost simultaneously and becomes associated with the protropocollagen constituents. This would tend to prevent cross-linking reactions of the active sites of the protropocollagen constituents with other proteins in the cytoplasm. The first-order granules aggregate to form the second-order clusters, and these, in turn, to form the third-order granule. The formation of aggregates could be visualized in simplified form as a series of conveyor belts: A conveyor (transport) belt of relatively high speed transports newly synthesized proteins along the RER. It deposits them on another conveyor belt operating at a lower speed. At the point where the faster belt impinges on the slower one, the newly synthesized proteins would be expected to become heaped up to form second-order granules. This transfer point corresponds to the border of the Golgi apparatus. As the molecules accumulate and reach a certain size, the second-order granule stops growing because too much energy is required (and not enough available) for more molecules to be added to the outside of the granule and new second-order granules are formed. The second-order granules are separated from each other by inert protein molecules also synthesized in association with ribosomes. This inert material is the matrix of the third-order granule. Now the first- and second-order granules, thus enclosed and stabilized, move through the cytoplasm and cell processes and, in response to unknown stimuli, dump their load of granules extracellularly through an opening in the cell surface. In the different conditions of the new environment, every first-order granule or (pro)tropocollagen molecule unrolls to its fullest extent and becomes attached end to end and side by side to others of similar morphology. The mechanisms of such aggregations are being studied in great detail by biochemists and molecular biologists. Whether the granule represents a tropocollagen molecule or one of its constituent helices is unknown.

The speculative secretory process proposed above accounts (1) for the fixation and staining properties of the possible precursor granules of fibrogenic cells, (2) for their synthesis at the far end of a series of events, which begin with DNA molecules in the nucleus and end with their synthesis in relation to ribosomal aggregates, (3) for their distribution (in cartilage cell) at the border region of the Golgi apparatus where it impinges on the RER, (4) for the absence of fibers in the cytoplasm, (5) for the absence of third-order granules from extracellular spaces, and (6) for the appearance of fibers on or near the surface of the cells of origin by molecular reorganization and aggregation.

It is worth pointing out that in cartilage, chondrocytes impinge directly on the extracellular matrix, with no intervening space. This close contact has been illustrated in Chapter 5 (this volume) and will be treated in greater detail in the next chapter, which deals with the chemical organization of cartilage matrix. A space appears in all electron micrographs of all researchers who have used fixatives in solution to preserve epiphyseal plate; the space is lacking only in the work of Durning (4), who fixed his specimens by freezing and drying, like the authors of this article. It has been known since the end of the last century that pericellular spaces could be produced at will by variations in tonicity of the fluid in which cartilage is immersed. It requires only a little solution of the protein component (at least) of the protein–polysaccharide complex during immersion fixation, and subsequent reprecipitation, to account for the amorphous, granular, and fibrillar components described in the pericellular space. See Schaffer (15) for early references.)

It should be emphasized that in the fibrogenic cells, as in hepatic and pancreatic acinar cells, protoplasm comprises a two-phase system (7). Staining with platinum tetrabromide emphasizes two regions of cartilage cells. The first corresponds with the RER, whose extensive cisternal walls are thick, with thinner walls extending between the two leaves of the flat cisternae. The second is extraordinarily pale-staining. It comprises numerous submicroscopic vacuoles or compartments separated by thin walls. These extend to the cell surface, and their pattern is not dissimilar to that of the extracellular matrix. A possible relation of both intra- and extracellular pattern to the origin and organization of the protein–polysaccharide complex of cartilage matrix will be discussed in the next chapter.

Summary

Certain distinctive granules occur in the cytoplasm of chondrocytes of the epiphyseal plate, in osteoblasts of rat parietal bone (as well as in osteocytes and osteoclasts), and in fibroblasts of rat tail tendon. They have not been observed in any other cell type. In the cartilage cells, the number of granules falls sharply during and preceding the periods of accelerated growth of the epiphyseal plate. These changes are correlated with changes in distribution of nucleic acids, including DNA and RNA. The granules in cartilage cells are denser in or near the periphery of the Golgi apparatus. When observed in the RER, the granules always are in the intercisternal space, never within the cisternae. The granules have not been observed in the extracellular space. The granules are believed to be a possible precursor of tropocollagen. A model is proposed which may account for most observations from the origin of the granules to the formation of collagen fibrils.

References

1. Anderson, C. E., and Parker, J. (1968). Electron microscopy of the epiphyseal cartilage plate. A critical review of electron microscopy observations on enchondral ossification. *Clin. Orthop.*, **58**, 225–241.
2. Bloom, W., and Fawcett, D. W. (1968). "A Textbook of Histology," 9th ed. Saunders, Philadelphia, Pennsylvania.
3. Bondareff, W. (1957). Submicroscopic morphology of connective tissue ground substance with particular regard to fibrillogenesis and aging. *Gerontologia* **1**, 222–233.
4. Durning, W. C. (1958). Submicroscopic structure of frozen-dried epiphyseal plate and adjacent spongiosa of the rat. *J. Ultrastruct. Res.* **2**, 245–260.
5. Fitton Jackson, S. (1964). Connective tissue cells. *In* "The Cell" (J. Brachet and A. E. Mirsky, eds.), Vol. 6, pp. 387–520. Academic Press, New York.
6. Fitton Jackson, S., and Smith, R. H. (1957). Studies on the biosynthesis of collagen. I. The growth of fowl osteoblasts and the formation of collagen in tissue culture. *J. Biophys. Biochem. Cytol.* **3**, 897–912.
7. Gersh, I., Isenberg, I., Bondareff, W., and Stephenson, J. L. (1957). Submicroscopic structure of frozen-dried liver specifically stained for electron microscopy. II. Biological. *Anat. Rec.* **128**, 149–169.
8. Godman, G. C., and Porter, K. R. (1960). Chondrogenesis studied with the electron microscope. *J. Biophys. Biochem. Cytol.* **8**, 719–760.
9. Greep, R. O. (1966). "Histology," 2nd ed. McGraw-Hill, New York.

10. Heller-Steinberg, M. (1951). Ground substance, bone salts and cellular activity in bone formation and destruction. *Amer. J. Anat.* **89**, 347–379.
11. Hirschman, A., and McCabe, D. M. (1969). Staining of intracellular granules in fresh epiphyseal cartilage by cationic dyes. *Calcif. Tissue Res.* **4**, 260–268.
12. Molnar, Z. (1959). Development of the parietal bone of young mice. I. Crystals of bone mineral in frozen-dried preparations. *J. Ultrastruct. Res.* **3**, 39–45.
13. Revel, J.-P., and Hay, E. D. (1963). An autoradiographic and electron microscopic study of collagen synthesis in differentiating cartilage. *Z. Zellforsch. Mikrosk. Anat.* **61**, 110–144.
14. Ross, R., and Benditt, E. P. (1961). Wound healing and collagen formation. I. Sequential changes in components of guinea pig skin wounds observed in the electron microscope. *J. Biophys. Biochem. Cytol.* **11**, 677–700.
15. Schaffer, J. (1930). Knorpelgewebe. *In* "Handbuch der mikroskopischen Anatomie des Menschen" (W. H. W. v. Möllendorff, ed.), Vol. 2, Part 2, pp. 210–390. Springer, Berlin.
16. Wassermann, F. (1956). The intercellular components of connective tissue: origin, structure and interrelationship of fibers and ground substance. *Ergeb. Anat. Entwicklungsges.* **35**, 240–333.

7

Morphochemical Study of the Matrix of Epiphyseal Plate and Joint Cartilage and the Origin of Protein–Polysaccharide Complex

Isidore Gersh

The polysaccharides of hyaline cartilage matrix are bound to protein, and together form a series of protein–polysaccharide complexes (P-PCs) (3, 27, 32, 41, 47, 48, 58), also called glycosaminoglycans and chondromucoproteins. While there is no agreement on the exact structure of these molecules, it seems clear that the P-PCs are of very high molecular weight, and consist of a protein backbone, to which are attached many polysaccharide chains, each a polymer of some fifty monomers. The P-PC molecules occupy an enormous volume, and are thought to be closely attached to collagen. The origin and disposition of the protein and polysaccharide moieties of the P-PC in cartilage and their relation to each other and to collagen are the major subjects of this chapter. These subjects were studied chiefly in the epiphyseal plate. The matrix of joint cartilage was also studied to ascertain whether the findings on the matrix of epiphyseal plate were equally applicable to the former.

Cartilage matrix has been subdivided into three microscopic zones: the capsule immediately adjacent to the cartilage cells, the pericellular territory, and the interterritorial region. In freeze-dried material, the surface of cartilage cells abuts directly on the matrix, with numerous delicate processes extending into it for short distances. At the interface, the matrix is identical with that of the pericellular and interterritorial regions, and there is no indication of any structure or space between the matrix and the cell surface *(11)*. Early workers had shown that such spaces could be produced at will by hypertonic fluids and by fixative solutions. (See this volume, Chapter 5 for a discussion of the capsule as artifact.)

The pericellular and interterritorial regions appear in the light microscope to be composed of tightly packed compartments (see Figs. 5.37 and 5.42, this volume,

Chapter 5). In the interterritorial regions of the longitudinal septa, the compartments are somewhat elongated in the main axis, while in the transverse septa the compartments are less prominently oriented. In the pericellular territories, the compartments tend to be somewhat larger and more spherical. The dimensions of the compartments and of their walls were estimated from densitometric tracings to be, respectively, approximately 3000 Å, and somewhat less than 380 Å. The problem concerning the disposition of protein and polysaccharide moieties in cartilage matrix can now be restated more specifically: How are these components and collagen distributed as between the walls and contents of the microscopically visible matrix compartments, and how do they originate?

Methods

Collagen

Morphological Studies. Thin slices of the proximal epiphyseal plate of rat tibia were frozen ultrarapidly and dried at a temperature lower than $-40°C$, as described in Chapter 2 (Vol. I). Specimens were postfixed in 95% alcohol, and infiltrated with and embedded in water-soluble Durcupan. Some ultrathin sections were mounted on grids and examined in the electron microscope without staining, while others were stained on the grid with uranyl acetate.

Collagenase Used as an Aid in the Identification of Collagen. Freeze-dried specimens of the epiphyseal plate were postfixed *in vacuo* with vapors of FFSulfone at 50°C as described in Chapter 2 (Vol. I). After excess vapors were removed, specimens were infiltrated with 95% alcohol in which they remained for 1 day. All specimens were passed through 80, 60, and 40% alcohol (30 minutes in each change) and briefly immersed in buffer (0.0067 M phosphate buffer, pH 7.4, plus 0.45% NaCl). Some specimens were transferred to a solution of 0.1 ml of Worthington's purified collagenase solution in 5 ml of the buffer. The collagenase solution was claimed to be free of proteolytic activity as tested with casein. Other specimens remained in the buffer. After incubation at 37°C for 2 hours, the specimens were dehydrated (2 hours each in 40, 50, 60, 70, 80, and 95% alcohol) and embedded in water-soluble Durcupan. Some sections were mounted directly on grids, while others were stained with uranyl acetate.

Polysaccharide Component

Identification by Adding Mass. Di- and multivalent cations are known to combine with sulfated polysaccharides, especially when they are highly polymerized. After freezing and drying, specimens were treated with such cations in vapor form *in vacuo*, and sites of increased density in the matrix in electron micrographs were noted. While this method cannot be claimed to be specific, it is clear that in cartilage matrix, most of the anions are sulfates with relatively few carboxyls and phosphates *(27, 47).* The reagents used and the conditions of use are listed in Table 7.1. Technical details are given in Chapter 2 (Vol. I), except in the case of benzethonium acetate. Details on this are found in Table 7.1.

Methods

TABLE 7.1
CATIONIC REAGENTS USED TO INCREASE MASS OF SULFATED POLYSACCHARIDES AND THEREBY INCREASE CONTRAST IN ELECTRON MICROGRAPHS OF CARTILAGE MATRIX

Reagent	Temperature (°C)	Reaction time (hours)	Temperature for evacuation of excess (°C)
Ba acac · $2H_2O$	40	5	50
Ca acac	40	5	50
La acac	27	5	35
Co tfac	27	5	35
Hg hfac	0	5	27
B A (benzethonium acetate)	27	5	27

TABLE 7.2
TREATMENT OF SECTIONS WITH VARIOUS SOLUTIONS TO TEST SOLUBILITY OF MATERIAL IN LMCWs, INCLUDING TEMPERATURE AND EXPOSURE TIME

Reagent and Conditions	Decrease of contrast in LMCW	Solubility of "purified" P-PC according to Schubert (47)
1. Control (no treatment) (Fig. 7.12)	No	
2. 0.15 N KCl, 0°C, 10 minutes	Yes	Yes
3. 0.1 N NaOH, 24°C, 10 minutes	Yes	Yes
4. 0.3 M NH_2OH, pH 7, 24°C, 2 hours (Figs. 7.15 and 7.16)	Yes	Yes
5. 0.5 N KCH_3COO, 0°C, 2 hours (Fig. 7.13)	Yes	Yes
6. 4.0 N KCH_3COO, 0°C, 1 hour (Fig. 7.14)	No	No

Benzethonium resembles cetylpyridinium in that it is a heavily charged ammonium derivative, but has a larger molecular weight. The acetate was prepared from the commercially available chloride by my colleague, Dr. R. M. Iyengar, who described the method of preparation as follows:

> Dowex 1–chloride anion exchanger was converted to the hydroxide by repeated batch-washing with 2 N NaOH. The resin was packed into a column, washed exhaustively with distilled water followed by 2 N acetic acid. The excess acid was removed from the column by repeated washing with distilled water. Five milliliters of an aqueous solution of benzethonium chloride was then passed slowly through the column, and 100 ml. of eluant were collected. The water was removed by distillation *in vacuo*, and the residual benzethonium was collected as a waxy powder.

Concentrated alcoholic ferric chloride solution was also effective in adding mass to the walls of the large matrix compartments. Technical details are given in the next chapter.

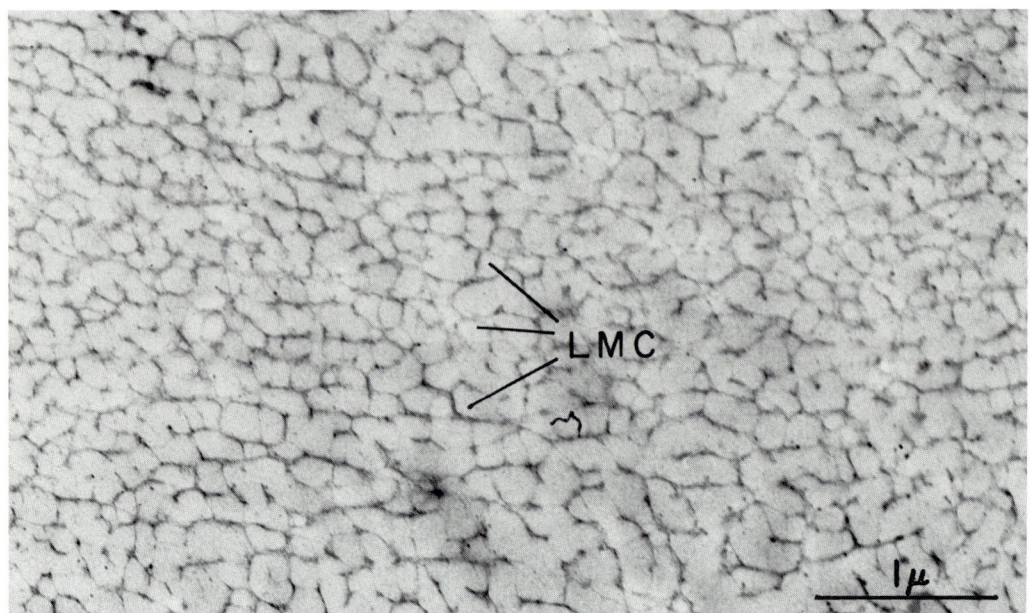

Fig. 7.1.

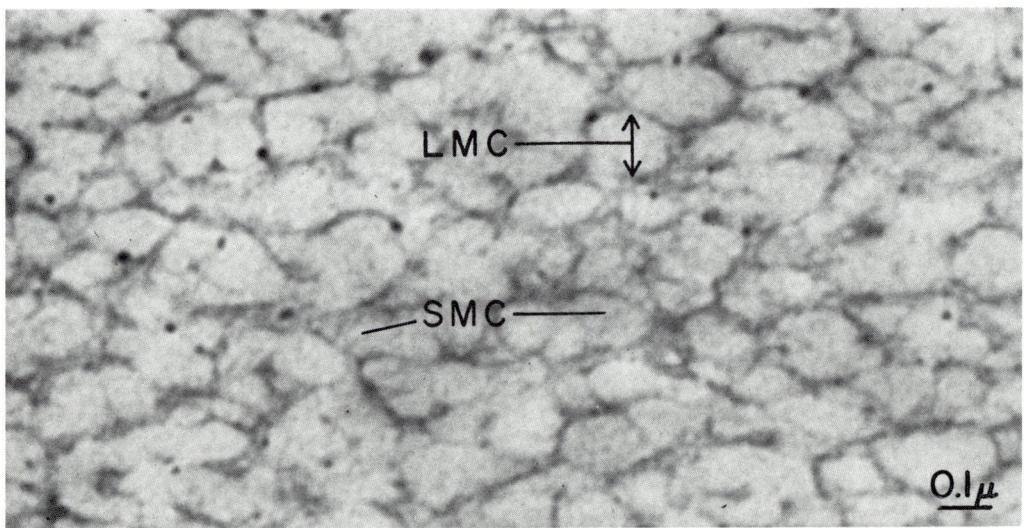

Fig. 7.2.

Figs. 7.1 and 7.2. Electron micrographs of an exceptionally thin section, postfixed in 95% alcohol, unstained. The large matrix compartments (LMC) are surrounded by walls about 380 Å thick. These enclose numerous small matrix compartments (SMC) of more uniform diameter, 500–600 Å, enclosed by thin walls with a maximum thickness of about 100 Å. Figure 7.1, × 2400; Fig. 7.2, × 70,000.

Methods

Identification by Solution Studies. P-PCs are readily extractable, and their solubilities have been of great value in their purification *(48).* After freezing and drying, specimens were postfixed in 95% alcohol and embedded in water-soluble Durcupan. Sections were treated with certain solvents and then studied for changes in density, which could be attributed to solution of the P-PCs. The solubility experiments were made on freeze-dried specimens of epiphyseal plate after postfixation in 95% alcohol. Sections of specimens, postfixed in alcohol and stained with uranyl acetate, served as a base line. Other sections of the same block were immersed in a single drop of the test solution, the solution was washed off with alcohol, and the sections were then stained with uranyl acetate. Comparable prints of both control and extracted section were made to ascertain whether the material in the walls of the large matrix compartments was removed. The conditions and results are given in Table 7.2, p. 151.

CELLULAR ORIGIN OF P-PC

A variety of methods for staining proteins was used, which consisted mainly of platinum tetrabromide in 95% alcohol and various organometallic compounds in vapor form, some in sequence. The most informative sequences thus far are HEF-DOD, and Hg hfac followed by FFDNB, under the conditions described in Chapter 2 (Vol. I). Also helpful in this connection was the use of benzethonium acetate.

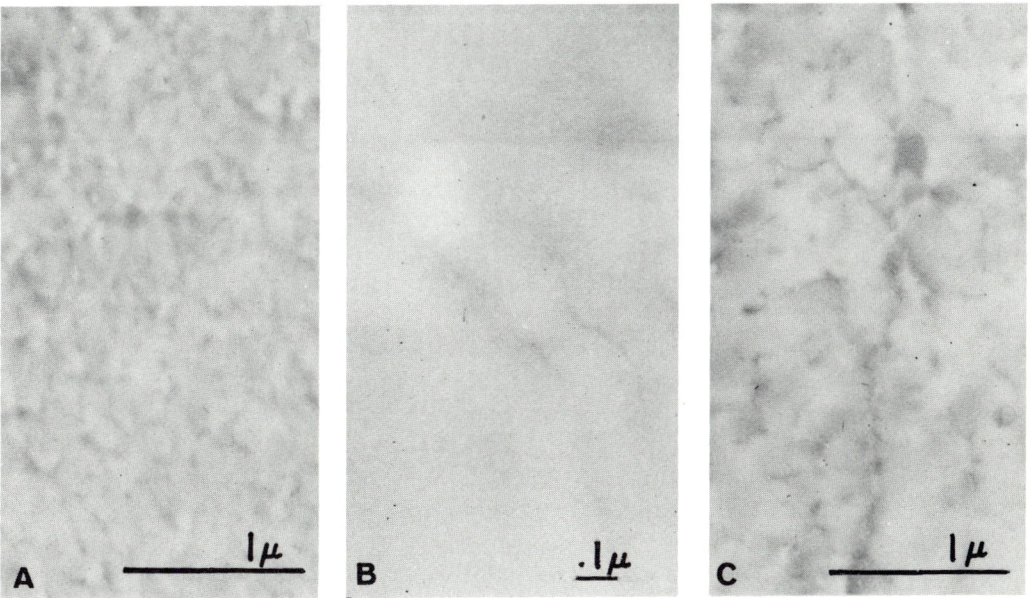

FIG. 7.3. Cartilage matrix of freeze-dried epiphyseal plate, postfixed in 95% alcohol. The LMCWs are denser than the contents in some parts, but not visible in others. These three examples show the range of variation, as a base line control, for preparations stained with cations (in Figs. 7.7–7.11). (A) (left) and (C) (right) × 24,000; (B) (middle) × 50,000.

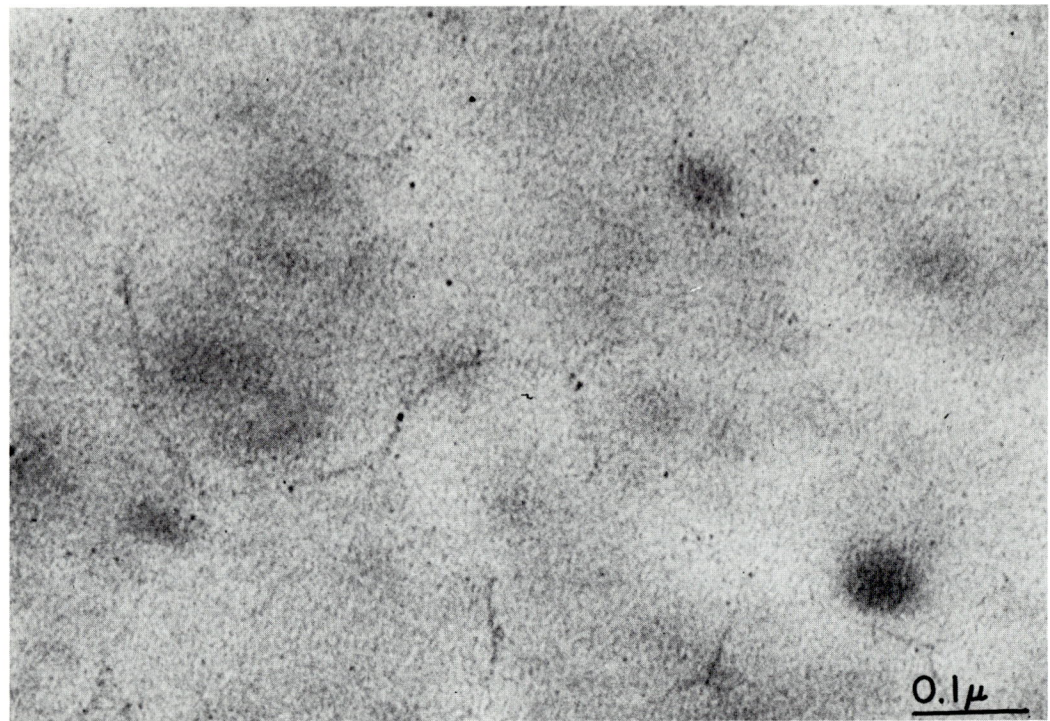

Fig. 7.4. Higher magnification, to show the rather uniform diameter (about 40 Å) of the collagen fibrils in the walls of the large matrix components. The fibers have a periodicity of about 200 Å. × 140,000.

Observations

In one series of unusually thin preparations of epiphyseal plate, the walls of the matrix compartments appeared with exceptional clarity (Fig. 7.1). At higher magnification, the compartments were found to be more finely divided into still smaller units whose walls were more or less uniform (Fig. 7.2). The submicroscopic compartments are approximately 500–600 Å in diameter, and their walls have a maximum apparent thickness of about 100 Å. The two classes of compartments will be referred to hereafter as large matrix compartments (LMC) and small matrix compartments (SMC), and their walls as LMCW and SMCW. The LMC is microscopic, while the SMC is submicroscopic, as are also the LMCWs and the SMCWs. The term "microscopic" refers to objects resolvable with the light microscope; objects not resolvable in this way are regarded as submicroscopic.

The location of collagen in the walls of the LMCs was clearly shown in electron micrographs of epiphyseal plate postfixed in 95% alcohol, without staining. In nearly all electron micrographs of sections of such specimens, the walls of the compartments are visible in the matrix, and appear of unequal density (Figs.

Observations

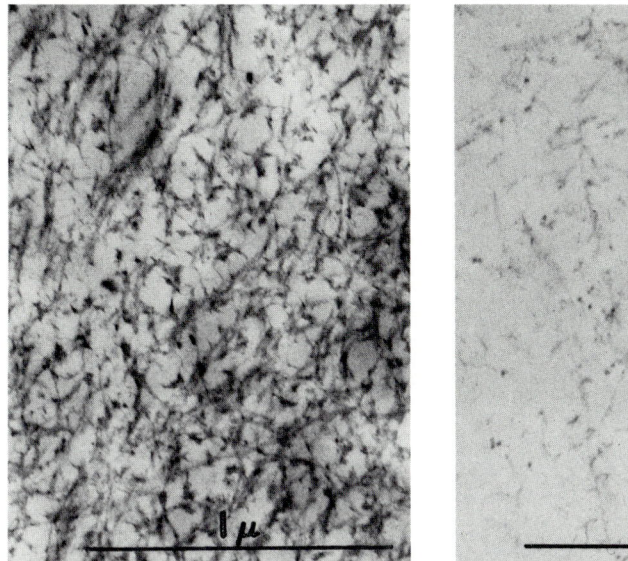

FIGS. 7.5 and 7.6. Specimens were treated with buffer (Fig. 7.5) or with collagenase dissolved in the same buffer (Fig. 7.6). Sections were stained with uranyl acetate. In the control specimen, numerous straight segments of collagen fibrils are present in the matrix, while in the collagenase-treated specimens, a few shortened traces of collagen fibrils may be seen. × 40,000.

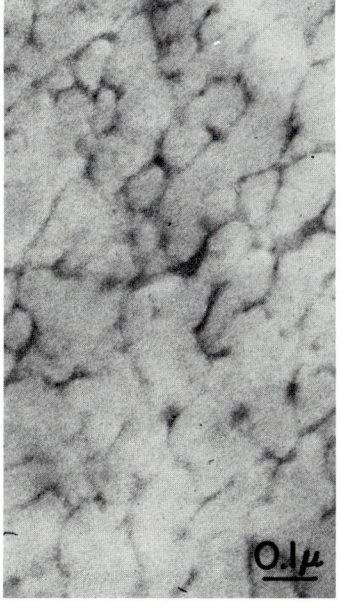

FIG. 7.7. Staining of LMCWs with vapors of hydrated barium acetylacetonate. Some SMCWs are also stained. × 70,000.

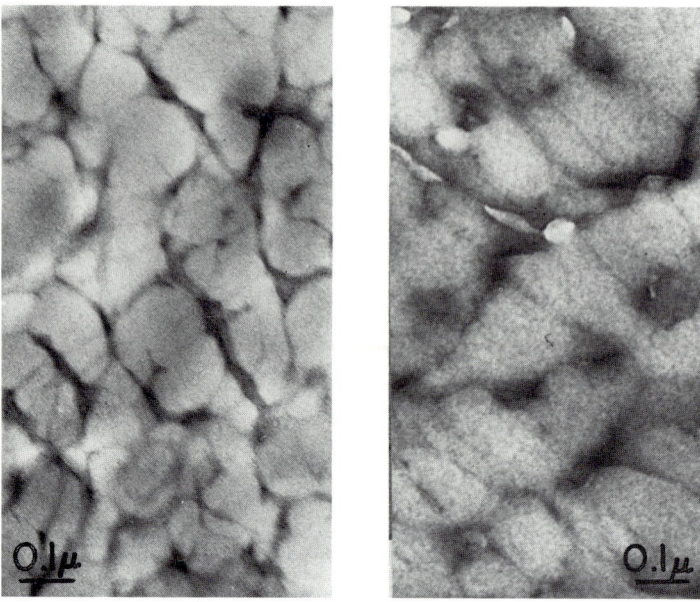

FIGS. 7.8 (LEFT) and 7.9 (RIGHT). Staining of LMCWs and SMCWs *in vacuo* with vapors of calcium acetylacetonate. Figure 7.8, × 70,000; Fig. 7.9, × 70,000.

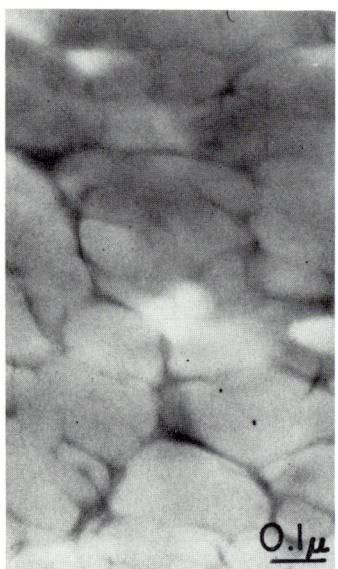

FIG. 7.10. Staining of LMCWs with vapor of cobalt (III) trifluoroacetylacetonate. The LMCWs are clearly stained. × 70,000.

Observations

7.3A–C). When the denser regions are studied at higher magnifications, fibrils are observed which have a maximum diameter of about 40 Å and a periodicity of about 200 Å (Fig. 7.4). The fibrils are also visible in stained sections (see Fig. 7.16). In joint cartilage (p. 173), most fibrils are of larger diameter.

An experiment using collagenase confirmed this identification of collagen fibrils. Thin slivers of frozen-dried epiphyseal plate were postfixed in FFSulfone and treated with collagenase or an appropriate buffer free of enzyme. Collagen fibrils were visible in sections of unstained buffer controls, but were not apparent in sections of enzyme-treated specimens. The stained sections were confirmatory in that numerous collagen fibrils were observed in controls (Fig. 7.5), while only a few pale staining short fibrils were apparent in enzyme-treated specimens (Fig. 7.6). Unlike all other fibrils described above, these were straight, and lacked the delicate curvatures of fibrils observed in all specimens not treated with aqueous reagents. This point was noted by Durning *(11)*, and will be discussed on p. 173.

Localization of Polysaccharide Component of the P-PC

All the reagents tested (Table 7.1) increased contrast preferentially of both LMCWs and SMCWs, with the exception of Ba acac which failed to enhance

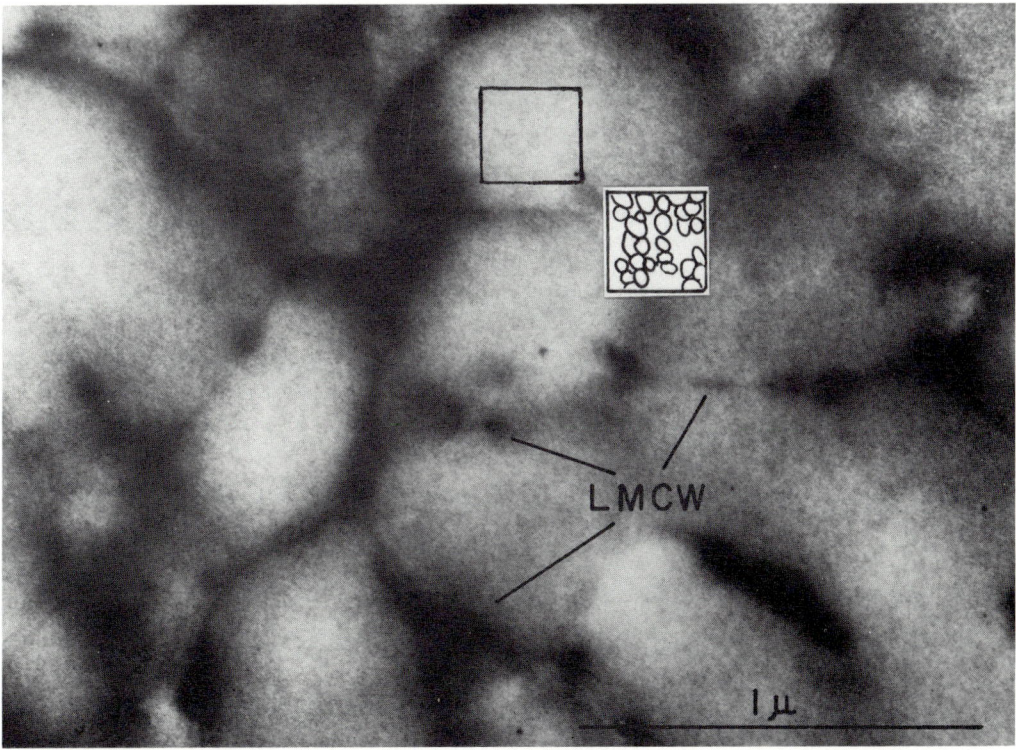

Fig. 7.11. Staining of SMCWs with vapors of benzethonium acetate (indicated in the insert). The walls are thicker than usual. × 54,000.

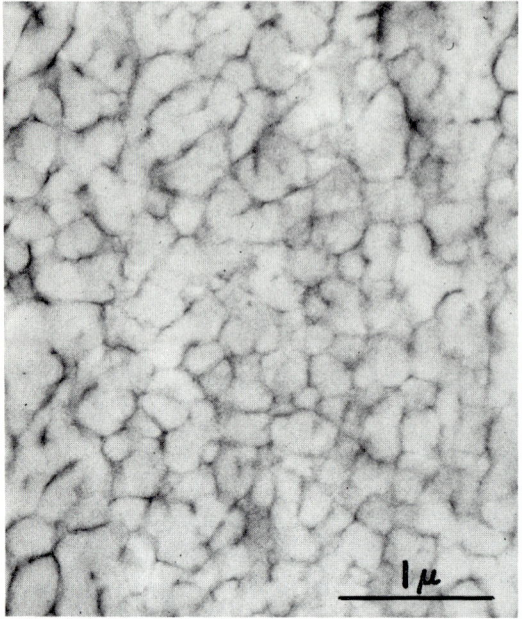

Fig. 7.12.

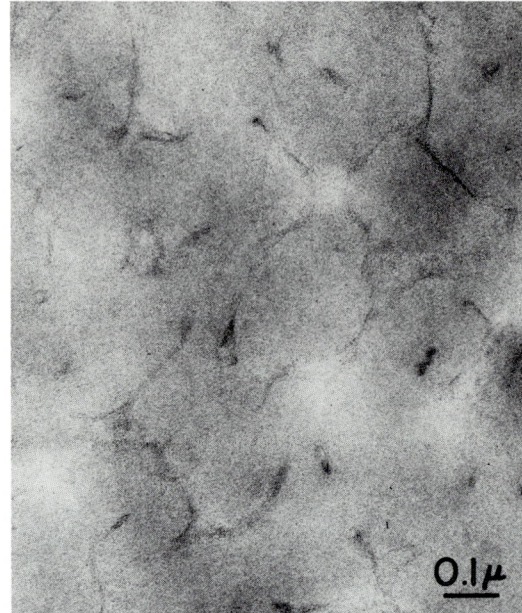

Fig. 7.13.

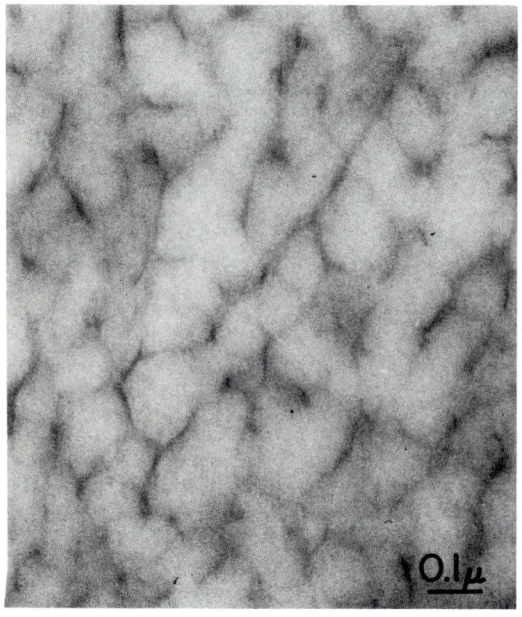

Figs. 7.12–7.14. Cartilage matrix postfixed in alcohol and stained with uranyl acetate (Fig. 7.12) as a control for the effects of 0.5 N potassium acetate (Fig. 7.13) and 4 N potassium acetate (Fig. 7.14). All sections were stained with uranyl acetate. In comparison with the uniform, rather dense staining of the LMCWs of the controls, the walls are paler and more irregular after extraction with dilute potassium acetate solution. The denser, thin lines in the LMCWs are probably collagen fibrils. The more concentrated potassium acetate solution leaves the LMCWs relatively unaltered, and SMCWs can bee seen. Figure 7.12, × 20,000; Figs. 7.13 and 7.14, × 70,000. Please note differences in magnificaton in figures of control versus extracted specimens.

Fig. 7.14.

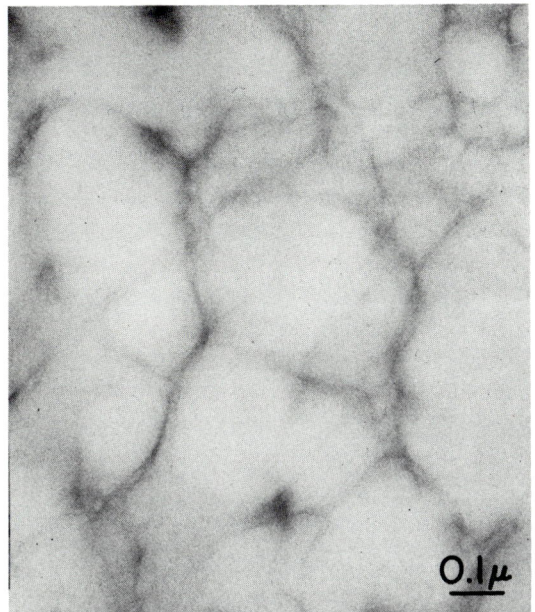 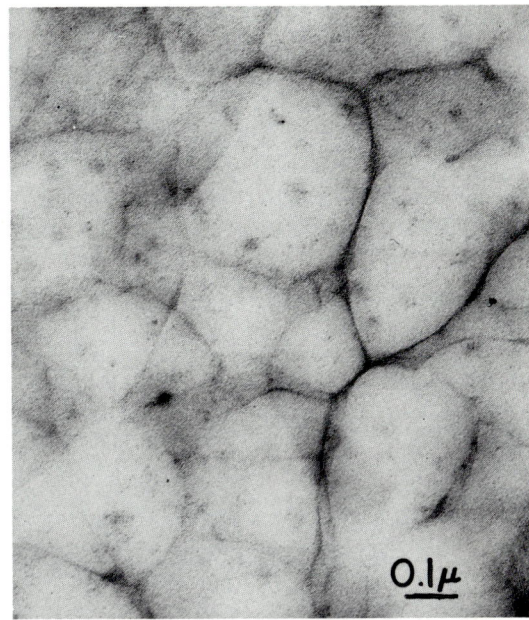

Figs. 7.15 (LEFT) and 7.16 (RIGHT). Cartilage matrix postfixed with alcohol and stained with uranyl acetate as a control (Fig. 7.15) for the extraction of the section with 0.3 M hydroxylamine (Fig. 7.16). The hydroxylamine has extracted the contents of the LMCWs except for some fine fibrils. $\times$ 70,000.

contrast of the latter (Figs. 7.7–7.11). These observations indicate that sulfated polysaccharides are present in the walls of the large and small matrix compartments. This was confirmed by the solubility studies, the results of which are summarized in Table 7.2, at least as concerns the LMCWs. It appears from the observations that the major components which contribute to contrast in the LMCWs have the same solubilities as purified P-PC. It should be noted that when the material of the LMCWs was dissolved, there remained always some short, sharply outlined lines corresponding with the collagen fibers of control specimens. Examples of the evidence on which these statements are based are presented in Figs. 7.12–7.16.

Cellular Origin of P-PC

In the preceding chapter, three examples were shown of the appearance of cartilage cells after postfixation and staining with platinum tetrabromide (this volume, Chapter 6, Figs. 6.8–6.10). It is clear from these electron micrographs and from others such as Fig. 7.17, prepared in a similar way, that the contents of the intercisternal spaces closely resemble the compartmentalized structure of the cartilage (extracellular) matrix. This could be observed also in cartilage cells postfixed and stained with many vapor reagents. In all such preparations, cell processes

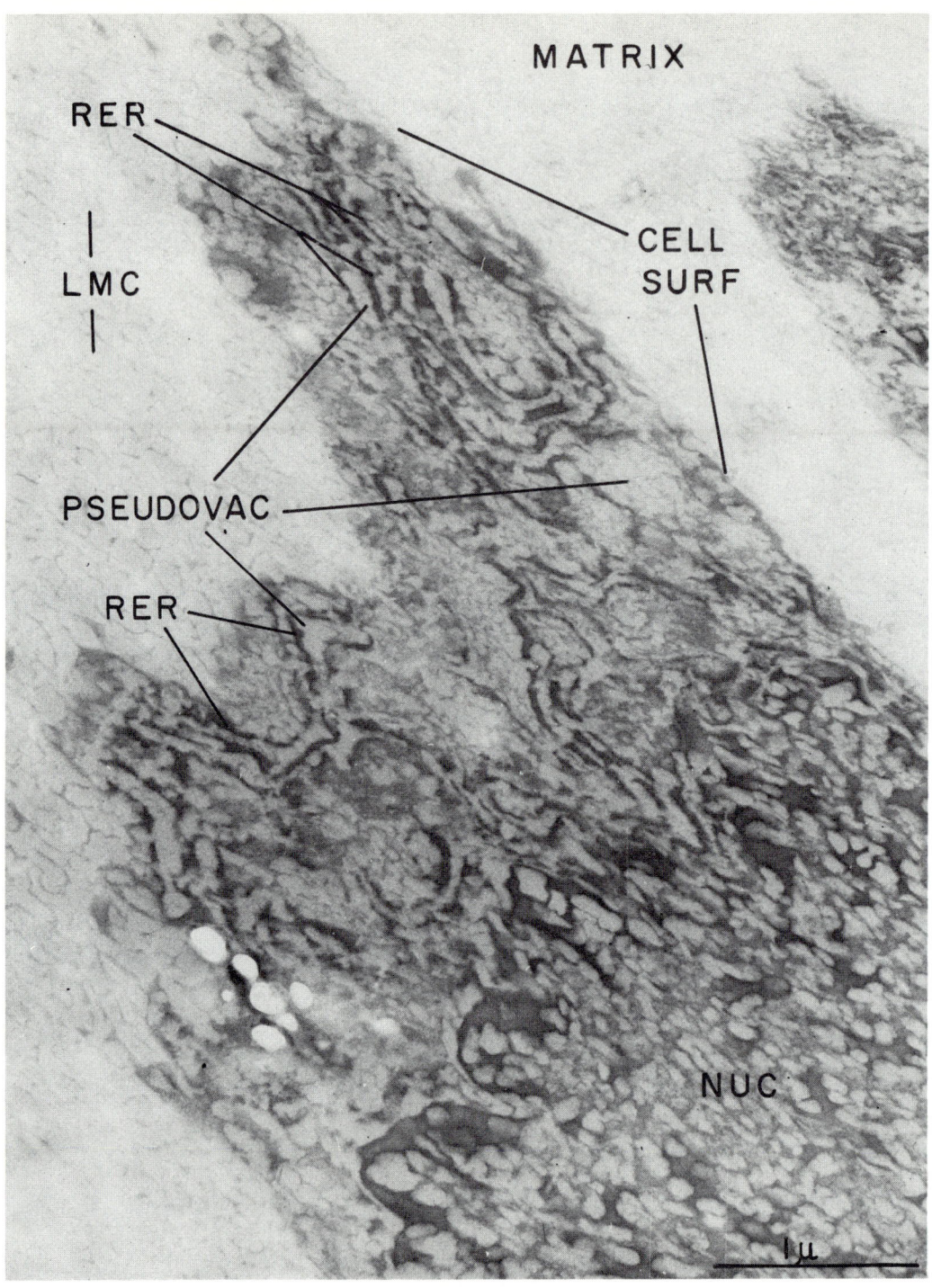

Fig. 7.17. Cartilage cell of epiphyseal plate postfixed and stained with platinum tetrabromide. The cell abuts directly on the cartilage matrix (MATRIX), with no space intervening between it and the cell surface (CELL SURF). Pseudovacuoles (PSEUDOVAC) characterize both cytoplasm (CYT) and nucleus (NUC). The walls of large matrix compartments are stained (LMC). × 30,000.

Observations

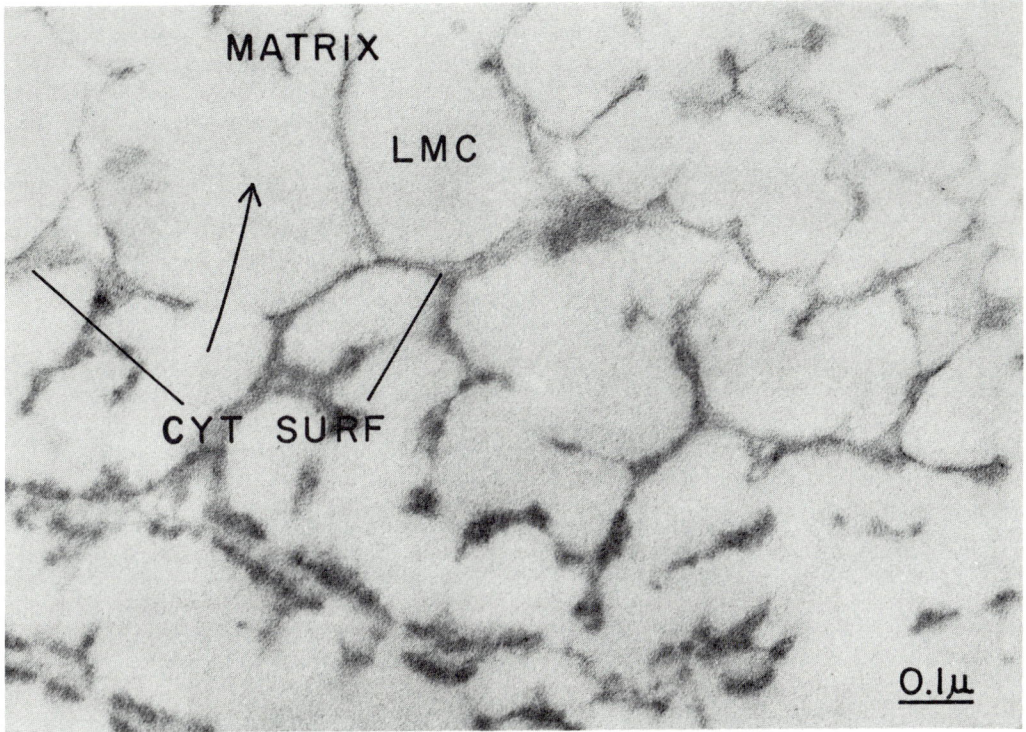

Fig. 7.18.

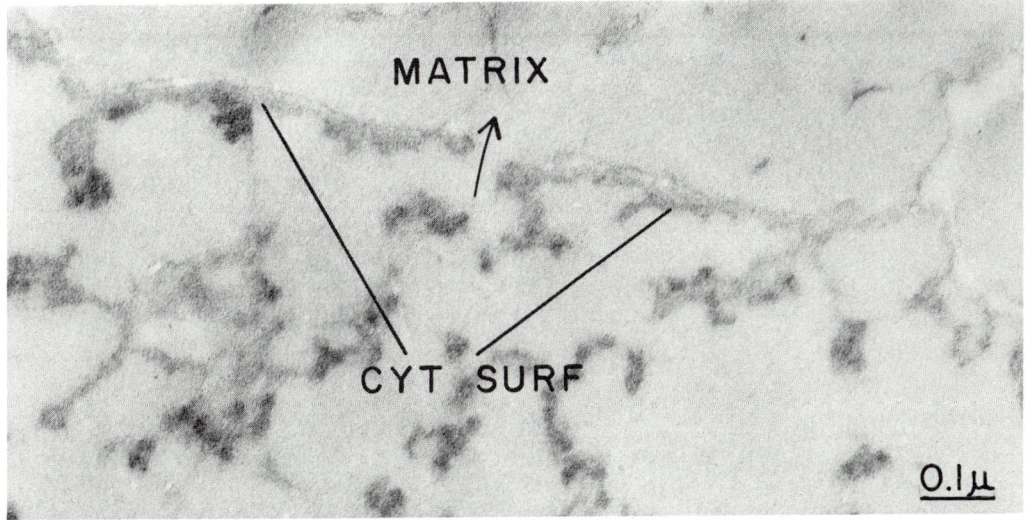

Fig. 7.19.

Figs. 7.18–7.20. Cartilage cells of epiphyseal plate postfixed and stained with alcoholic platinum tetrabromide to show continuity of large matrix compartments (LMC) of cartilage matrix (MATRIX), with cytoplasm through a gap in the surface of the cytoplasm (CYT SURF), where indicated by the curved arrow. × 100,000.

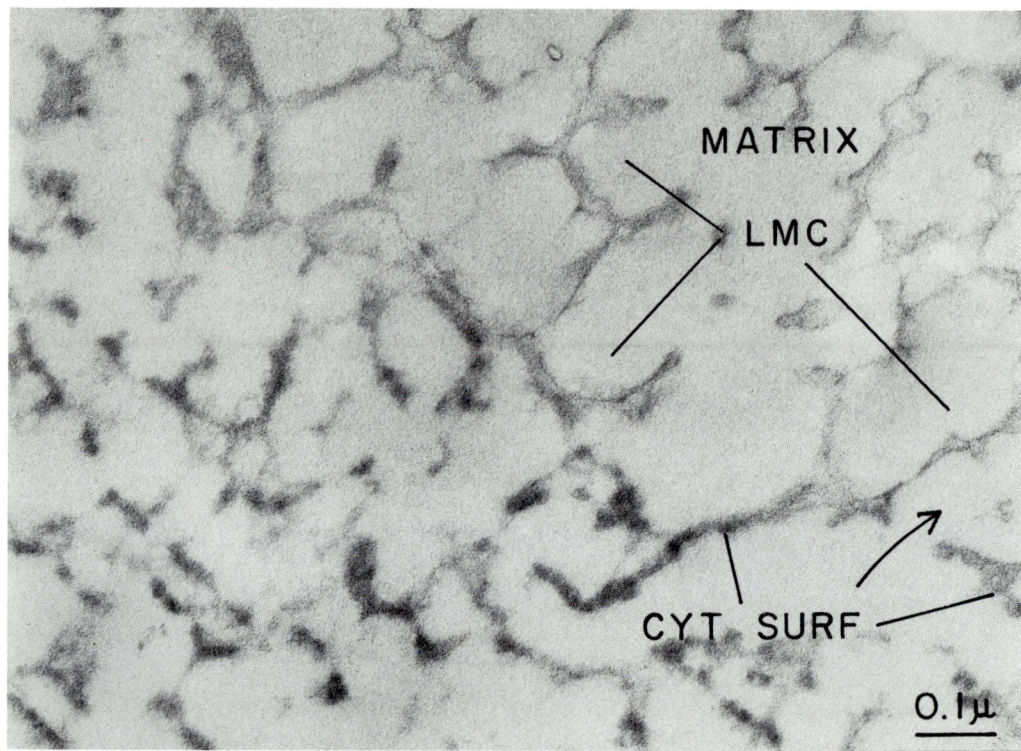

Fig. 7.20. See p. 161 for complete legend.

are present, and these frequently seem to enclose cartilage matrix, in a manner which might be taken as evidence of secretion. Such a deduction is, however, of dubious value, since the appearances might be accounted for on the basis of plane of section. More significant evidence bearing on the mode of secretion is the occurrence of gaps in the surface of the cell. It should be recalled that the outer layer of protoplasm is irregularly thin and sharply outlined over the whole surface of the cell. Sometimes, especially on the side facing the longitudinal matrix, the cell surface is broken, and there is a narrow orifice of about 1000 Å between the matrix and the pseudovacuolar content of the cytoplasm. Three examples of such orifices are shown in Figures 7.18–7.20, of the many observed. These cannot be interpreted to be the result of the vagaries of plane of section, or as a result of mechanical or chemical manipulation, since the cells were in the inner portions of the specimen, and not on the surface. I see no reason to regard these gaps in the cell surface as anything but real.

While the evidence presented suggests that secretion of P-PC takes place from the surface of cells through minute, temporary apertures, proof for the mechanism requires that the contents of the cytoplasmic submicroscopic pseudovacuoles be identified as equivalent to the contents of the extracellular large matrix compartments. This was studied in preparations of epiphyseal plate postfixed and stained in three different ways: (1) vapors of HEFDOD with subsequent section staining

Observations 163

with uranyl acetate (Figs. 7.21 and 7.22); (2) vapors of Hg hfac followed by vapors of FFDNB, with subsequent section staining with uranyl acetate (Figs. 7.23 and 7.24); (3) vapors of benzethonium acetate (BA) alone (Figs. 7.25 and 7.26). In all preparations, the walls of the cytoplasmic submicroscopic pseudovacuoles were stained like the walls of the large matrix compartments. Moreover, the pseudovacuoles in all three preparations were composed of smaller spherical units which resembled closely those of the small matrix compartments. It seems highly likely that the cytoplasmic submicroscopic "vacuole" could be considered equivalent to a large cytoplasmic compartment which in turn is equivalent to a large matrix compartment of extracellular cartilage matrix.

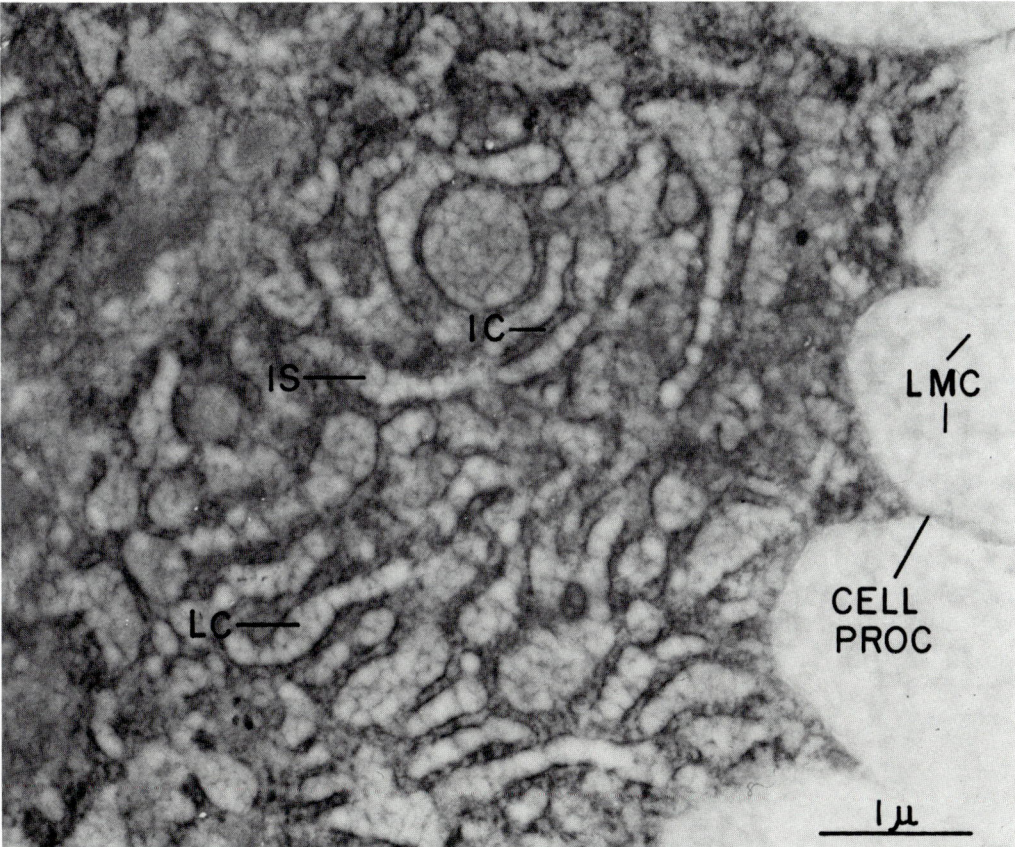

Fig. 7.21. Cartilage cell of epiphyseal plate, postfixed with vapors of HEFDOD, followed by section staining with uranyl acetate. The RER is dark, with very narrow intracisternal spaces (IC). The intercisternal spaces (IS) are paler and clearly marked off into subunits which are not very different from the extracellular large matrix compartments (LMC). Delicate cell processes (CELL PROC) extend into the matrix. LC, large compartment (cytoplasmic). × 20,000. G2 print. See Preface for definition of term.

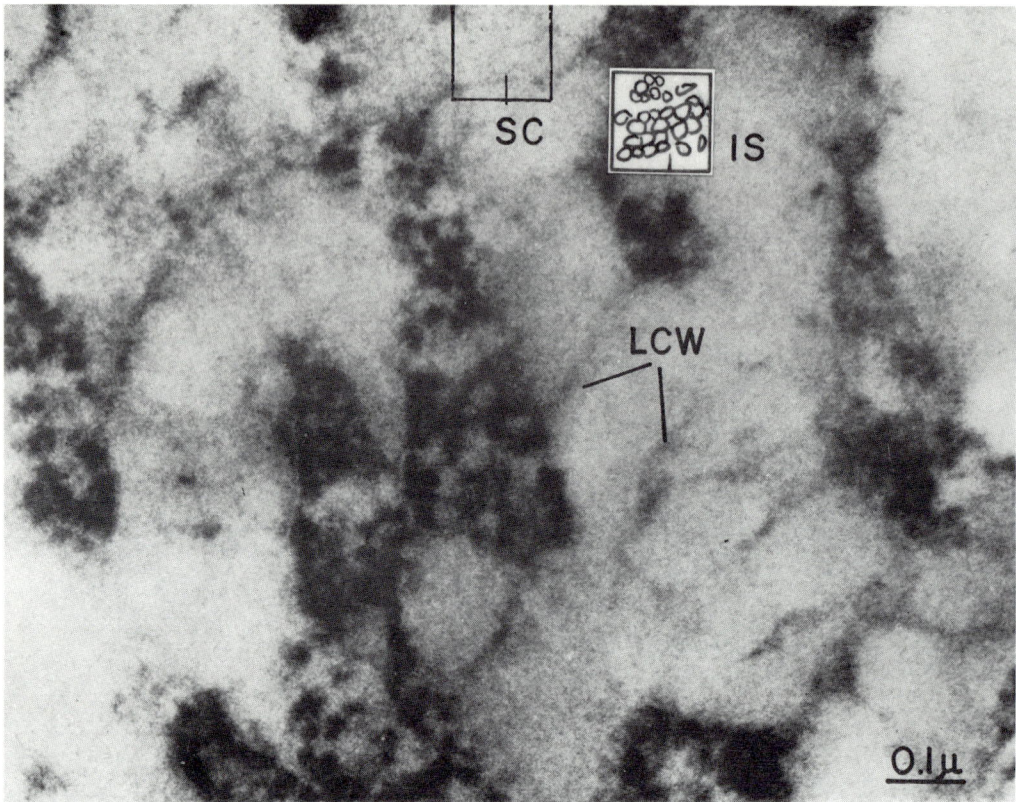

Fig. 7.22. Prepared as in Fig. 7.21. The contents of the intercisternal spaces (IS) comprise numerous weakly outlined spaces (SC), which fill the large compartments (LC). The small compartments are indicated in the insert. The large compartments of the intercisternal space are believed to be the same as the large matrix compartments, and the smaller pale staining globules within the large compartments are believed to be the same as the small matrix compartments. The small compartments of the intercisternal space are believed to be the same as the walls of the corresponding compartments of the matrix. The peripheral cytoplasmic compartments are transposed extracellularly to the matrix via small, transient openings in the cell surface. The very dark granules are ribosomes. × 105,000. G2 print.

Fig. 7.23. (A and B). Cartilage cell of epiphyseal plate postfixed *in vacuo* successively with vapors of Hg hfac and FFDNB, followed by section staining with uranyl acetate. The structure of the RER, the intracisternal space (IC) and the intercisternal space (IS) is essentially the same as in Fig. 7.21. The matrix is fragmented, probably because of its brittleness, which is a consequence of the method of postfixation. × 36,000. G2 print. (B) Enlargement of a portion of part of A. The contents of the large matrix components were shattered from the section during the sectioning. The small matrix compartments (SMC) are well outlined. They resemble very closely the small compartments (SC) of the cytoplasm. The very dark granules are ribosomes. × 70,000. G2 print.

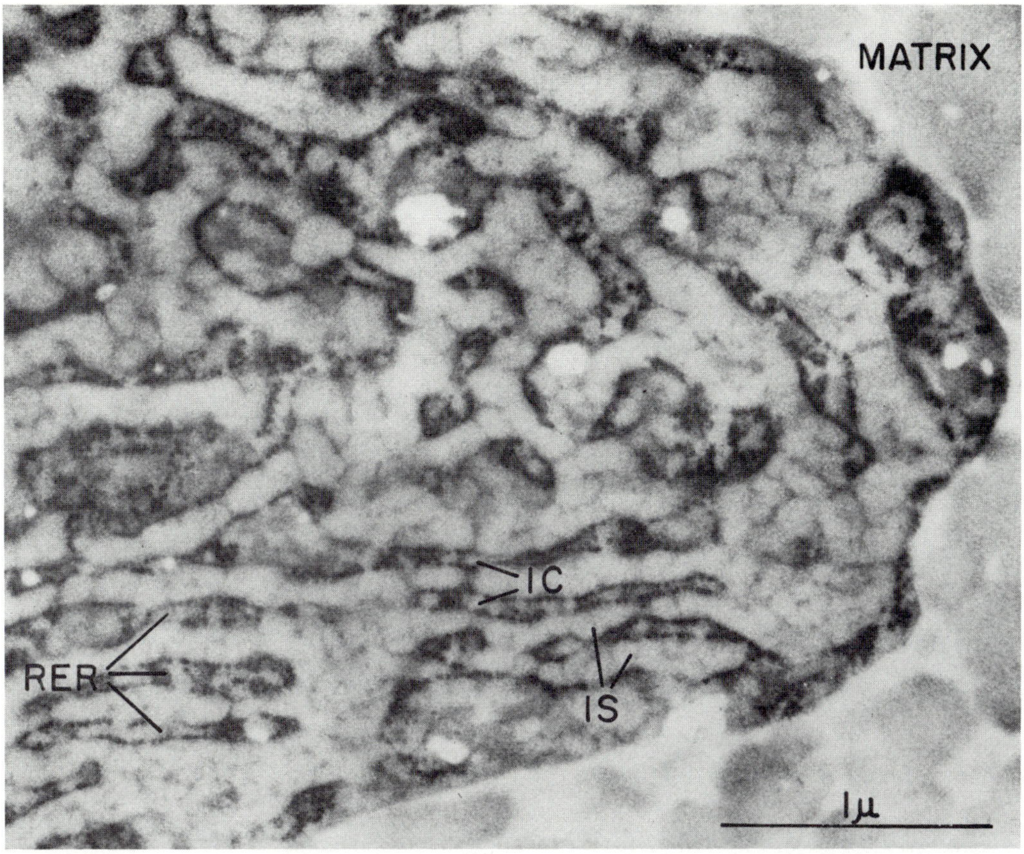

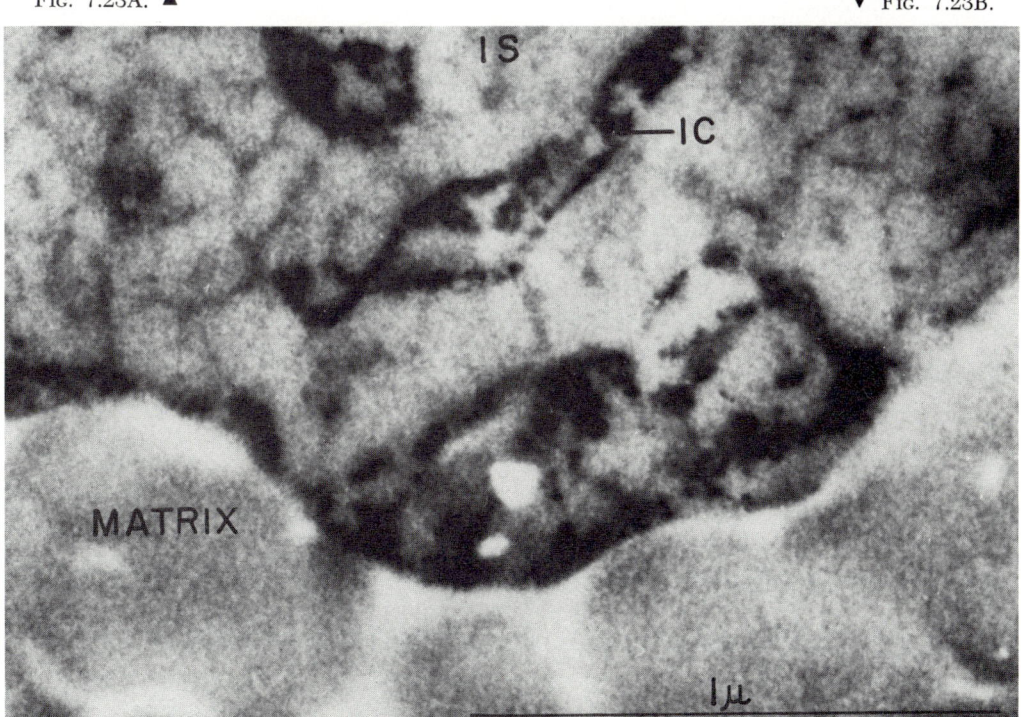

Fig. 7.23A. ▲   ▼ Fig. 7.23B.

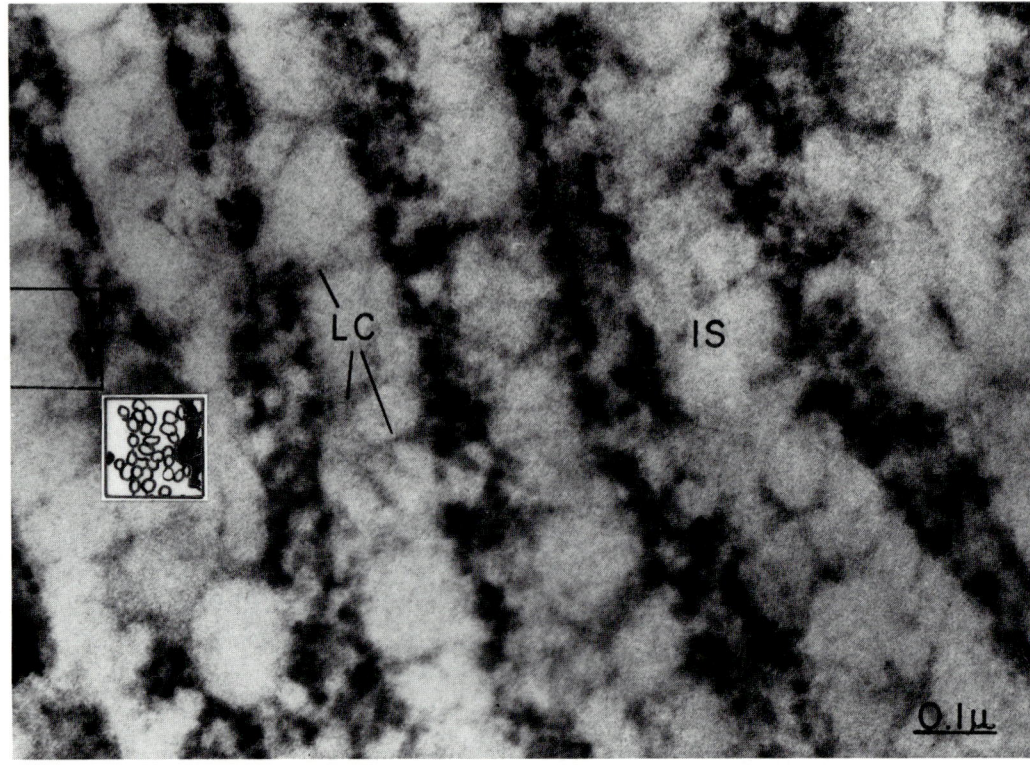

Fig. 7.24. Prepared as in Fig. 7.23. The structure of the large (LC) and small (SC) compartments of the intercisternal space (IS) is essentially the same as in Fig. 7.17. The small compartments are indicated in the insert. × 105,000. G2 print.

Discussion

Chemical Organization of the Cartilage Matrix of Epiphyseal Plate

Cartilage matrix is organized as large matrix compartments (LMC) which can be observed with the light microscope. These may correspond with similar structures photographed by Bütschli in 1898 *(4, 5)* and drawn by Nowikoff *(31)* in 1908 and by Ruppricht in 1910 *(42)*. These are the second-order compartments, and are in turn subdivided into small, first-order, matrix compartments (SMC), which are submicroscopic. The LMCs differ somewhat in their morphology, depending on their relations to the chondrocytes. The LMCs adjacent to cells are more spherical and slightly larger than those farther removed from the cells. The interterritorial LMCs are slightly elongated. The walls of the large and small matrix compartments bind certain bi- and multivalent cations and the contents of the walls have certain solubilities. These are the same as those attributable to sulfated polysaccharides of protein–polysaccharide complexes extracted from cartilage. It

Discussion

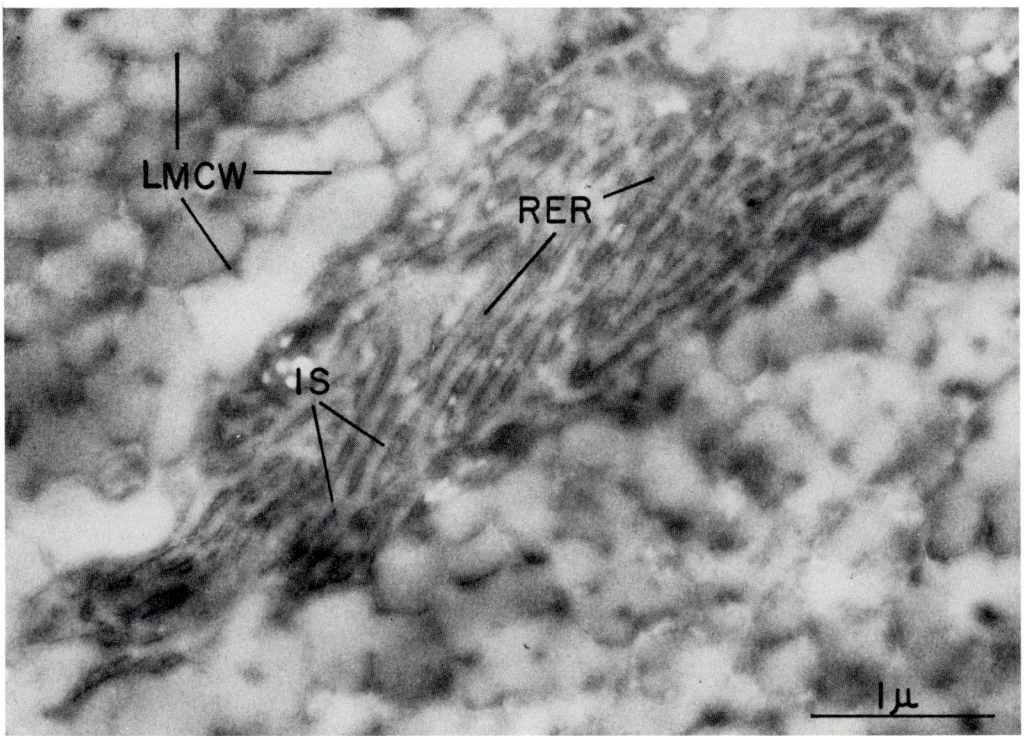

Fig. 7.25. Cartilage cell of epiphyseal plate, postfixed and stained with benzethonium acetate alone. The walls of the large matrix compartments (LMCW) are deeply stained, showing that they contain highly charged, negative, reactive groups, probably mostly chondroitin sulfate. The contents of most of the large matrix compartments are also dense, though in many of the contents are unstained, probably because they were chipped out during sectioning. The density of the regions described should be compared with control areas of alcohol-fixed, unstained control specimens in Fig. 7.3 (A) and (C) at the same magnification. In the cartilage cell, the RER is very dense, probably because of combination of RNA with vapors of benzethonium acetate. The intercisternal space (IS) is much paler. × 24,000. G2 print.

is, therefore, suggested that the walls of both large and small compartments contain or consist largely of the polysaccharide chains of the P-PC. The small compartments would then contain the protein moiety, but whether this exists as a monomer, dimer, or polymer could not be ascertained (40). Whether additional proteins are trapped among the polysaccharide polymer chains as suggested (15, 32, 33), or are integrally involved in the aggregation of P-PC (19, 43), could not be ascertained. It is possible that some polysaccharide is included in the protein moiety, but not detected, either because the former is not accessible to the reagent or because the resolution (sensitivity) is not sufficiently fine.

Restricted to the LMCWs are fibrils, with a periodicity of about 200 Å which are digestable by collagenase. These are probably collagen fibrils. Their maximum

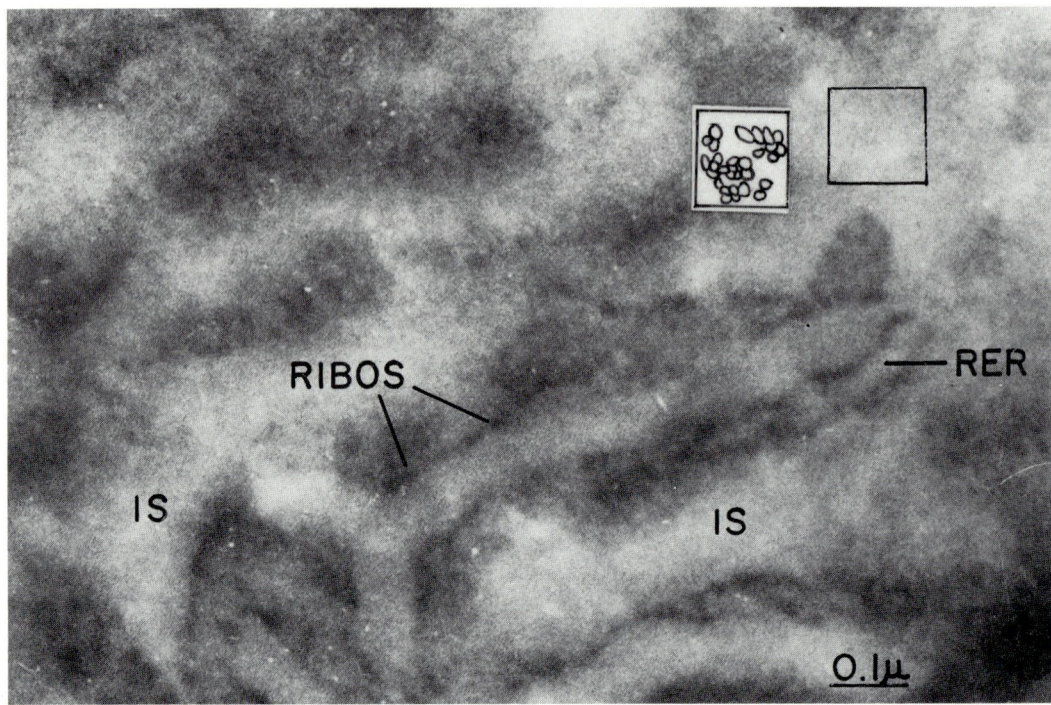

Fig. 7.26. Prepared as in Fig. 7.25. In addition to the dense staining of the ribosomes (RIBOS) of the RER, the intercisternal space (IS) is filled with unstained globules surrounded by a stainable layer. The pale globules (arrow heads) with their stained outer parts correspond with the same structures in the large compartments of the intercisternal space (IS) stained with HEFDOD (Fig. 7.22), with Hg hfac + FFDNB (Fig. 7.24), and with the extracellular small matrix compartments (Fig. 7.11). × 105,000. G2 print.

diameter was found to be about 40 Å, which is approximately the equivalent of three troprocollagen molecules lying side by side. Tightly packed, such fibrils, if round, would comprise seven molecules. Though smaller fibrils were not observed in electron micrographs, one could assume that fibrils containing fewer molecules of tropocollagen down to the single macromolecule also occur in LMCWs. The finding that collagen fibrils are confined to the same regions where sulfated polysaccharides seem to be located (i.e., the LMCW) is of interest as all biochemical models associate the polysaccharide moiety most closely with collagen molecules or fibrils.

The spatial relations of the polysaccharide and protein moieties of P-PC and of collagen and their organization in the first (SMC)- and second (LMC)-order compartments are represented diagrammatically in a two-dimensional drawing of a model of cartilage matrix (Fig. 7.27).

Collagen fibrils frequently appear in published electron micrographs as straight

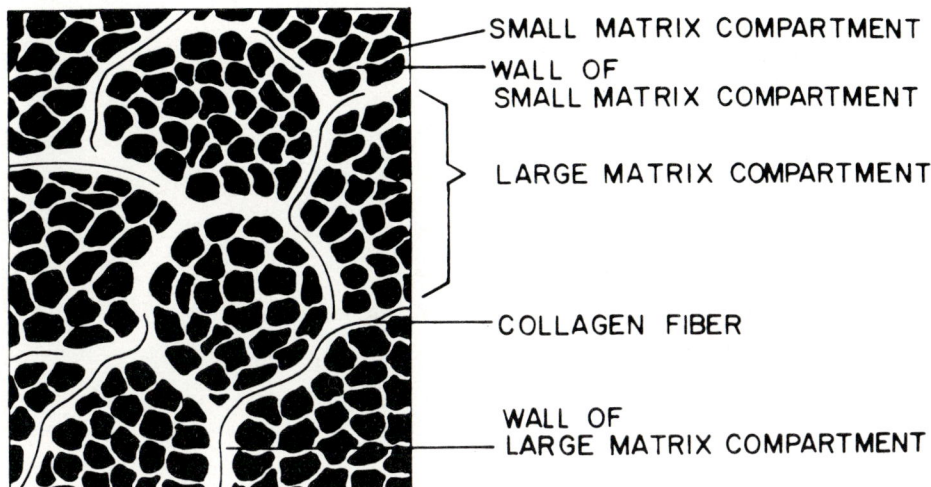

Fig. 7.27. Diagram of submicroscopic composition and structure of hyaline cartilage of epiphyseal plate. Large matrix components are enclosed in walls which contain polysaccharide components (white) of protein–polysaccharide complexes and fine collagen fibrils. Coarser fibrils may also be present, as in joint cartilage, and cause or contribute to a marked irregularity in size and shape of the LMCs. Polysaccharide components (white) extend into the walls of the small matrix compartments. Protein components of the P-PCs are shown in black, and constitute the contents of the SMCs and the greater part of the LMCs.

segments generally surrounded by a pale region, as in the review by Cameron *(6)*, with little or no trace of the gentle bends and curves like those in Fig. 7.4. Collagen may appear as straight fibrils in frozen-dried specimens, but only after the specimen has been treated with water, as in Durning's work *(11)* and shown in Fig. 7.5. It is suggested that the clear areas in electron micrographs of cartilage matrix fixed by immersion in aqueous reagents are the result of solution of the polysaccharide component of the P-PC, accompanied by displacement of the protein component, or else of solution of some of the P-PC. It is also suggested that collagen fibrils in cartilage matrix are normally under tension, and that their course is shortened as the P-PC is dissolved.

The partial solution of P-PC in aqueous solutions and fixatives is necessarily accompanied by displacement of one or the other component (or both) in the remainder of the tissue block, and by precipitation of the displaced constituents away from their normal site. This is the lesson of Fischer *(14)*, Hardy *(18)*, and others of the last century. A further consequence of this process is that the soluble constituents tend to precipitate toward the outer surface of the block, where the reacting fixative is penetrating the specimen. This is especially prominent when the precipitating reagent is of colloidal dimensions, or forms amorphous precipitates which impede further penetration of the specimen by the fixative. Examples may be found in recent references *(12, 13, 21, 26, 44)*. For this reason, I believe the

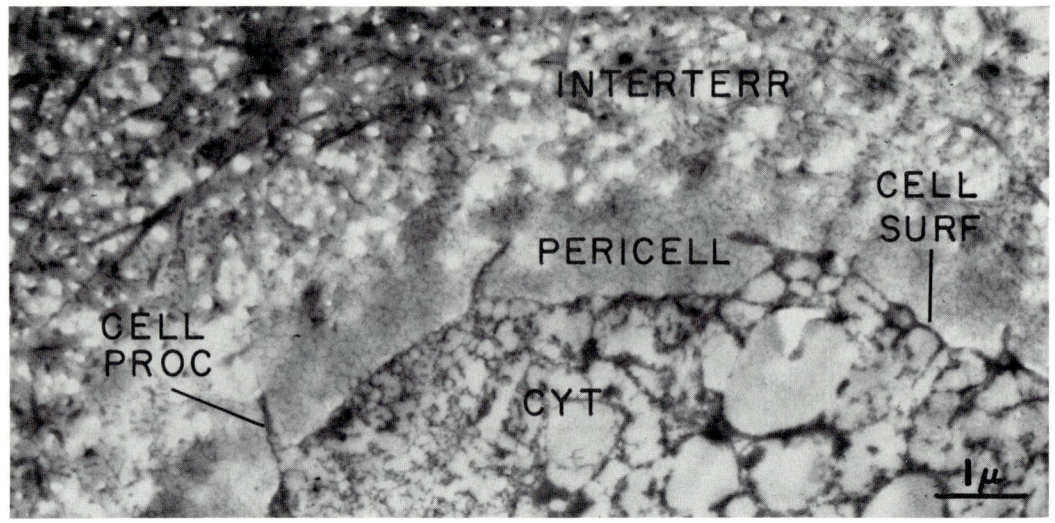

Fig. 7.28. Joint cartilage of dog, postfixed and stained en bloc with alcoholic platinum tetrabromide, and in sections with uranyl acetate. The cartilage cell and its processes (CELL PROC) are surrounded by a zone rich in tightly packed large matrix compartments whose walls are free of collagen fibrils (at this magnification) (PERICELL); the latter are apparent in the interterritorial zone (INTERTERR) where they are coarse and seem to form a tight feltlike structure. CELL SURF, cell surface. × 12,000.

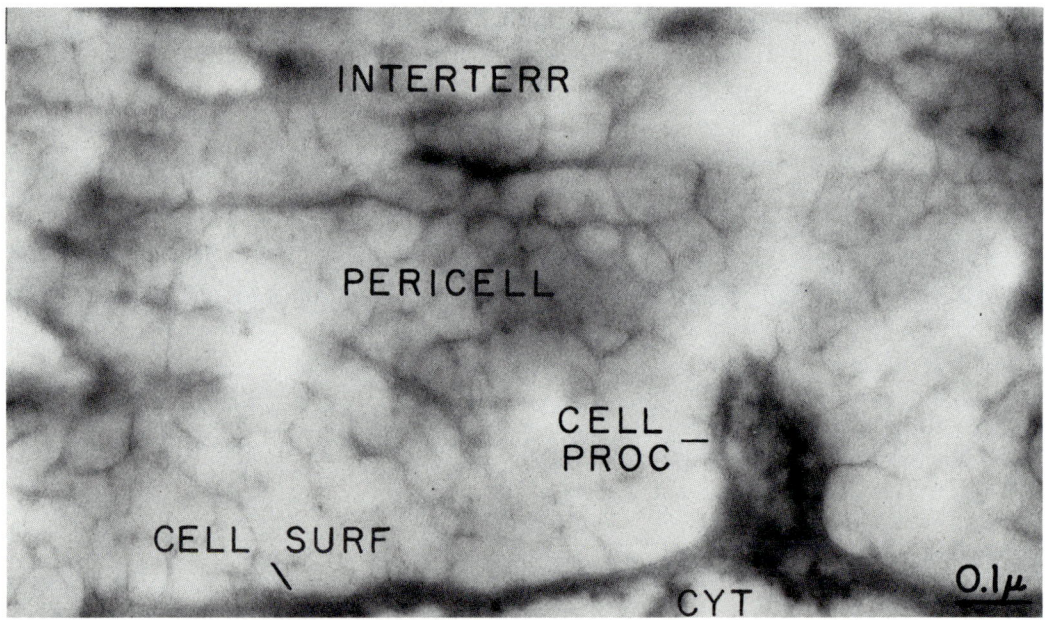

Fig. 7.29. Joint cartilage of same dog, prepared as above. The large and some small matrix compartments are outlined by, respectively, thick and thin walls. Collagen fibrils are not detectable in the pericellular matrix (PERICELL). The rather thick collagen fibrils in the interterritorial regions (INTERTERR) course in the walls in a somewhat parallel arrangement, and distort the shape of the large matrix compartments. Other abbreviations as noted earlier. × 70,000.

Discussion

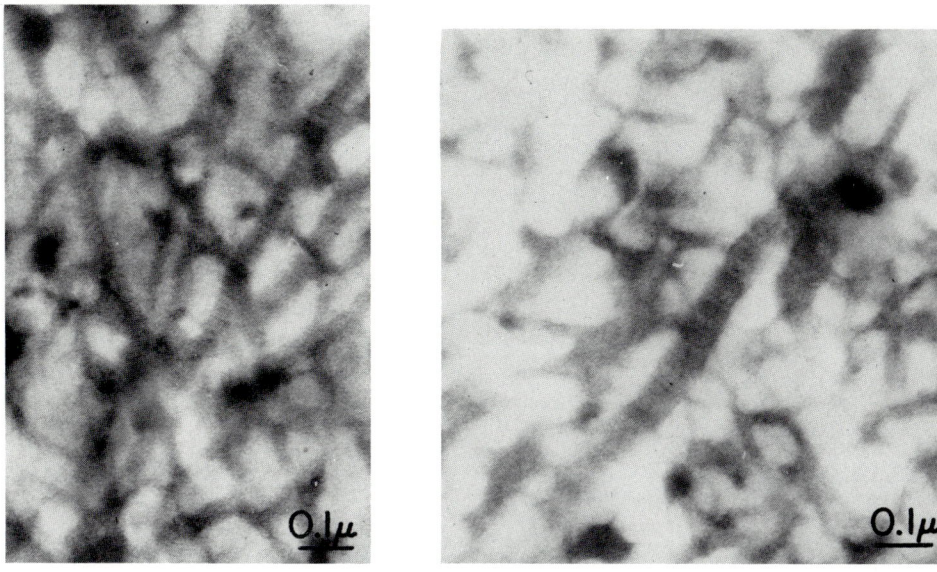

FIG. 7.30.   FIG. 7.31.

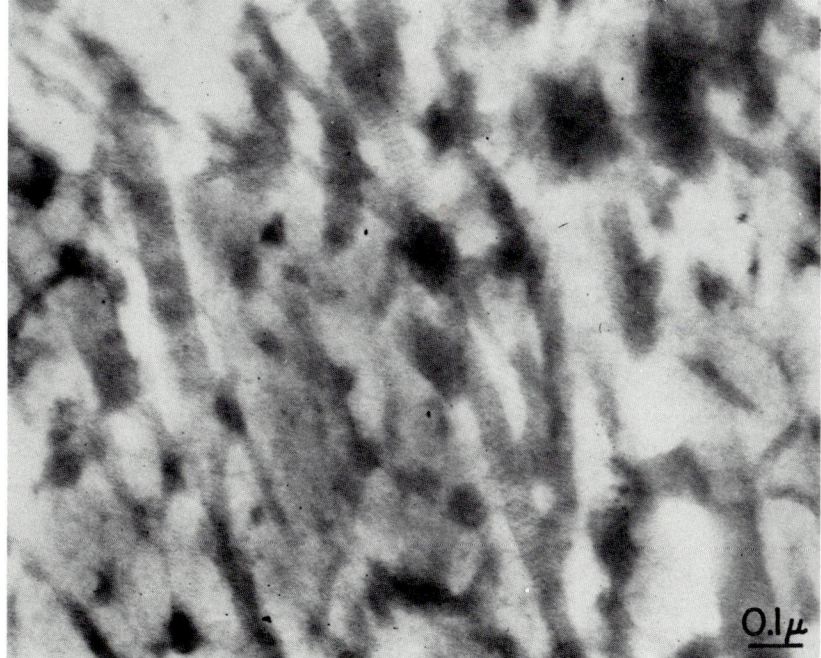

FIG. 7.32.

FIGS. 7.30–7.32. Joint cartilage of the same dog, prepared as in Fig. 7.28, to show the fine periodicity in the fibrils, the variability in their thickness, and the irregularity in the pattern of the LMCs. The SMCWs are rather uniformly delicate. × 70,000.

large number of reports in the literature on reagents which claim to visualize polysaccharides for electron microscopic study carry within them the seeds of disbelief that the sites visualized are the same as those occupied by the polysaccharides before fixation *(8, 11, 17, 22, 24–26, 28–30, 34, 36–38, 46, 49, 50, 52–55, 59)*. The methods used in this report avoid this difficulty, as the cations penetrate the specimen in vapor form and react in the absence of a fluid solvent. In any case, it should be emphasized that, regardless of the method of fixation, the chemical identification of the polysaccharide (or of the P-PC) is based on the single property of its strongly anionic nature. This is supplemented by use of the salting out and solubility properties of the P-PC which may be affected by the denaturation accompanying fixation *(45)*. The use of autoradiography for the submicroscopic localization of sulfated polysaccharides is inadvisable, in addition to the reasons given above, because the resolution is coarser than the dimensions of the structures (which range between approximately 100 and 500 Å).

Comments on the Origin of P-PC

Nearly all investigators who have published on this topic associate ribosomal activity of the RER, and the Golgi apparatus with the origin of P-PC, almost always on the basis of autoradiographical studies *(39)*. The only exceptions are Campo and Dziewiatkowski *(7)* who found that radioactive sulfate is rapidly bound as organic sulfate by cartilage cells, and that this is stored in the peripheral parts of the cells. The limited success in the fixation of P-PC in cartilage matrix must surely apply to its preservation in cells, and it seems unwise to rely exclusively on techniques which have this failing.

Fortunately, there are some pertinent biochemical observations which point to a simultaneous synthesis and comparable turnover rate of both protein and polysaccharide components of P-PC. These have been reviewed by Dorfman *(10)* and Priest *(35)*. Support for this thesis comes from studies of particulate fractions of cartilage cells *(20, 23, 40, 51, 56)*. The inference is that the mechanisms involved in the attachment of the polysaccharide component to the protein core as well as its enlargement may not be spatially far removed from the ribosome system, which is probably responsible for the synthesis of the protein core. The morphological evidence I have observed in freeze-dried specimens which bears on the subject is more in agreement with the biochemical findings than with most of the autoradiographical findings. In freeze-dried specimens, regardless of the stain employed, the space occupied between the double plates of the RER consists of a large, thin-walled submicroscopic "vacuole" which cannot be distinguished morphologically from the large matrix compartments adjacent to the cell, except for the absence of collagen fibrils in the former and their presence in the latter. The cytoplasmic components extend to the cell surface, where a thin film only a few Ångstroms thick separates them from the extracellular matrix. It is suggested that the thin-walled submicroscopic vacuoles in the periphery of the cartilage cells represent stored precursors of the LMCs, which are extruded from numerous gaps in the cell surface and which probably form briefly and intermittently. A related suggestion on the mode of release from the cell was made by Salpeter *(44)*.

Several deductions may be made if one assumes that the proposed method of secretion of cartilage matrix is real. The first is that all components of the P-PC are synthesized in association with the RER, as proposed by several biochemists. The second is that the flow of the secretory product during its synthesis is away from the Golgi apparatus and oriented toward the cell surface. The third is that the complex structure and organization of the large and small matrix compartments of cartilage matrix are prefabricated as such in the cytoplasm from self-assembled units. These may be the ones postfixed and stained by the vapor reagents.

The probable identification of the secretion precursor is of interest also in connection with the question of ice formation during the freezing and drying of the specimens. For, in this instance, not only is the "space" occupied by a cell protein, but the cell protein is highly structured, and composed of primarily one type of substance, the P-PC complex (see Vol. I, Chapters 1 and 2 for a detailed discussion).

Cartilage Matrix of Joint Cartilage

It was thought desirable to learn how the distribution of P-PC in cartilage matrix would be influenced by the presence of collagen fibers of larger diameter than those observed in epiphyseal plate. For this purpose, specimens were prepared from the proximal tibial joint cartilage of three dogs. The most interesting electron micrographs were of the joint cartilage of a dog with an estimated age of 12 years. Thin slivers were frozen ultrarapidly, dried at a temperature below $-30°C$, postfixed with 95% alcohol (1 day at $-30°C$, 1 day at room temperature), stained with alcoholic platinum tetrabromide for 5 hours, and washed with 95% alcohol for 18 hours. The specimens were embedded in water-soluble Durcupan, and sections were stained with uranyl acetate.

Cartilage cells here are enclosed by a narrow pericellular zone free of visible collagen fibrils (Figs. 7.28 and 7.29). The matrix consists of large and small compartments, the walls of the former varying somewhat in density and thickness (Fig. 7.28). The LMSs in the interritorial zone are irregular in size, shape, thickness, and density. The walls of the SMCs are thin and rather uniform. Distortions of the LMCWs in the interterritorial regions are associated with the presence in them of collagen fibers which run singly and apparently unbranched and mostly straight or very slightly curved. They may form an extremely dense three-dimensional felt, or they may run predominantly in one plane. The diameters of fibers vary from about 150 to 700 Å. Most of the fibers have a very fine periodicity (Figs. 7.30–7.32).

In short, the basic structure of cartilage matrix exemplified by epiphyseal plate may be observed in joint cartilage, but restricted to pericellular regions. Elsewhere in joint cartilage the basic structure of the LMCs is distorted by the occurrence in the LMCWs of periodically banded collagen fibers of rather large diameter.

The sharply delimited pericellular region of joint cartilage, at least in some zones, has been observed by others (*1, 52, 57*). All other descriptions are not as clear on this point because of the large shrinkage space present around cartilage cells *(2, 9, 16, 60)*. All these authors agree, however, that collagen fibrils near cells are of smaller diameter than those in interterritorial regions, and that the 640 Å periodicity is present only in the latter fibrils.

Summary

Cartilage matrix is organized as large matrix compartments (LMC) which are of light microscopic dimensions. These are subdivided into small matrix compartments (SMC) which are submicroscopic in size. The LMCs differ somewhat in their morphology, depending on their relations to chondrocytes, their position, and the presence of coarse striated collagen fibrils about 150 to 700 Å in diameter. These, as well as extremely fine ones (about 40 Å in diameter) run in the walls of the LMCs. Evidence is presented which shows that the polysaccharide component of the protein–polysaccharide complexes (P-PC) of cartilage matrix are an important part of the walls of the SMCs and the LMCs, and enclose the protein components of the P-PC. Thus the contents of the SMC, which constitute most of the LMC, are mainly the protein moiety of the P-PC. These relations are illustrated in a diagram shown as Fig. 7.27.

References

1. Barland, P., Janis, R., and Sandson, J. (1966). Immunofluorescent studies of human articular cartilage. *Ann. Rheum. Dis.* **25**, 156–164.
2. Barnett, C. H., Cochrane, W., and Palfrey, A. J. (1963). Age changes in articular cartilage of rabbits. *Ann. Rheum. Dis.* **22**, 389–400.
3. Brandt, K. D., and Muir, H. (1971). Heterogeneity of protein-polysaccharides of porcine articular cartilage. *Biochem. J.* **123**, 747–755.
4. Bütschli, O. (1898). "Untersuchungen über Strukturen insbesondere über Strukturen nichtzelliger Erzeugnisse des Organismus und über ihre Beziehungen zu Strukturen, welche ausserhalb des Organismus entstehen," pp. 337–344. W. Engelmann, Leipzig.
5. Bütschli, O. (1898). "Atlas zu den Untersuchungen über Strukturen," Plate 16, Figs. 4–7. W. Engelmann, Leipzig.
6. Cameron, D. A. (1963). The fine structure of bone and calcified cartilage. *Clin. Orthop.* **26**, 199–228.
7. Campo, R. D., and Dziewiatkowski, D. D. (1962). Intracellular synthesis of protein-polysaccharides by slices of bovine costal cartilage. *J. Biol. Chem.* **237**, 2729–2735.
8. Clark, A. E., and Curran, R. C. (1964). Staining reactions of collagen in "Epon" sections. *Nature (London)* **202**, 912–913.
9. Davies, D. V., Barnett, C. H., Cochrane, W., and Palfrey, A. J. (1962). Electron microscopy of articular cartilage in the young adult rabbit. *Ann. Rheum. Dis.* **21**, 11–22.
10. Dorfman, A. (1962). Biochemistry of connective tissue. In "The Biology of Connective Tissue Cells," pp. 34–44. The Arthritis and Rheumatism Foundation, New York.
11. Durning, W. C. (1958). Submicroscopic structure of frozen-dried epiphyseal plate and adjacent spongiosa of the rat. *J. Ultrastruct. Res.* **2**, 245–260.
12. Eisenstein, R., Sorgente, N., and Kuettner, K. E. (1971). Organization of extracellular matrix in epiphyseal growth plate. *Amer. J. Pathol.* **65**, 515–534.
13. Engfeldt, B., and Hjertquist, S.-O. (1967). The effect of various fixatives on the preservation of acid glycosaminoglycans in tissues. *Acta Pathol. Microbiol. Scand.* **71**, 219–232.
14. Fischer, A. (1899). "Fixirung, Färbung unde Bau des Protoplasmas." Fischer, Jena.

References

15. Fitton Jackson, S. (1965). Macromolecular order in the ground substance. *In* "Structure and Function of Connective and Skeletal Tissue." (S. Fitton Jackson, R. D. Harkness, S. M. Partridge, G. R. Tristram, eds.), pp. 156–160. Butterworth, London.
16. Ghadially, F. N., and Roy, S. (1970). "Ultrastructure of Synovial Joints in Health and Disease." Appleton-Century-Crofts, New York.
17. Godman, G. C., and Lane, N. (1964). On the site of sulfation in the chondrocyte. *J. Cell Biol.* **21**, 353–366.
18. Hardy, W. B. (1899). On the structure of cell protoplasm. *J. Physiol.* **24**, 158–210.
19. Hascall, V. C., and Sajdera, S. W. (1969). Protein–polysaccharide complex from bovine nasal cartilage. *J. Biol. Chem.* **244**, 2384–2396.
20. Horwitz, A. L., and Dorfman, A. (1968). Subcellular sites for synthesis of chondromucoprotein of cartilage. *J. Cell Biol.* **38**, 358–368.
21. Józsa, L., and Szederkényi, G. (1967). Über Verluste der Gewebsmukopolysaccharide während der Fixierung. *Acta Histochem.* **26**, 255–260.
22. Khan, T., and Overton, J. (1969). Staining of intercellular material in reaggregating chick liver and cartilage cells. *J. Exp. Zool.* **171**, 161–173.
23. Kleine, T. O., Kirsig, H. J., and Hilz, H. (1968). Untersuchungen zur Biosynthese der Chondroitinsulfat-Proteins. II. Über die Koordinierung von Protein-, Polysaccharid und Sulfatester-Synthese im Rippenknorpel von Kälbern. Hoppe-Seylers Z. *Physiol. Chem.* **349**, 1037–1048.
24. Luft, J. H. (1965). The fine structure of hyaline cartilage matrix following ruthenium red fixative and staining. *J. Cell Biol.* **27**, 61A.
25. Luft, J. H. (1971). Ruthenium red and violet. I. Chemistry, purification, methods of use for electron microscopy and mechanism of action. *Anat. Rec.* **171**, 347–368.
26. Luft, J. H. (1971). Ruthenium red and violet. II. Fine structural localization in animal tissues. *Anat. Rec.* **171**, 369–415.
27. Mathews, M. B. (1967). Biophysical aspects of acid mucopolysaccharides relevant to connective tissue structure and function. *In* "The Connective Tissue" (B. M. Wagner, and D. E. Smith, eds.), pp. 304–329. Williams & Wilkins, Baltimore, Maryland.
28. Matukas, V. J., Panner, B. J., and Orbison, J. L. (1967). Studies on ultrastructural identification and distribution of protein-polysaccharide in cartilage matrix. *J. Cell. Biol.* **32**, 365–377.
29. Myers, D. B., Highton, T. C., and Rayns, D. G. (1969). Acid mucopolysaccharides closely associated with collagen fibrils in normal human synovium. *J. Ultrastruct. Res.* **28**, 203–213.
30. Nicolson, G. L., and Singer, S. J. (1971). Ferritin-conjugated plant agglutinins as specific saccharide stains for electron microscopy: application to saccharides bound to cell membranes. *Proc. Nat. Acad. Sci. U.S.* **68**, 942–945.
31. Nowikoff, M. (1908). Beobachtungen über die Vermehrung der Knorpelzellen, nebst einigen Bemerkungen über die Struktur der "hyalinen" Knorpelgrundsubstanz. *Z. Wiss. Zool.* **90**, 205–257.
32. Partridge, S. M. (1968). The chondroitin sulfate-protein complex from bovine cartilage. *In* "The Chemical Physiology of Mucopolysaccharides" (G. Quintarelli, ed.), pp. 51–58. Little, Brown, Boston, Massachusetts.
33. Partridge, S. M., Whiting, A. H., and Davis, H. F. (1965). The presence of aggregates containing non-covalently linked protein in preparations of the chondroitin sulfate-protein complex from bovine cartilage. *In* "Structure and Function of Connective and Skeletal Tissue" (G. R. Tristram and S. Fitton Jackson, S. M. Partridge, and R. D. Harkness, eds.), pp. 160–164. Butterworth, London.
34. Pihl, E., Gustafson, G. T., and Falkmer, S. (1968). Ultrastructural demonstration of cartilage acid glycosaminoglycans. *Histochem. J.* **1**, 26–35.
35. Priest, R. E. (1967). Endocrine control of connective tissue metabolism. *In* "The Connective Tissue" (B. M. Wagner and D. E. Smith, eds.), pp. 50–60. Williams & Wilkins, Baltimore, Maryland.
36. Quintarelli, G., Zito, R., and Cifonelli, J. A. (1971). On phosphotungstic acid staining. I and II. *J. Histochem. Cytochem.* **19**, 641–647, 648–653.
37. Rambourg, A., Hernandez, W., and Leblond, C. P. (1969). Detection of complex carbohydrates in the Golgi apparatus of rat cells. *J. Cell Biol.* **40**, 395–414.

38. Revel, J.-P. (1964). A stain for the ultrastructural localization of acid mucopolysaccharides. *J. Microsc.* **3**, 535–544.
39. Revel, J.-P. (1970). Role of the Golgi apparatus of cartilage cells in the elaboration of matrix glycosaminoglycans. In "Chemistry and Molecular Biology of the Intercellular Matrix" (E. A. Balazs, ed.), pp. 1485–1502. Academic Press, New York.
40. Richmond, M. E., de Luca, S., and Silbert, J. E. (1972). Macromolecular assembly of chondroitin. *Biochem. Bophys. Res. Commun.* **46**, 263–268.
41. Rosenberg, L., Hellman, W., and Kleinschmidt, A. K. (1970). Macromolecular models of proteinpolysaccharide from bovine nasal cartilage based on electron microscopic studies. *J. Biol. Chem.* **245**, 4123–4130.
42. Ruppricht, W. (1910). Über Fibrillen und Kittsubstanz des Hyalinknorpels. *Arch. Mikrosk. Anat.* **75**, 748–771.
43. Sajdera, S. W., Hascall, V. C., Gregory, J. D., and Dziewiatkowski, D. D. (1970). The proteoglycans of bovine nasal cartilage: structure of the aggregate. In "Chemistry and Molecular Biology of the Intercellular Matrix" (E. A. Balazs, ed.), Vol. 2, pp. 851–858. Academic Press, New York.
44. Salpeter, M. M. (1968). H^3-proline incorporation into cartilage: electron microscope autoradiographic observations. *J. Morphol.* **124**, 387–400.
45. Sasai, Y. (1971). Effect of tissue fixation on the histochemical identification of acid mucopolysaccharides. *Tohoku J. Exp. Med.* **104**, 85–91.
46. Saunders, A. M., and Silverman, L. (1967). Electron microscopy of chondromucoprotein and the products of its digestion with hyaluronidase and papain. *Nature (London)* **214**, 194–195.
47. Schubert, M. (1964). Intercellular macromolecules containing polysaccharides. *Biophys. J. Suppl.* **4**, 119–138.
48. Schubert, M. (1966). Structure of connective tissues, a chemical point of view. *Fed. Proc.* **25**, 1047–1052.
49. Serafini-Fracassini, A., Wells, P. G., and Smith, J.-W. (1970). Studies on the interactions between glycosaminoglycans and fibrillar collagen. In "Chemistry and Molecular Biology of the Intercellular Matrix" (E. A. Balazs, ed.), pp. 1201–1215. Academic Press, New York.
50. Shea, S. M. (1971). Lanthanum staining of the surface coat of cells. Its enhancement by the use of fixatives containing Alcian blue or cetylpyridinium chloride. *J. Cell Biol.* **51**, 611–620.
51. Silbert, J. E., and DeLuca, S. (1969). Biosynthesis of chondroitin sulfate. III. Formation of a sulfated glycosaminoglycan with a microsomal preparation from chick embryo cartilage. *J. Biol. Chem.* **244**, 876–881.
52. Smith, J. W. (1970). The disposition of proteinpolysaccharide in the epiphysial plate cartilage of the young rabbit. *J. Cell Sci.* **6**, 843–864.
53. Smith, J. W., and Frame, J. (1969). Observations on the collagen and protein-polysaccharide complex of rabbit corneal strama. *J. Cell Sci.* **4**, 421–436.
54. Smith, J. W., Peters, T. J., and Serafini-Fracassini, A. (1967). Observations on the distribution of the protein-polysaccharide complex and collagen in bovine articular cartilage. *J. Cell Sci.* **2**, 129–136.
55. Szubinska, B., and Luft, J. H. (1971). Ruthenium red and violet. III. Fine structure of the plasma membrane and extraneous coats in amoebae *(A. proteus* and *Chaos chaos). Anat. Rec.* **171**, 417–441.
56. Telser, A., Robinson, H. C., and Dorfman, A. (1965). The biosynthesis of chondroitin-sulfate protein complex. *Proc. Nat. Acad. Sci. U.S.* **54**, 912–919.
57. Weiss, C., Rosenberg, L., and Helfet, A. J. (1968). An ultrastructural study of normal young adult human articular cartilage. *J. Bone Joint. Surg.* **50A**, 663–674.
58. Woodward, C. B., Hranisavljevic, J., and Davidson, E. A. (1972). Physical properties of cartilage proteoglycans. *Biochemistry* **11**, 1168–1176.
59. Yardley, J. H., and Brown, G. D. (1965). Fibroblasts in tissue culture. Use of colloidal iron for ultrastructural localization of acid mucopolysaccharides. *Lab. Invest.* **14**, 501–513.
60. Zelander, T. (1959). Ultrastructure of articular cartilage. *Z. Zellforsch. Mikrosk Anat.* **49**, 720–738.

8

The Movement of Ferrocyanide in Cartilage Matrix

Isidore Gersh

Many of the factors involved in transport by cartilage matrix were analyzed by Maroudas and her co-workers *(14–19, 21)*. These include the fixed charge density of cartilage matrix, nonuniform protein distribution, the exclusion effect of the macromolecular protein–polysaccharide complex (P-PC), and the size and degree of ionization of the substances being transported. In view of the importance which they attribute to fixed charges of macromolecules in permeability of cartilage to smaller molecules and ions, it is difficult to understand how the English workers could ignore completely until very recently *(18)*, the theoretical and experimental results of a group of three American researchers who have published on this topic from 1952 *(4)* to the present time. Their denial completely of the existence of a two-phase system in cartilage matrix is unfortunately based on cartilage slices from which the protein or polysaccharide, which could be the basis of a two-phase system, was leached out by soaking in iso- or hypotonic solutions of sodium chloride. Regardless of terminology, the fact is that ferrocyanide is not distributed uniformly in cartilage matrix, as it would have to be in a homogeneous structure on a "colloidal" scale. The English group also suggested that the diameter of the flow channels and their tortuosity were important. In one of the earlier papers, they *(17)* had already shown the effect of agitation of the fluid surrounding the cartilage, and noted that the diffusion of dyes through cartilage matrix was not dependent on the activity of living cells *(17)*. Maroudas also found that molecules of the size of hemoglobin (about 50 Å) are about the largest to penetrate normal articular cartilage, at least to some degree. This estimate of the maximum "pore" size in cartilage is not far removed from that made by McCutchen on the basis of some simplifying assumptions, i.e., 62 Å *(20)*. She wondered how the much larger P-PC could be distributed uniformly through the matrix after secretion by the chondrocyte, a problem which will be discussed later in this chapter.

The kinetic factors mentioned above must operate within the limits imposed by the ordered arrangement in space of the protein and polysaccharide moieties of P-PC and other proteins, including collagen. It was thought that a histochemical study of transport of ferrocyanide in cartilage might give information bearing on some of these factors. The choice of ferrocyanide was determined by the fact that the test for it is sensitive and precise, and that it lends itself well to electron microscopic study because of the large molecular weight and the correspondingly high contrast of Prussian blue when present in sufficiently large amounts. In addition, it was hoped to integrate earlier work on sites of distribution of ferrocyanide in other connective tissues with the newer findings on cartilage.

As is well known, cartilage is avascular, except for vessels which pass through it without branching internally into capillaries, for the most part. (See references *(23, 24)* and this volume, Chapter 10.) The significance of the chemical anatomy of the P-PC of cartilage and of the physical parameters is that together they determine the permeability of cartilage matrix to metabolites and nutrients, all of which must reach their destination by diffusion. This is emphasized by recent studies with fluorescent dyes and radioactive sulfate *(1, 3, 9, 11, 12)* and in a recent review on this topic *(10)*.

Methods

Three types of experiments were performed on cartilage of mice, rats, and rabbits:

1. Three-week-old mice were injected intraperitoneally with 0.5 ml of a freshly prepared 20% solution of sodium ferrocyanide. After 5 minutes, the mouse was decapitated and the proximal end of the tibia was exposed. A thin longitudinal slice of the proximal epiphyseal plate was separated from the diaphysis and epiphysis, placed on a small square of aluminum foil, and frozen ultrarapidly. (Details are given in Vol. I, Chapter 2.) It was dried at a temperature lower than $-40°C$, and flooded rapidly *in vacuo* with a stream of a 60% solution of hydrated ferric chloride in 95% alcohol chilled to about $-30°C$. After a few minutes, the vacuum was broken, the specimen was removed from the foil and transferred to a large volume of 95% alcohol, where it remained at room temperature. After 2 days, the specimen was cut into smaller slivers which were then infiltrated with water-soluble Durcupan. Thin slices of epiphyseal plate from another mouse of the same age not injected with ferrocyanide were treated in the same way and served as controls.

2. The proximal epiphyseal plate of 23-day-old rats was removed and treated in one of three ways: (a) Controls were immersed for 15 minutes in normal sheep serum at 27°C, washed briefly in fresh serum, blotted, cut in thin slices on a cold plate at 0°C in a moist box, frozen ultrarapidly, dried, and tested with alcoholic ferric chloride as above for viewing with the electron microscope. (b) Experimentals were immersed for 5 minutes in normal sheep serum, and transferred for 10 minutes to serum containing 4% of sodium ferrocyanide, both at

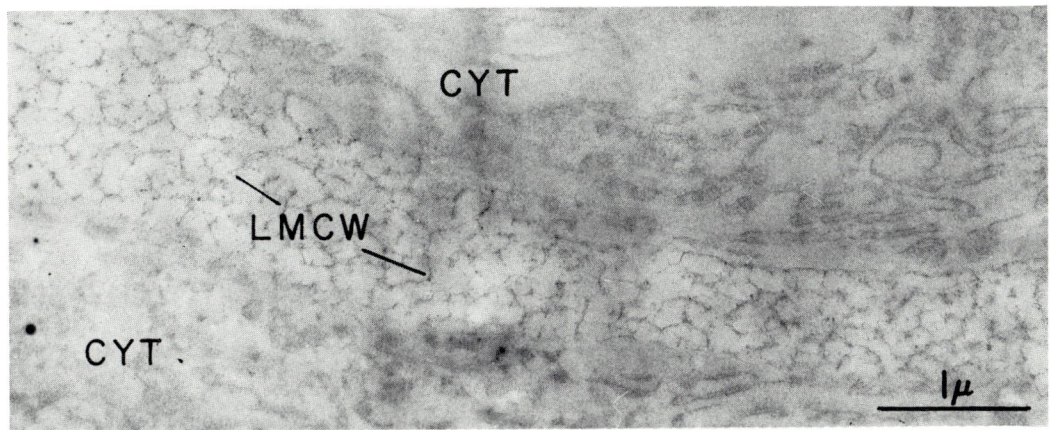

Fig. 8.1.

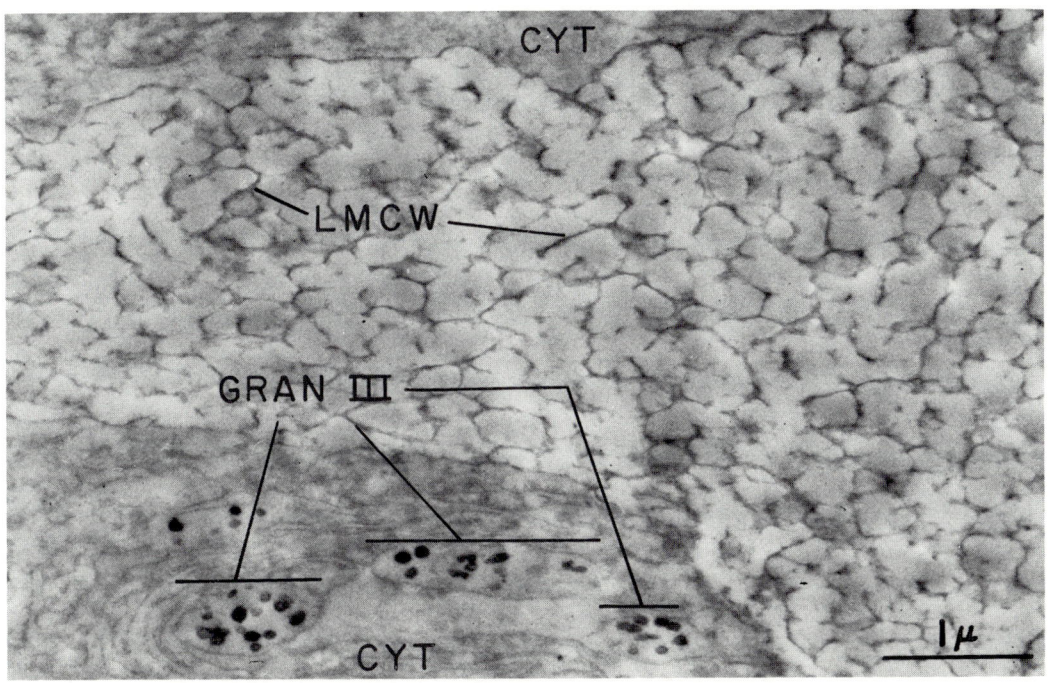

Fig. 8.2.

Figs. 8.1 and 8.2. Epiphyseal cartilage of mouse injected intraperitoneally with a large dose of sodium ferrocyanide (Fig. 8.2) and of control untreated mouse (Fig. 8.1). The walls of the large matrix compartments (LMCW) of the control preparation are stained with ferric ions, but are of lower contrast than the LMCWs of the mouse injected with ferrocyanide. The increased contrast is probably caused by Prussian blue. Incidentally, third-order granules (GRAN III) identical with those in rat chondrocytes are visible in one cell. CYT, cytoplasm. (See also Chapter 6, this volume.) × 20,000.

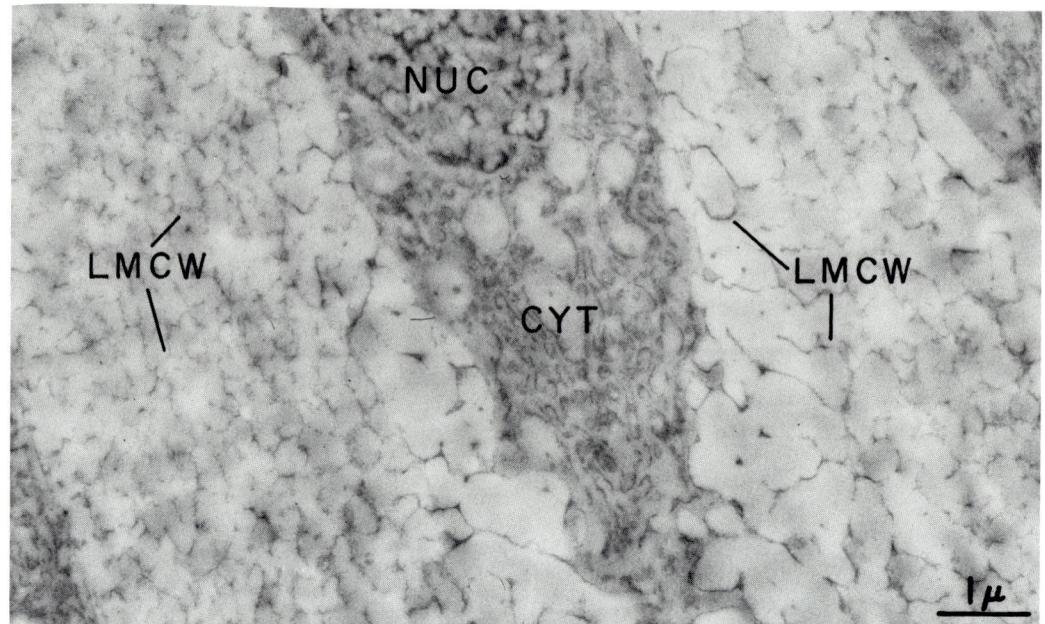

Fig. 8.3.

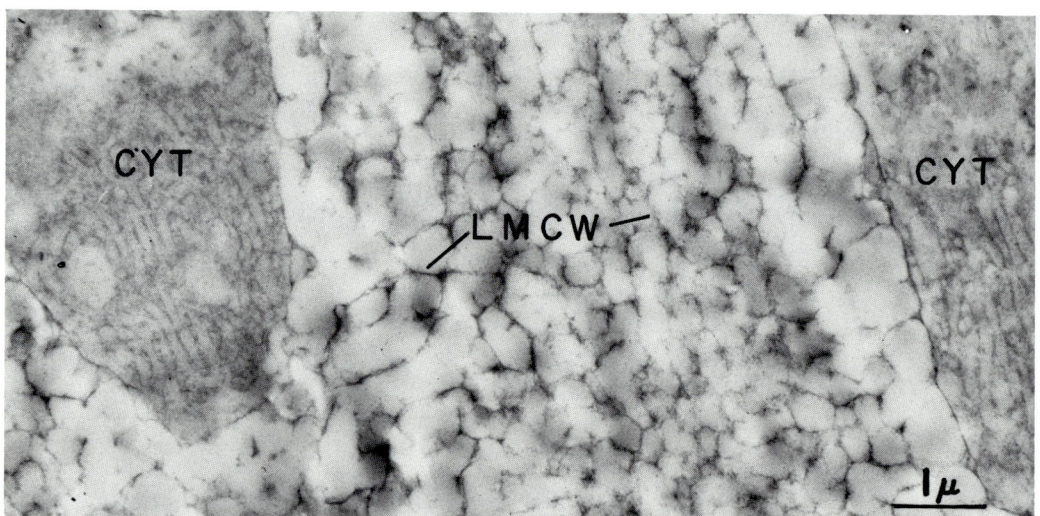

Fig. 8.4.

Figs. 8.3–8.5. Epiphyseal plate of 3-week-old rats after immersion in sodium ferrocyanide at room temperature (Fig. 8.4) and 0°C (Fig. 8.5) as compared with a control not immersed in ferrocyanide (Fig. 8.3). The LMCWs of both ferrocyanide-treated specimens are darker than in the control specimen, because of the presence of ferric ferrocyanide. All abbreviations as given earlier. × 12,000.

Methods

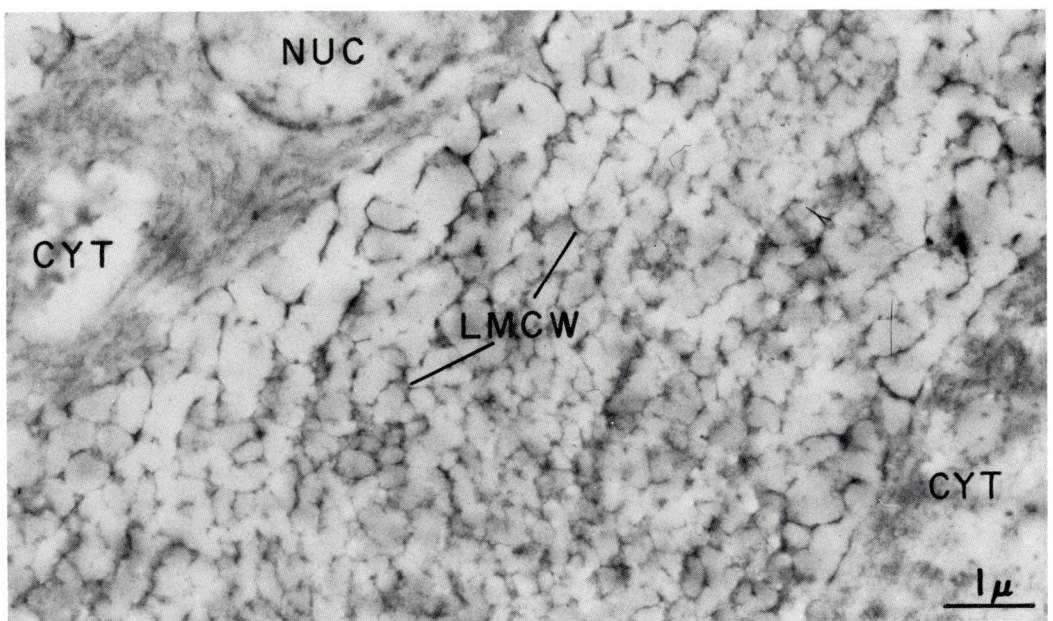

Fig. 8.5.

27°C. The specimens were removed, washed briefly with normal sheep serum, frozen ultrarapidly as in (a), dried, and tested in the same way with ferric chloride. (c) Other experimentals were immersed as in (b) above, but at 0°C, briefly washed with normal serum at the same temperature, freeze-dried, and tested with alcoholic ferric chloride as usual. The sheep serum, which happened to be available, was used to reduce the chances that changes in tonicity in the matrix might take place through a Donnan effect if an isotonic Ringer's solution had been used.

3. (a) The knee joint cavity of an anesthetized, 4-week-old rabbit was injected with 0.5 ml of a very dilute solution of hyaluronidase in saline, and then irrigated with a freshly prepared 20% solution of sodium ferrocyanide. After 10 minutes, the proximal tibial joint was exposed and thin slices of the joint cartilage were frozen-dried in the usual manner. They were then tested for ferrocyanide with ferric chloride as in (1) and (2) above. (b) The proximal tibal joint cartilage of an uninjected rabbit of the same age was cut into thin slices which were immersed in a freshly prepared 20% solution of sodium ferrocyanide for 10 minutes. They were frozen ultrarapidly, dried, and tested with ferric chloride as above. (c) As controls, the proximal tibial joint cartilage of another uninjected rabbit of the same age was cut into thin slices in a moist box, frozen ultrarapidly, dried, and treated with ferric chloride as in (a) and (b) above. Ultrathin sections of all specimens were prepared for study with the Hitachi HU-11A electron microscope. Electron micrographs were enlarged × 2 in printing.

Observations

In all cartilage controls, not exposed to ferrocyanide, the walls of the large matrix components (LMCW) are stained by the ferric ion (Figs. 8.1, 8.3, and 8.6). In all experimental cartilage specimens, exposed to ferrocyanide, the LMCWs are of higher contrast than in their corresponding controls, the increased contrast

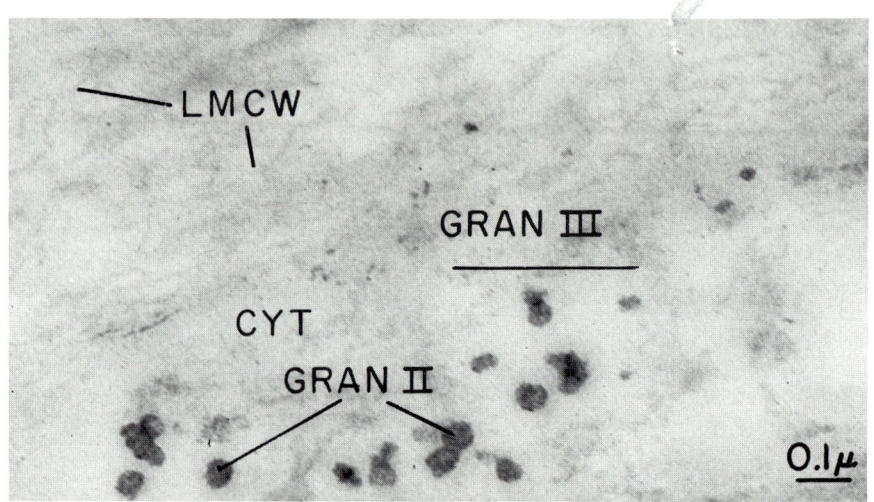

Fig. 8.6 ▲   ▼ Fig. 8.7

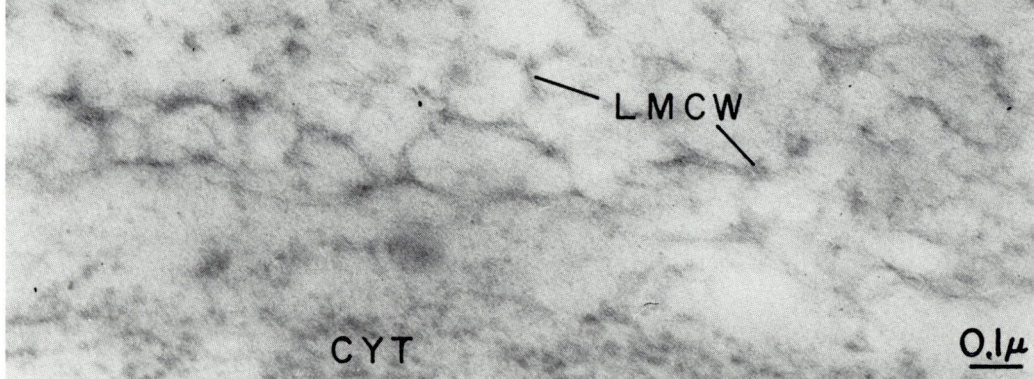

Figs. 8.6–8.8. Proximal tibial joint cartilage of 4-week-old rabbits. Figure 8.7 is of cartilage from a joint perfused with a solution of sodium ferrocyanide. Figure 8.8 is of cartilage slice immersed in a solution of sodium ferrocyanide. Both should be compared with Fig. 8.6, which is of control, untreated cartilage. The LMCWs of Figs. 8.7 and 8.8 have more contrast than those of the control specimen in Fig. 8.6, indicating that ferrocyanide (viewed as Prussian blue) is present in the walls of the LMCWs. In Fig. 8.8, numerous SMCWs also have high contrast, probably because of the presence of ferrocyanide in them. Second- and third-order granules identical with those in rat chondrocytes described in Chapter 6 (this volume) are visible in the cytoplasm of the cartilage cell of Fig. 8.6. All abbreviations as previously given. × 70,000.

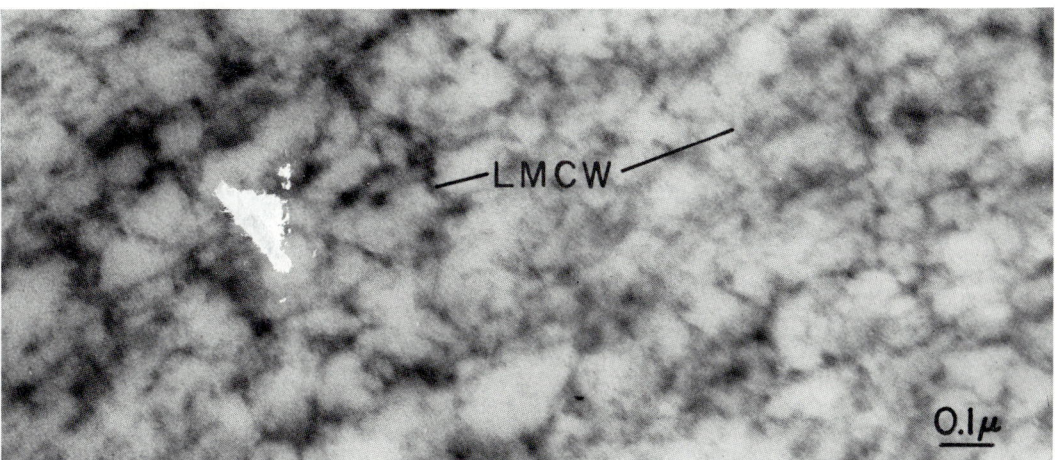

Fig. 8.8.

being attributable to ferrocyanide precipitated as Prussian blue (Figs. 8.2, 8.4, 8.5, 8.7, 8.8). Contrast was highest in the rabbit tibial cartilage immersed in the highest concentration of ferrocyanide (Exp. 2b, Fig. 8.8); it is probable that ferrocyanide in this specimen was present also in the walls of the small matrix compartments (SMCW). The results were the same whether ferrocyanide was administered intraperitoneally, by perfusion of a joint cavity, or by immersion of the specimen. Increased contrast attributable to ferrocyanide in LMCW of the mouse was approximately the same in specimens exposed at room temperature and at 0°C. The last observation largely excludes enzymic or other cellular mechanisms, as suggested by Maroudas et al. (17).

Discussion

It appears that ferrocyanide transport is restricted largely to LMCWs, but may also involve SMCWs. It should be noted that both structures are composed chiefly of the polysaccharide components of P-PC. It is suggested that the diffusion of most nutrients and metabolites is similarly restricted to the same structures in cartilage matrix. What seems to be essential is a three-dimensional mechanism for a rapidly alternating or cyclic process which has a progressive or directional change in the fixed charges. The long, sulfated polysaccharide chains attached to the protein core of the P-PC would seem to be particularly deformable in response to charge effects. It is also possible that more free water is present in the walls of the LMCs and the SMCs, which alone could limit diffusion to their walls.

If the LMCWs and SMCWs are regarded as "diffusion channels" in cartilage matrix, their degree of tortuosity can be estimated from electron micrographs. Their real diameter cannot be so treated, since the effective space occupied by the polysaccharide chains and by the collagen fibrils is unknown. In view of the location of collagen fibers in the LMCWs, it seems obvious that the reduced permeability of cartilage associated with collagen *(19, 21)* is attributed primarily to the reduction in the effective thickness of the LMCWs caused by the space occupying collagen fibers. The actual thickness of the walls (380 and 100 Å, respectively) is larger than the space available in them for diffusion. The penetrability of large molecules, such as hemoglobin, would determine the actual "clear" channel. Even so, this dimension is probably variable depending on the nature of and the degree to which the polysaccharide chains are deformable and responsive to (sub) micro environmental changes. Such adjustments, while feasible within certain limits in the case of globular proteins, are more probable for extended proteins like P-PC.

It was suggested in Chapter 5 (this volume) that cartilage matrix must be in a state of flux, adjusting to differing growth rates in different parts of, for example, the epiphyseal plate. Such accommodations probably take place also in nongrowing cartilage in the ordinary turnover of P-PCs, which is rather appreciable when studied biochemically *(13)*. The continual shift of macromolecules in cartilage matrix, which is proposed here, moderates somewhat the seemingly sharp boundaries of first- and second-order matrix compartments visualized in the model presented in the preceding chapter. The contrast between movement of P-PC and of hemoglobin molecules in cartilage matrix is the contrast between the continually shifting readjustments of the structural components of cartilage matrix (P-PC macromolecules) and the movement of metabolites and nutrients through and between them.

The remainder of this section is devoted to a reinterpretation of the results of studies of ground substance of bone and connective tissue, that have been published from my laboratories.

Durning *(7)* and Molnar *(22)* described the osteoid of frozen-dried specimens in terms of density, after staining developing bone with alcoholic platinum tetrabromide. The dense walls which enclose the less dense submicroscopic spaces of osteoid are remarkably similar to the LMCWs of cartilage matrix described in the preceding chapter. Collagen fibers appear in the walls as the osteoid matures, much as they appear in joint cartilage, though more highly ordered and more numerous. It is interesting that Chase [cited by Gersh *(8)*] found that ferrocyanide is localized specifically in the denser walls of osteoid. Interpreting these morphological findings histochemically, I would suggest now that the denser walls of the osteoid compartments, referred to in earlier articles as the more dense phase, contain the polysaccharide component of the P-PC of osteoid, and that the less dense phase of the compartments includes the protein moiety, perhaps with some polysaccharide. Collagen fibrils present in the walls of the osteoid become highly ordered and prominent as the osteoid matures. As in calcifying, calcification is accompanied by loss of P-PC, which is catabolized.

Freeze-dried rat tail tendon has been studied after numerous methods of postfixation and staining (reference 2, and this volume, Chapter 4). Following all methods of preparation, a characteristic feature was densely stained walls enclosing less dense "submicroscopic vacuoles." This was especially prominent adjacent to fibroblasts, and was interpreted as a two-phase system. Reinterpreted histochemically, I suggest now that the denser walls represent the peripheral polysaccharide components of the P-PC of developing collagenous connective tissues. The protein component, perhaps with some polysaccharide, would be present in the less dense "submicroscopic vacuole," or it may have some other distribution.

The distribution of ferrocyanide in the ground substance of loose connective tissue could not be studied in rat tail tendon for technical reasons, but was studied instead in the diaphragm, both with the light microscope and with the electron microscope (5, 6). In certain conditions, ferrocyanide could be observed with the light and electron microscope to be distributed as "vacuoles" which could be resolved into submicroscopic structures. I suggest a reinterpretation of these findings on the basis of the findings in cartilage. I think now that the ferrocyanide vacuoles represent portions of the P-PC referred to in the previous paragraph, which have become disaggregated, and water-rich, and more positively charged.

Summary

A histochemical test for ferrocyanide was used to ascertain the pathways for diffusion of this ion through cartilage matrix. It was found in three series of different experiments that ferrocyanide was always found in the walls of large matrix compartments, and sometimes in the walls of the small matrix compartments. These narrow channels are regarded as the sites where nutrients and metabolites move in their passage between cartilage cells and the peripheral parts of the avascular cartilage.

References

1. Bergman, B. (1968). Diffusion of some cationic and anionic dyes through mandibular disk and cartilage *in vitro*. *Acta Odontol Scand.* **26**, 103–110.
2. Bondareff, W. (1957). Submicroscopic morphology of connective tissue ground substance with particular regard to fibrillogenesis and aging. *Gerontologia* **1**, 222–233.
3. Brower, T. D. (1969). Normal articular cartilage. *Clin. Orthop.* **64**, 9–17.
4. Catchpole, H. R., Joseph, N. R., and Engel, M. B. (1952). The action of relaxin on the pubic symphysis of the guinea pig, studied electrometrically. *J. Endocrinol.* **8**, 377–385.
5. Chase, W. H. (1959). Extracellular distribution of ferrocyanide in muscle. *A.M.A. Arch. Pathol.* **67**, 525–532.
6. Dennis, J. B. (1959). Effects of various factors on the distribution of ferrocyanide in ground substance. *A.M.A. Arch. Pathol.* **67**, 533–549.
7. Durning, W. C. (1958). Submicroscopic structure of frozen-dried epiphyseal plate and adjacent spongiosa of the rat. *J. Ultrastruct. Res.* **2**, 245–260.
8. Gersh, I. (1959). Aging and ground substance of connective tissue. "VA Prospectus-Research in Aging," pp. 5–28. VA Prospectus. Veterans Administration, Washington, D. C.
9. Greenwald, A. S., and Haynes, D. W. (1969). A pathway for nutrients from the medullary cavity to the articular cartilage of the human femoral head. *J. Bone Joint Surg.* **51B**, 747–753.
10. Hamerman, D. (1970). Synovial joints: aspects of structure and function. *In* "Chemistry and Molecular Biology of the Intercellular Matrix" (E. A. Balazs, ed.), Vol. 3, pp. 1259–1277. Academic Press, New York.
11. Hodge, J. A., and McKibbin, B. (1969). The nutrition of mature and immature cartilage in rabbits. *J. Bone Joint Surg.* **51B**, 140–147.

12. Honner, R., and Thompson, R. C. (1971). The nutritional pathways of articular cartilage. *J. Bone Joint Surg.* **53A**, 742–748.
13. Mankin, H. J., and Lippiello, L. (1969). The turnover of adult rabbit articular cartilage. *J. Bone Joint Surg.* **51A**, 1591–1600.
14. Maroudas, A. (1968). Physicochemical properties of cartilage in the light of ion exchange theory. *Biophys. J.* **8**, 575–595.
15. Maroudas, A. (1970). Distribution and diffusion of solutes in articular cartilage. *Biophys. J.* **10**, 365–379.
16. Maroudas, A. (1970). Effect of fixed charge density of the distribution and diffusion coefficients of solutes in cartilage. *In* "Chemistry and Molecular Biology of the Intercellular Matrix" (E. A. Balazs, ed.), Vol. 3, pp. 1389–1401. Academic Press, New York.
17. Maroudas, A., Bullough, P., Swanson, S. A. V., and Freeman, M. A. R. (1968). The permeability of articular cartilage. *J. Bone Joint Surg.* **50B**, 166–177.
18. Maroudas, A., and Evans, H. (1972). A study of ionic equilibria in cartilage. *Connect. Tissue Res.* **1**, 69–77.
19. Maroudas, A., and Muir, H. (1970). The distribution of collagen and glycosaminoglycans in human articular cartilage and the influence on hydraulic permeability. *In* "Chemistry and Molecular Biology of the Intercellular Matrix" (E. A. Balazs, ed.), Vol. 3, pp. 1381–1387. Academic Press, New York.
20. McCutchen, C. W. (1962). The frictional properties of animal joints. *Wear* **5**, 1–17.
21. Muir, H., Bullough, P., and Maroudas, A. (1970). The distribution of collagen in human articular cartilage with some of its physiological implications. *J. Bone Joint Surg.* **52B**, 554–563.
22. Molnar, Z. (1959). Development of the parietal bone of young mice. I. Crystals of bone mineral in frozen-dried preparations. *J. Ultrastruct. Res.* **3**, 39–45.
23. Spira, E., and Farin, I. (1967). The vascular supply to the epiphyseal plate under normal and pathological conditions. *Acta Orthop. Scand.* **38**, 1–22.
24. Wilsman, N. J., and Van Sickel, D. C. (1972). Cartilage canals, their morphology and distribution. *Anat. Rec.* **173**, 79–94.

9

Relation of the Walls of Large Matrix Compartments of Epiphyseal Cartilage to the Formation of Calcium Crystals

Isidore Gersh

The literature on the mechanism of calcification has been exhaustively and critically reviewed by Glimcher and Krane *(24)*. They presented the theoretical basis and some evidence to support the thesis that the initiation of apatite crystals takes place by heterogeneous nucleation, the necessary space and substrate for nucleation being provided by the "holes" in collagen fibrils which occur periodically because of the precise packing of tropocollagen molecules. They say (p. 239) ". . . the formation of the first solid phase is dependent on a cooperative event involving the interactions between a number of mineral ions, ion clusters, matrix-bound mineral components and the surfaces within the holes of the collagen (or other tissue components) than on any one single interaction." Subsequent growth of crystals involves the transport of calcium and phosphate ions to the site, the state of polymerization of the protein–polysaccharide complexes (P–PC), and the presence of a local reserve of calcium and phosphate in the form of a noncrystalline precipitate. While the theory and observations may explain many aspects of calcification, the authors cannot exclude the possibility that other sites may act as nucleation centers.

Most recent research on calcification has been on bone and teeth, with a lesser amount on tendon and reconstituted collagen. Relatively little research has been published on calcification in cartilage. The uniform finding is that calcium crystals occur in cartilage matrix in bunches and aggregates, with no clear relations to collagen fibers or any other structure *(13)*.

The restrictions imposed by technical limits are more fully appreciated in this aspect of electron microscopy than in any other, as, for example, the solubility of calcium phosphate compounds, crystalline and noncrystalline, in the solvent fixative, the water in the sectioning boat, and even in the solvent of the section staining solution, especially because of the enormously large volumes of water used at each step relative to the amount of calcium in the section (24). In addition, there is a possibility that the calcium salts may recrystallize in a site significantly removed from their original location. Moreover, there is an intrinsic uncertainty in locating the original site of nucleation in that it may be anywhere along the length of the fully developed crystal. Finally, there is in cartilage, particularly, the matter of solution, reprecipitation, and displacement of residual P–PC, as well as the shifting of the course of collagen fibrils during fixation in aqueous solutions (see this volume, Chapter 8).

The methods I have used in this study avoid most, but not all, of these difficulties, in that at least the organic substrate of cartilage matrix is retained more or

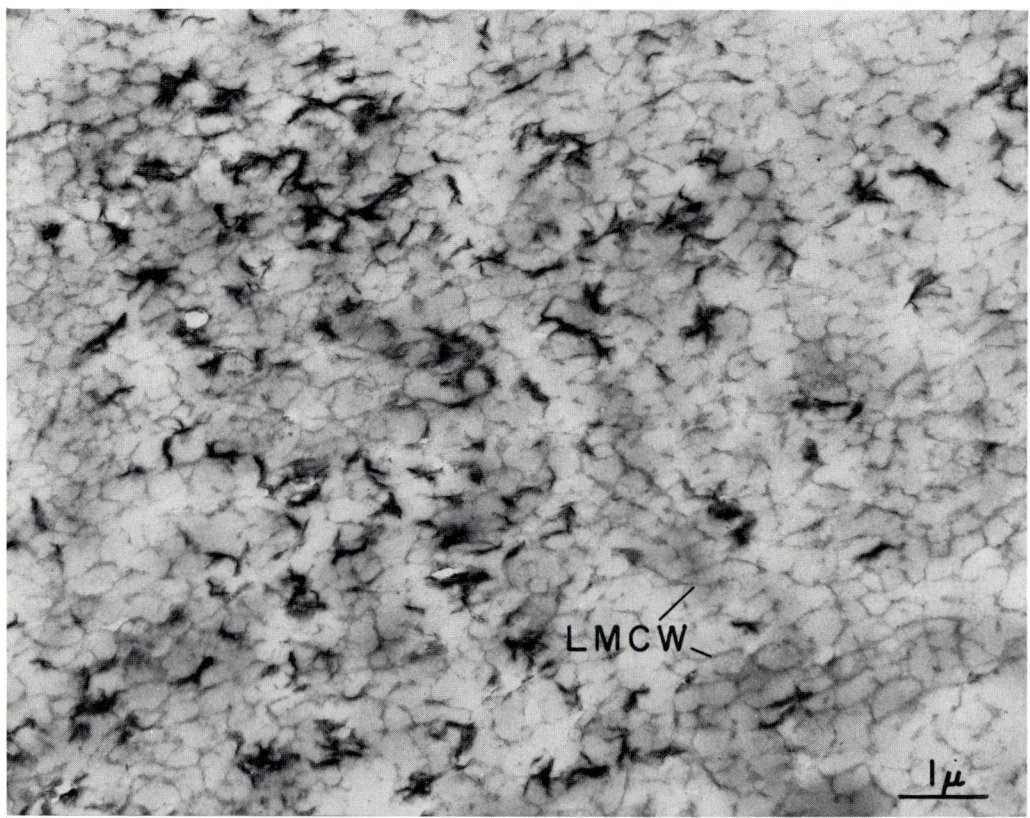

FIG. 9.1. Proximal tibial epiphyseal cartilage of 3-week-old rat at level of zone of provisional calcification. Specimen had been treated with alcoholic ferric chloride. The smaller crystals (whose high contrast is intrinsic and unrelated to the ferric chloride) occur frequently in the large matrix compartment walls (LMCW). These are stained by the ferric chloride throughout all levels of the epiphyseal plate. × 12,000.

less undisturbed and unextracted and in that more calcium is retained by the method of fixation.

Method

Thin slices of epiphyseal plate of the rat were frozen ultrarapidly in the usual way (see this volume, Chapters 5–7). The specimens were dried at a low temperature by sublimation of water, and were postfixed with 95% alcohol, with alcoholic ferric chloride (as for ferrocyanide, this volume, Chapter 8), or with vapor reagents for the detection and the making insoluble of calcium compounds. Sections were cut with the ultramicrotome; much of the time the section-floating water was nearly saturated with calcium phosphate. Sections were removed from the surface of the water and dried as rapidly as possible. The addition of calcium phosphate

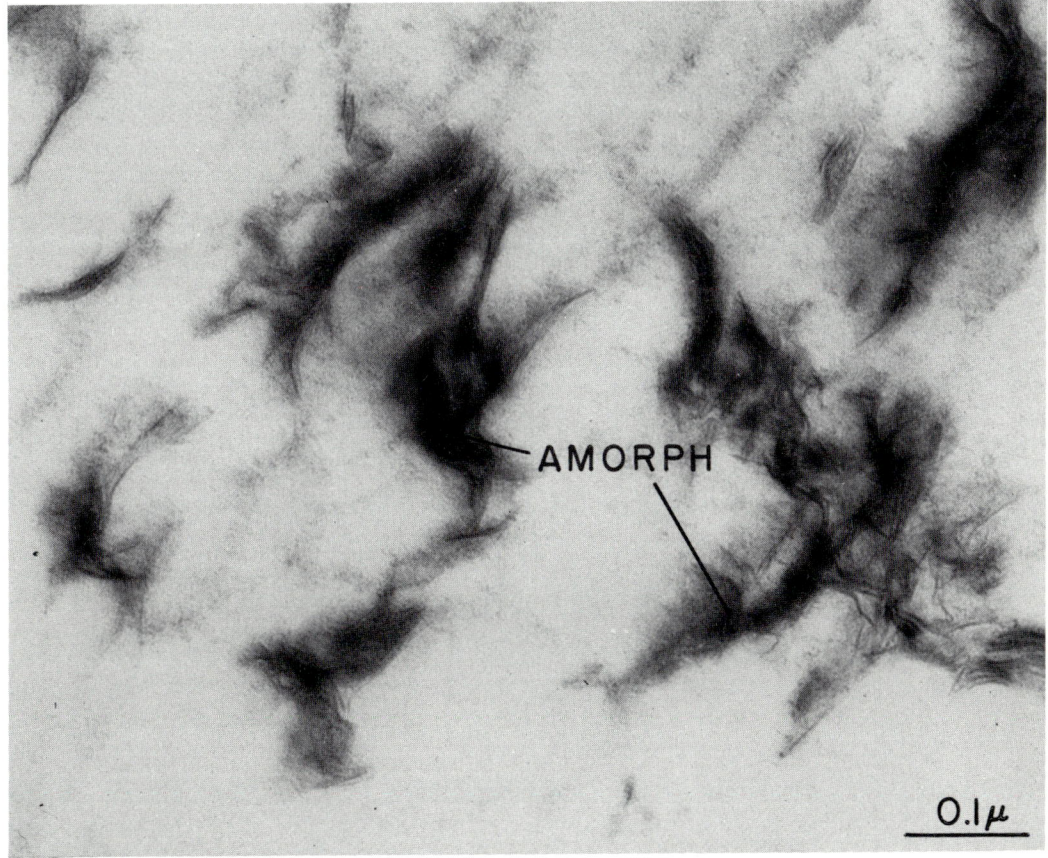

Fig. 9.2. As in Fig. 9.1. The crystals are enclosed in a homogeneous, less dense amorphous mass (AMORPH). × 165,000.

seemed to preserve more of the dense material after postfixation in alcohol; it made no appreciable difference after the use of the vapor reagent, which preserved the dense material even without calcium phosphate. The vapor reagent is a carmine red, crystalline derivative of dipivaloylmethane, heptafluorodimethyloctanedione (HEFDOD) prepared by Pierce Chemical Company, Rockford, Illinois, through the cooperation of Mr. Roy Oliver. Dipivaloylmethane and suitable derivatives chelate with di- and polyvalent cations (43) (the chief of which in cartilage matrix would be expected to be calcium) and combine with proteins, probably mainly through their amine groups. HEFDOD has the advantage for electron microscopy over the nonfluorinated compound of having more mass, and, hence, improving contrast where it reacts. The mass is in the same range as that of 4-iodo-2,2,6,6-tetramethyl-3,5-heptadione (ITH), which is a white crystalline compound also prepared by Pierce Chemical Company. The results with both diones were nearly identical. However, the cutting qualities of various tissues with HEFDOD are so far superior, that work with ITH was discontinued.

Details of postfixation with alcohol and with alcoholic ferric chloride are given in Chapter 2 (Vol. I) and Chapter 8 (this volume). Postfixation with vapors of HEFDOD took place *in vacuo* for 4 to 8 hours at 0°C in the presence of water vapor. Evacuation of excess vapors was for 2 hours at 0°C and then overnight at room temperature. Specimens were infiltrated *in vacuo* with 95% alcohol, and then embedded in water-soluble Durcupan in the usual way. Sections were studied with an Hitachi EMU-11A electron microscope operating at an accelerating voltage of 50 kv. Electron micrographs were enlarged in printing × 2 (Fig. 9.1; see also Figs. 9.4 and 9.9) or × 5.

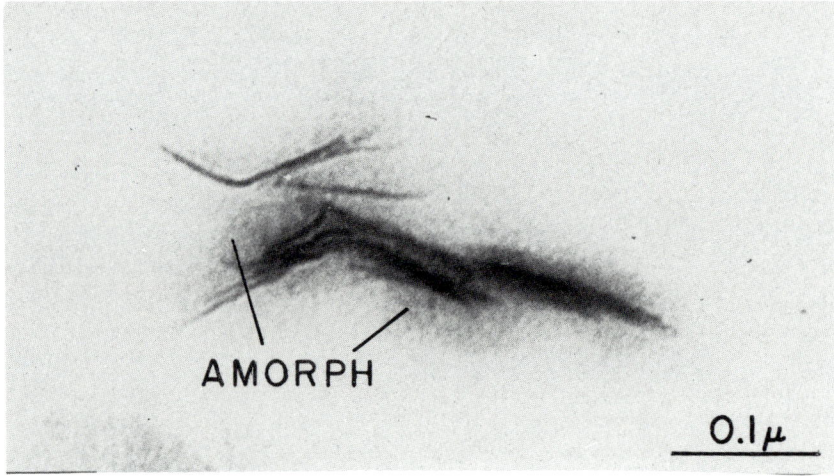

FIG. 9.3. A similar specimen postfixed with 95% alcohol. The crystals are enclosed in a very fine web, which represents the residue of the amorphous calcium deposits (AMORPH). × 200,000.

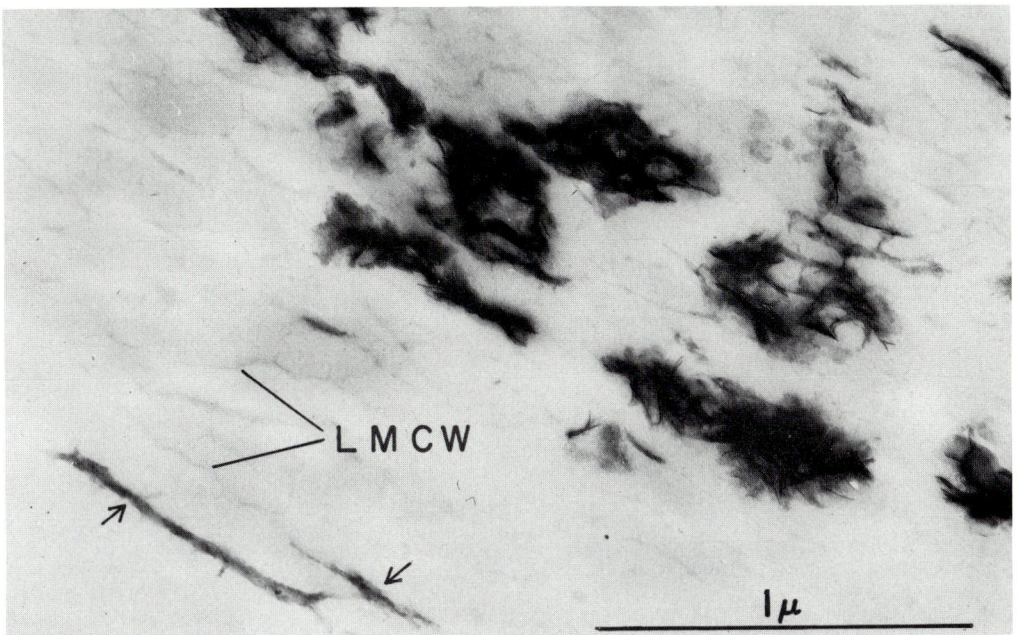

Fig. 9.4. A similar specimen of epiphyseal plate postfixed *in vacuo* with vapors of the chelating reagent HEFDOD (see p. 190). The LMCWs are stained. In addition, there are some noncrystalline regions of high density (arrows). The crystalline masses contain and are surrounded by amorphous dense material. × 50,000.

Observations

The first observations on calcium crystals in cartilage were made on sections of freeze-dried specimens of epiphyseal plate postfixed with alcoholic ferric chloride. In the electron micrographs, there are remarkable numbers of apatite crystals in an early stage of formation and aggregation which clearly lie in the course of the walls of large matrix compartments (LMCWs) as illustrated in Fig. 9.1. The crystals lie in an amorphous material of high contrast (Fig. 9.2). To rule out the possibility that the amorphous material represented calcium phosphate of apatite crystals, which had been dissolved and reprecipitated because of the acidity of the solution, freeze-dried specimens were postfixed in alcohol. Numerous examples were found also in this material of crystals lined up along the LMCWs. On enlargement of the electron micrographs, the crystals were found to be surrounded by a network of fine lines which were about 10 Å thick (Fig. 9.3). The crystals were not uniformly dense, and it is possible that the inequalities reflect the unit structure of apatite crystals *(41)*, or that they were caused by fragmentation during sectioning.

In specimens postfixed with vapors of HEFDOD, numerous crystals were again found to lie in the course of the LMCWs. Even at low magnifications, nearly all crystals were found to lie in a very broad region of high density probably representing in part amorphous calcium phosphate (Fig. 9.4). At higher magnification, the large amount of the amorphous material relative to the crystalline material is very impressive (Figs. 9.5–9.7). A very unexpected finding was the appearance of small regions of amorphous dense material in the walls of the LMCWs with no crystals visible. Sometimes these extend into the SMCWs. In such regions, all the LMCWs are rather well stained (Fig. 9.8). In cartilage matrix treated with vapors of HEFDOD, LMCWs are denser than in alcohol-fixed specimens. (Compare Figs. 9.8 and 9.9 with Figs. 7.3A–C and 7.4, in Chapter 7, this volume.)

In some ultrathin sections, which were transparent when viewed on the surface of the sectioning trough (i.e., 300–400 Å or less in thickness) and relatively undistorted, the improved resolution made it possible to detect numerous aggregates otherwise obscured by the dense, amorphous background in thicker sections

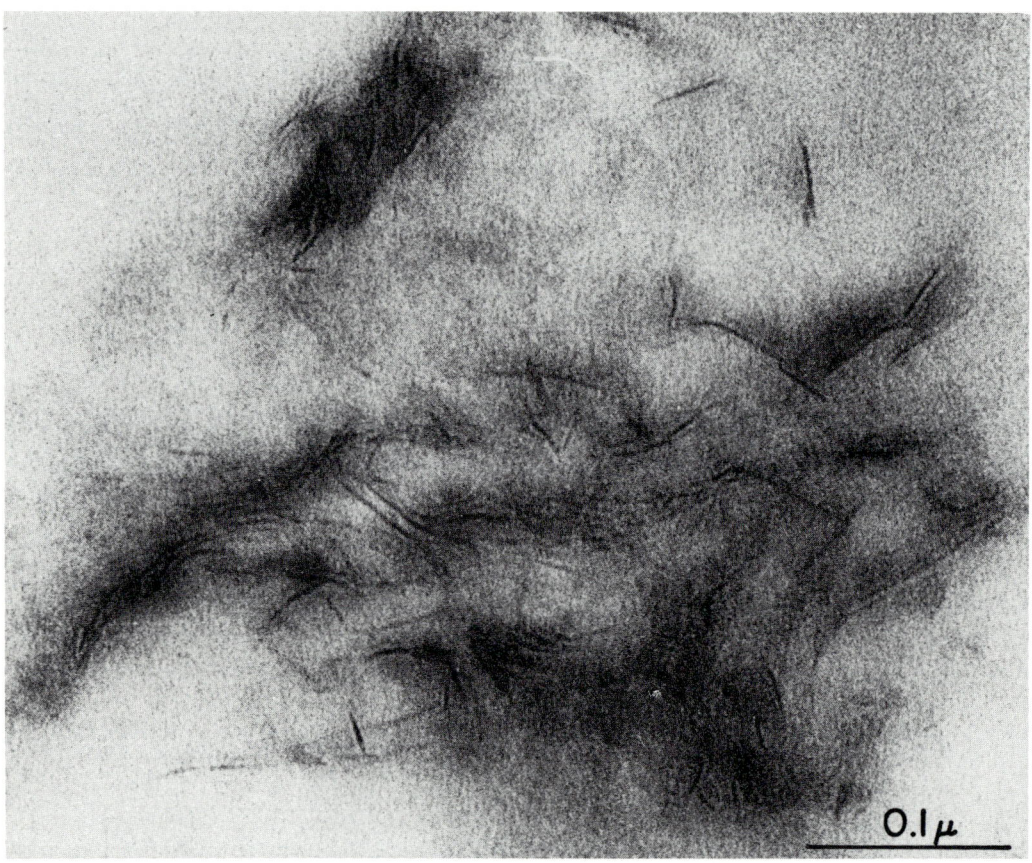

Figs. 9.5–9.7. Same section, at higher magnification, to show some examples of crystals and the amorphous material around them (AMORPH). Figures 9.5 and 9.7, × 270,000; Fig. 9.6, × 200,000.

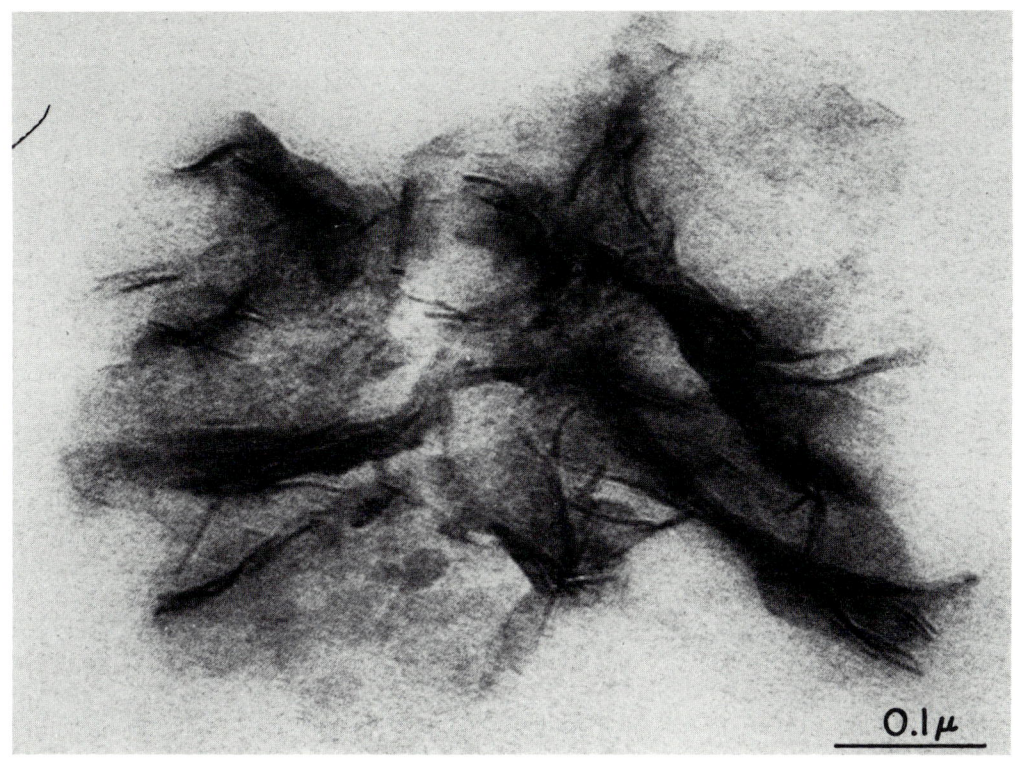

FIG. 9.6. ▲   ▼ FIG. 9.7.

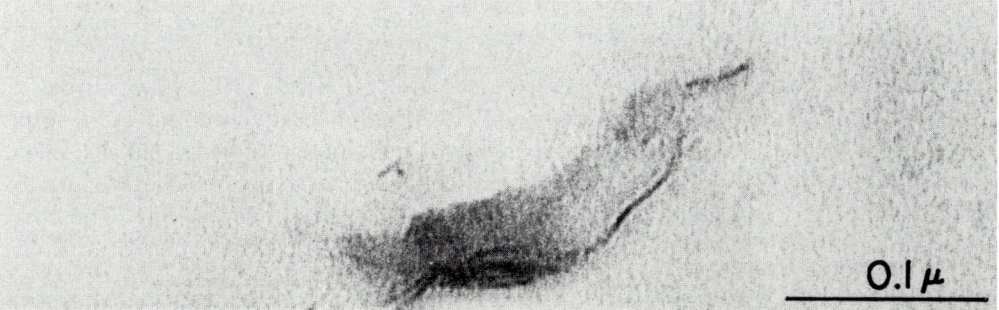

(500–600 Å). Some of these appear crystalline and may represent crystals in process of growth (Figs. 9.10–9.12). Some of them might conceivably represent crystal nuclei, which seem large enough to be resolvable (32). In these exceedingly thin sections, there seems to be a larger number of crystalline structures (which seem to lie closer to each other) as compared with those in thicker sections (500–600 Å).

As stated above, the reagent HEFDOD reacts with bi- and polyvalent cations as well as with proteins. Two experiments were performed to distinguish between these substances in the crystal masses of cartilage matrix. The first experiment

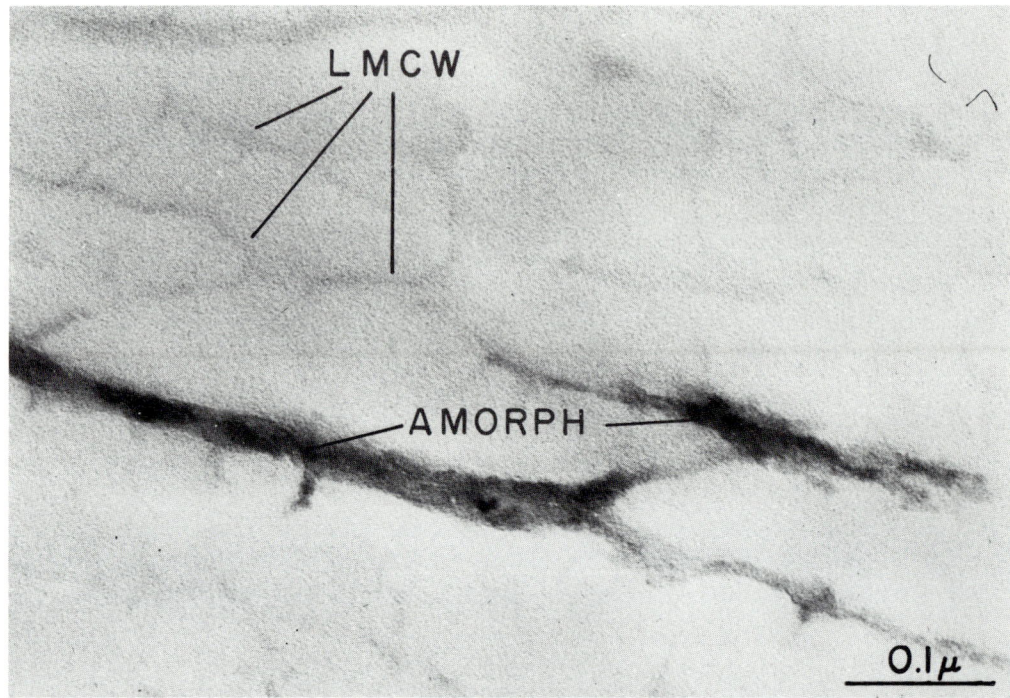

Fig. 9.8. Same section, at high magnification to show several regions of dense amorphous material (AMORPH) lying in LMCWs. The amorphous calcium extends partly into the walls of the small matrix compartment walls. × 200,000.

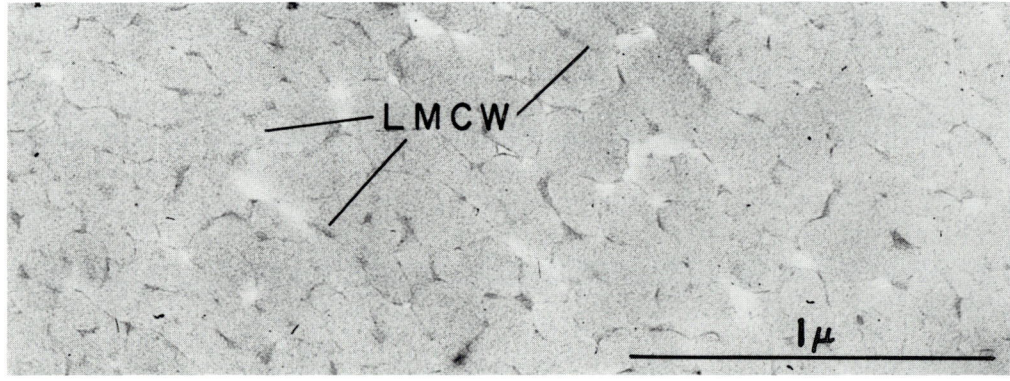

Fig. 9.9. Another specimen treated similarly from a noncalcifying zone of the epiphyseal plate, to show the LMCWs stained with the vapor chelating reagent for di- and polyvalent cations including calcium, HEFDOD. This reagent also combines with and stains proteins. × 48,000.

was a comparison of the effects on density in electron micrographs of specimens which are treated with vapor reagents which react almost exclusively with proteins (and not with calcium) with density of specimens treated with a vapor reagent which reacts with proteins and with calcium. Slivers of epiphyseal plate were frozen ultrarapidly and dried in the usual way. Some were overexposed to vapors of FFDNB and FFSulfone (10 hours at 0°C *in vacuo*). Others were exposed to vapors of HEFDOD for 4 hours at 0°C. Specimens were infiltrated with 95% alcohol and embedded in water-soluble Durcupan. The results were as follows: After exposure to the protein reagents, the crystalline aggregates were smaller, the crystals noticeably less numerous and less sharp (Fig. 9.13), and there was less amorphous dense material associated with them, as compared with specimens treated with HEFDOD (Figs. 9.10–9.12). These observations serve to identify the amorphous, dense material as probably not protein in nature, but as a bi- or polyvalent cation, probably calcium.

This conclusion was fortified by the results of the second experiment. Here ultrathin sections of specimens postfixed with vapors of HEFDOD were cut about 500 to 600 Å thick, mounted on grids, and dried. They were then immersed for 10 minutes in a drop of (1) a 0.1 M solution of sodium oxalate, or (2) a 0.1 M solution of the sodium salt of EDTA, with immersion of other grids in distilled water as a control. After oxalate treatment for 10 minutes no crystals could be observed.

After EDTA treatment for 10 minutes, few crystals and some amorphous material remained (Fig. 9.14). Spaces formerly occupied by crystals in the undecalcified controls (Fig. 9.15) appear empty. On further treatment for 20 minutes, the dense material is completely removed, and the site occupied by it appears less dense than noncalcified matrix in the same section (Fig. 9.16). The absence of dense residual regions indicates that the amorphous densities of electron micrographs are to be attributed virtually entirely to calcium compounds.

Discussion

I offer an interpretation of these findings which involves certain assumptions. The first assumption is that many of the crystals observed are "new" crystals, which arose *de novo* rather than as seeds from other crystals. The second is that when the latter appears unaccompanied by crystals, in a limited part of a section (as in Fig. 9.8), the material is what it seems to be, and is not a section of amorphous material which encloses crystals at another level above or below the plane of section examined. The third assumption is that the amorphous calcium concentrate may contain nuclei of crystals or fragments of crystals indistinguishable as such. The electron micrographs of crystals in "transparent" sections (Figs. 9.10–9.12) may be pertinent in this connection. The fourth assumption is that most of the dense material of high contrast in the matrix treated with vapors of HEFDOD represents calcium, and that other di- or polyvalent cations are present, if at all, in negligible amounts.

Published evidence suggests that the process of crystal formation begins with the trapping of calcium by the polysaccharide component of the LMCWs. This is illustrated by electron micrographs showing amorphous dense regions in the LMCWs stained by vapors of HEFDOD. The calcium is probably bound to P–PC, but some of it may be in transit through the cartilage matrix, like ferrocyanide (Chapter 8, this volume). Calcium probably combines with free sulfates of chondroitin (and perhaps other polysaccharides) and acts as a cross-linking agent *(9, 23, 28, 29, 31, 57)*. Whether the P–PC is normal or degraded at the critical site is not certain. [The possible involvement of P–PC in calcification has been analyzed by Campo *(14)*.] These findings must be juxtaposed to those of Eanes and Posner and their colleagues *(21, 22, 52)*, who find that as much as 60% of the calcium occurs in an amorphous state. It is easy to imagine that this (high) value must include the whole range of microsites from those where all the calcium may be amorphous to those where only 10 or 20% of the calcium may be amorphous.

These interpretations support the following quotations: " . . . calcification in its broadest sense encompassing all phases of mineral deposition and resorption, must involve in the first instance, transport processes which bring such ions and

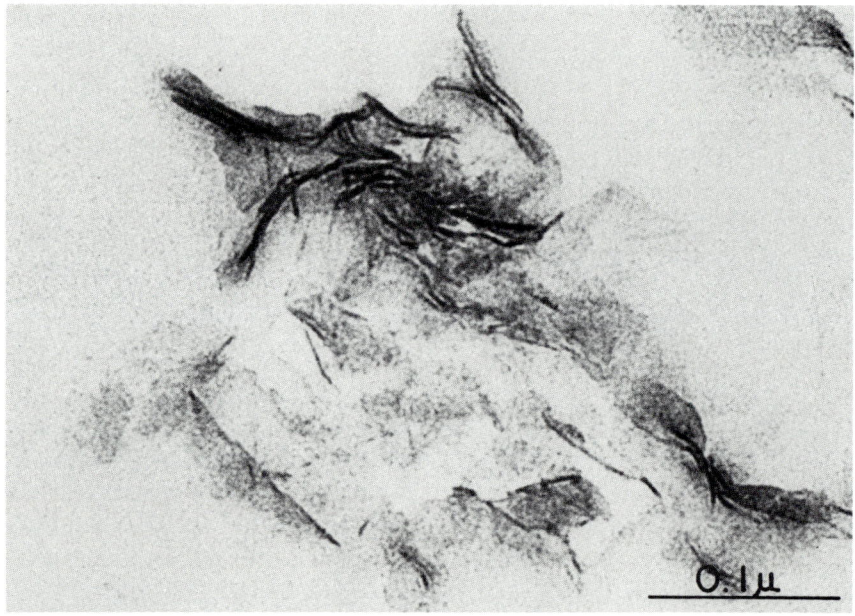

Fig. 9.10.

Figs. 9.10–9.12. Proximal tibial epiphyseal cartilage of 3-week-old rat at level of zone of provisional calcification. Specimen treated with vapor of HEFDOD, which reacts with calcium and other bi- and polyvalent cations as well as proteins. Sections was 300–400 Å thick or less. Crystals are surrounded by amorphous, dense material which contains many granules of minute dimensions which are about the size or slightly larger than nuclei apatite of crystals. × 250,000.

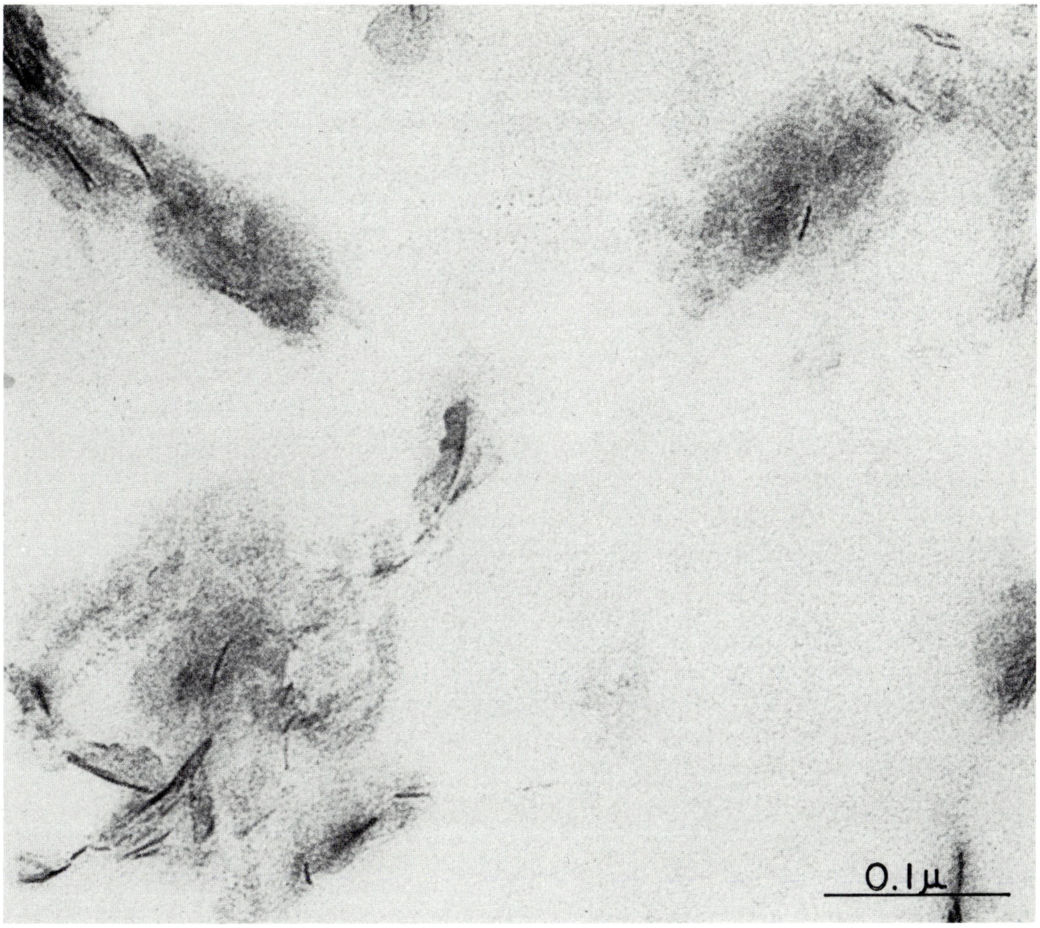

FIG. 9.11. ▲

▼ FIG. 9.12.

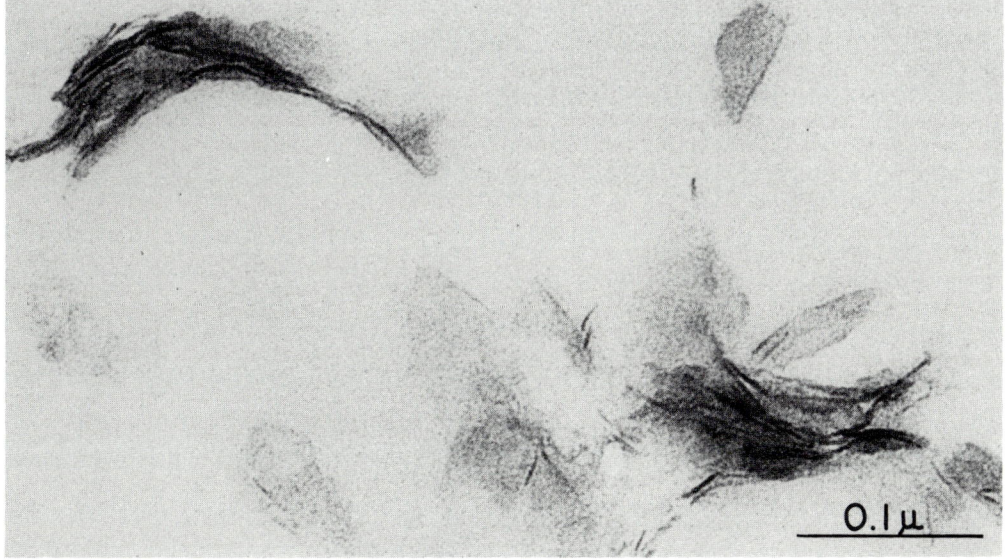

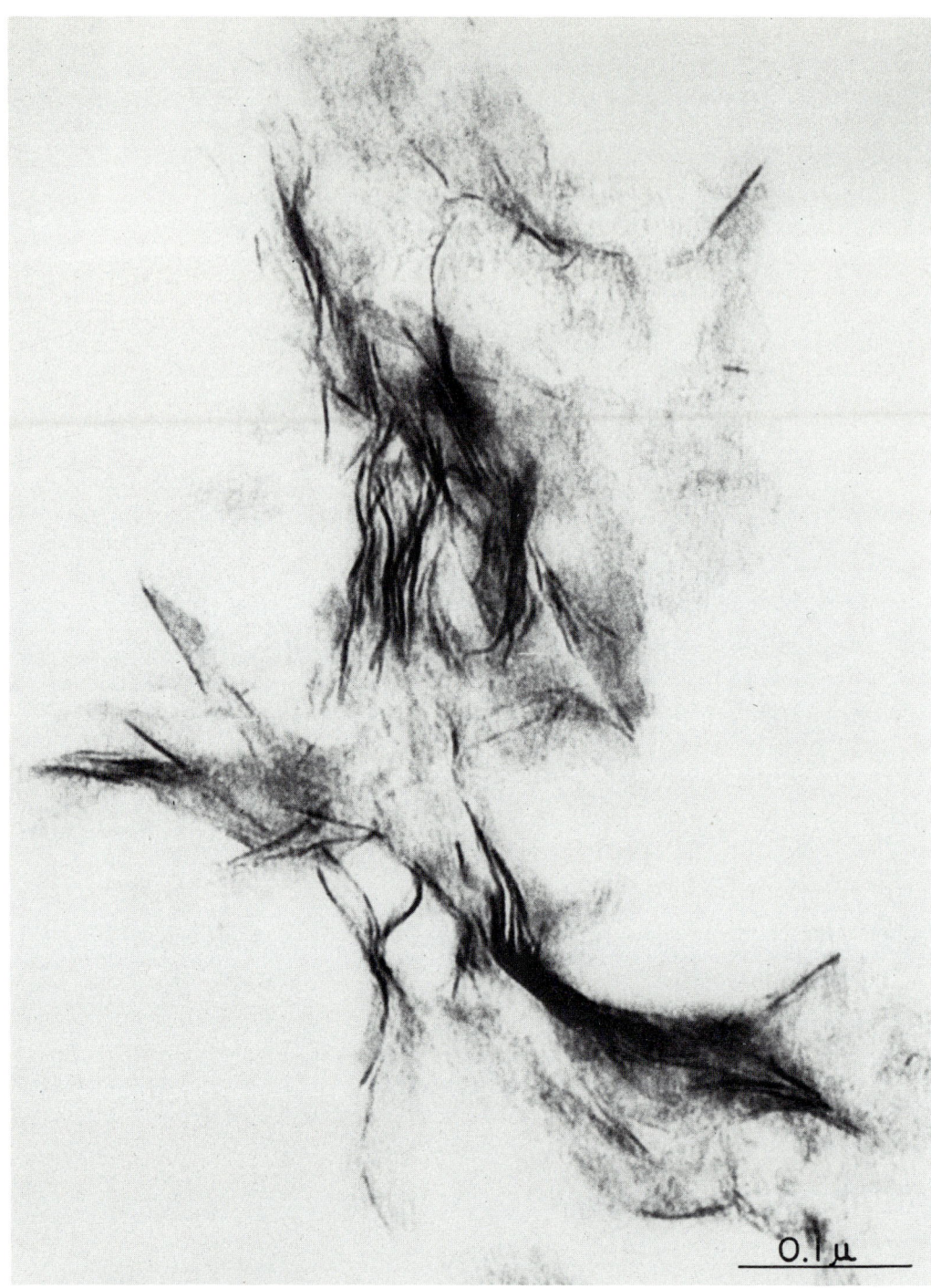

Fig. 9.13. Epiphyseal plate from the same animal and region as shown in Fig. 9.10. Specimen treated with vapors of FFDNB and FFSulfone, which reacts almost exclusively with proteins. Section was 300–400 Å thick. Crystals are surrounded by amorphous, dense material which is present in smaller amount than in Figs. 9.10–9.12. The crystals seem to be thicker, in part, and in process of "fading" into the amorphous background by dissolution. The amorphous material lacks the granular component present in the preceding three figures. × 250,000.

Discussion

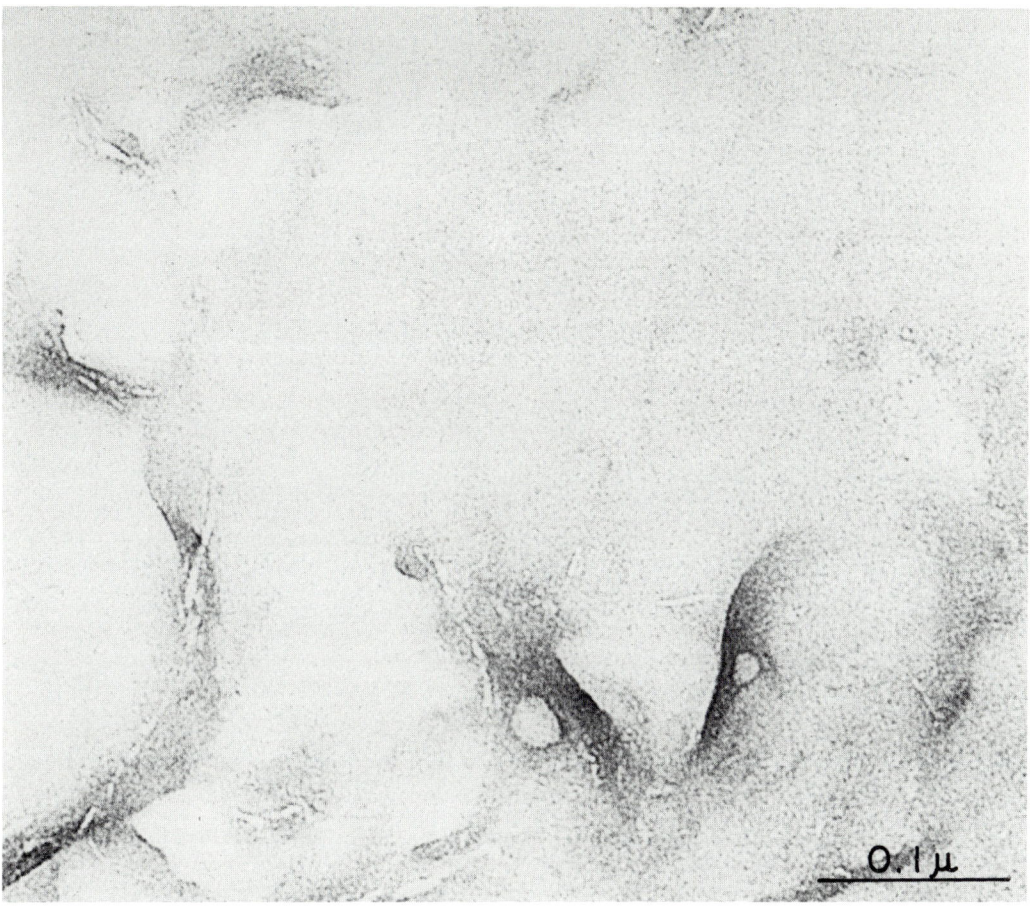

Fig. 9.14. Ultrathin section, from same specimen as used in Figs. 9.10–9.12, mounted on grid and treated for 10 minutes with 0.1 M solution of sodium salt of EDTA. The section was then washed in water, dried, and observed with the electron microscope. The density of the calcified region is greatly reduced as compared with water-treated controls. Spaces formerly occupied by crystals are very prominent. This is an intermediate stage of decalcification. × 250,000.

clusters to the interstices of the collagen fibrils" (or other sites of nucleation) (24) (p. 224); and "degradative activity, particularly that directed against the polypeptide trunk, might serve to release calcium locally and promote foci of metastatic calcification" (36). Also, the interpretation takes account of the reports in the literature that the amount of P–PC is reduced in regions of calcification. The brief observations in this chapter which bear on this topic (Figs. 9.14 and 9.16) are in support of this view. [See below and review by Campo (14).]

It is of critical importance to learn what happens to the P–PC at the very moment when amorphous calcium is deposited. The literature on this subject, with one important exception (27) indicates that there is less P–PC in calcifying regions than in hypertrophic, but uncalcified zones (14, 15, 26). It is clear that cartilage may contain hydrolytic enzymes capable of degrading P–PC to smaller units. Biochem-

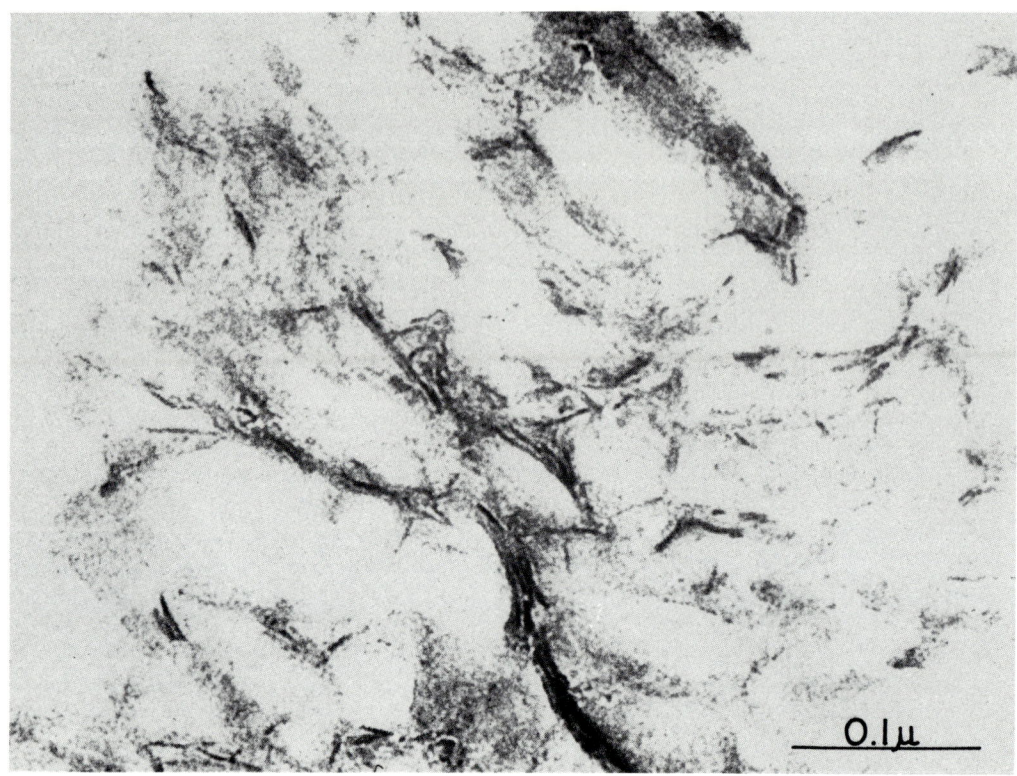

Fig. 9.15. Ultrathin section treated as a control for Fig. 9.14. The section mounted on the grid was immersed for 10 minutes in water instead of EDTA, washed like that in Fig. 9.14, dried, and observed with the electron microscope. There are not as many crystals as in untreated sections, but it is obvious that much amorphous and crystalline calcium persists. The result is essentially the same after extraction with water for 30 minutes. × 250,000.

ical evidence suggests that such enzymes are of lysosomal origin, and that cathepsin D may be one of the primary hydrolytic enzymes *(1, 2, 18, 20, 25, 30, 33–35, 42, 44, 54, 55)*. Granules rich in acid phosphatase have been described in chondrocytes of epiphyseal plate *(49, 51)*. Thus, it is possible that amorphous calcium is deposited when the calcium, which is bound to P–PC, is released locally (at sites where the protein is degraded enzymatically), in a grossly supersaturated solution. This proposal is supported by *in vitro* experiments on nucleation and growth of calcium crystals in relation to macromolecular P–PC and its digestion products *(16, 48)*.

This proposed system automatically includes ". . . a mechanism for producing really high concentrations of calcium and phosphate as well as providing for their effective release at or near a nucleation site" *(56)*. This may be a precondition in cartilage for the autocatalytic conversion of amorphous to crystalline calcium *(22)*. At the same time, crystal growth in cartilage, at least, becomes independent of the

Discussion

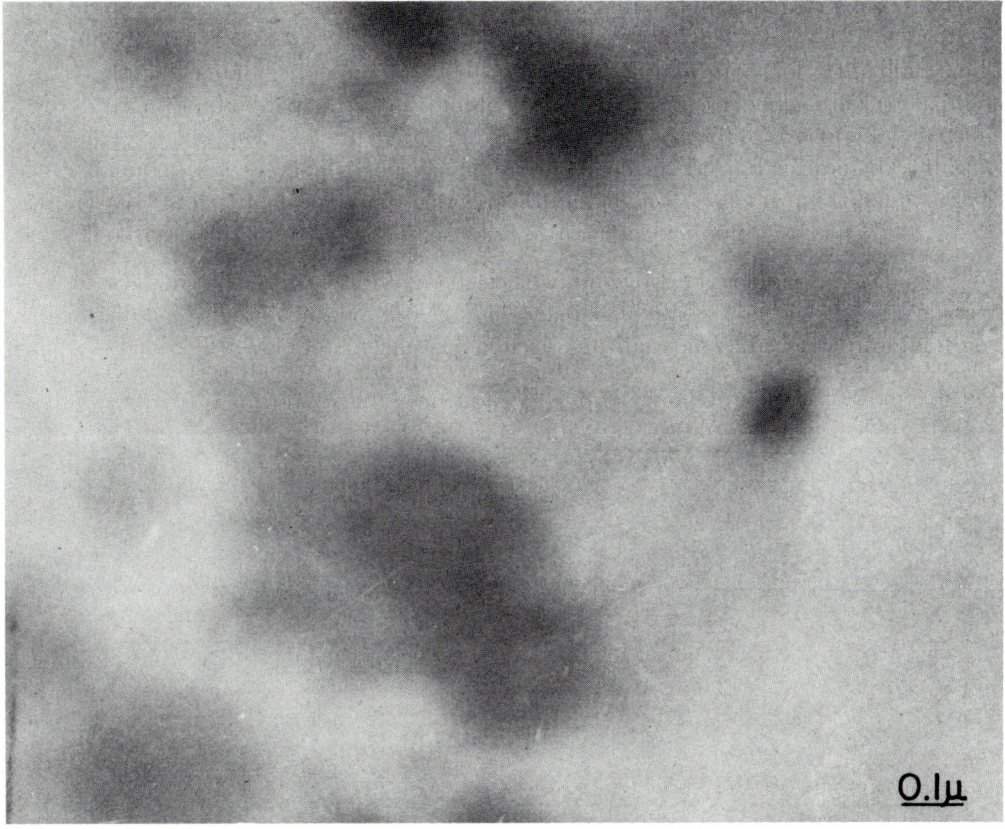

Fig. 9.16. Ultrathin section prepared as in Fig. 9.14, but treated with EDTA for 30 minutes instead of 10 minutes, to show a terminal stage of decalcification. The electron micrograph was printed as a negative in order to show the clear spaces of the cartilage matrix (black), which correspond with regions occupied by amorphous and crystalline calcium before treatment with EDTA. This serves (1) to identify the amorphous and crystalline material of untreated specimens as probably containing calcium, and (2) to show that the organic residue of regions occupied by crystal clusters is greatly reduced in regions of calcification as compared with uncalcified matrix. × 83,000.

rate of diffusion of calcium, which may be a limiting factor in other situations (32).

In the end, nucleation and growth of crystals take place. While there is good evidence that nucleation takes place in "holes" in collagen fibers, the evidence does not require that calcification should be restricted to them. In fact, crystal formation in tooth enamel proves that collagen is not a necessary requirement for nucleation. Any submicroscopic region would be suitable if it has the correct protected microenvironment conducive to nucleation and crystal growth. The only collagen fibrils observed in epiphyseal plate in the rat are about 40 Å in diameter (this volume, Chapter 6). These are, at the same time, the smallest fibrils capable of furnishing "holes" if the tropocollagen molecules are tightly packed as six molecules enclosing a seventh. If crystal formation were limited to collagen fibrils with "formal" holes, then it seems clear that in epiphyseal plate, nucleation sites exceed the number of collagen fibrils 40 Å in diameter, and that one must attribute some (or all) of nucleation to other structures, perhaps those formed in association with P–PCs.

If smaller fibrils may induce nucleation, then I have no basis for a judgment as to the comparative frequency of nucleation sites and collagen fibrils, since they are not detectable in electron micrographs when they are smaller than about 40 Å in specimens prepared as described earlier in this chapter. It is even conceivable that smaller collagen fibrils of five or even three tropocollagen molecules might supply the surface sufficient for nucleation. It is also conceivable that particular accumulations of polysaccharide chains (alone, cross-linked with calcium, or attached to collagen fibrils) might also form, for a sufficiently long time, a free space suitable for nucleation and crystal growth. The same particular accumulations could also result in a suitable space on the protein moiety of the P–PC.

By its very nature, when it exceeds a certain size, a crystal diffuses the basic information on the exact site of origin of the site of nucleation to the whole region included by or on the surface of the crystal. For this reason it is not possible to state categorically what is the site of nucleation in cartilage. I should point out that a similar situation was described in osteoid *(19, 40)*. There, an amorphous material resembling that in cartilage is deposited in the dense walls of the osteoid. These contain collagen fibrils and also a matrix which transports ferrocyanide *(17)*. As reinterpreted (this volume, Chapter 8), these denser walls in osteoid contain the polysaccharide component of the P–PC of osteoid, as do the LMCWs of cartilage matrix.

Numerous reports have appeared recently on the occurrence in cartilage matrix of a large variety of globules associated with early mineralization *(3–8, 10–12, 28, 37–39, 45, 46, 49, 50, 53, 54)*. They have been studied histochemically and it has even been claimed that they have been separated for chemical analysis. I have not seen such bodies in freeze-dried specimens of cartilage after treatment with numerous vapor reagents, or have they appeared after treatment of sections with chelating agents. It is, of course, possible that I have been unable to fix and stain them satisfactorily. It is also possible that they are a consequence of reprecipitation during fixation in fluid fixatives, and that they represent some calcium deposit which is bound, even after the use of decalcifying reagents, to the reprecipitated material. A similar interpretation is offered for the appearance of the lamina limitans *(47)*.

The appreciation of the technology of specimen preparation is perhaps more sophisticated in the study of calcification than in any other field. The study of the inorganic constituents of cells and tissues with the electron microscope is beset with difficulties, the chief one being their diffusivity during fixation and each subsequent manipulation involving liquids. A second important difficulty is the solution of the reaction product in the various solvents used. In studies on crystalline calcium, a third difficulty is the danger of solution and recrystallization of calcium compounds. The methods I have used reduce, but do not eliminate these possible sources of error. The avoidance of fluid fixatives undoubtedly reduces this error. Further security is achieved by the use of the vapor chelating reagent *in vacuo*. This success is marked by an increased amount of noncrystalline calcium in the LMCWs and between and surrounding the apatite crystals. There remain possibilities of diffusion and extraction by 95% alcohol and the components of the water-soluble Durcupan mixture, and by the water in the section floating boat. The last difficulty was reduced by cutting many excess sections and allowing them to float in the boat, by adding small crystals of calcium phosphate to the boat, and by removing sections rapidly for viewing directly after cutting and drying them. These technical dodges seemed to improve the results somewhat after postfixation with alcohol, but were without noticeable effect after postfixation with the vapor reagent. To the extent that the fluids employed in this study do actually extract calcium and phosphate the possibilities of error through recrystallization remain.

Summary

With the aid of a vapor reagent which reacts with bi- and polyvalent cations as well as proteins, calcium has been found in the walls of large matrix compartments, where it may exist as amorphous pools. Crystal formation takes place in the same sites, which contain collagen fibers and the polysaccharide moiety of protein–polysaccharide complex. It has not been possible to ascertain which of these components is primarily involved in crystal formation. The earliest and smallest crystals are surrounded by a region of high density which is largely amorphous. Some aggregates which may be crystal nuclei or growing crystals are also included in these amorphous regions. The findings are integrated into current hypotheses on apatite formation in hard tissues.

References

1. Ali, S. Y. (1964). The degradation of cartilage matrix by an intracellular protease. *Biochem. J.* **93**, 611–618.
2. Ali, S. Y., and Evans, L. (1969). Studies on the cathepsins in elastic cartilage. *Biochem. J.* **112**, 427–433.
3. Ali, S. Y., Sajdera, S. W., and Anderson, H. C. (1970). Isolation and characterization of calcifying matrix vesicles from epiphyseal cartilage. *Proc. Nat. Acad. Sci. U. S.* **67**, 1513–1520.
4. Anderson, H. C. (1969). Vesicles associated with calcification in the matrix of epiphyseal cartilage. *J. Cell Biol.* **41**, 59–72.
5. Anderson, H. C., and Matsuzawa, T. (1970). Membranous particles in calcifying cartilage matrix. *Trans. N. Y. Acad. Sci.* **32**, 619–630.
6. Appleton, J. (1970). Ultrastructural observations on early cartilage calcification. The use of chromium sulphate in decalcification. *Calcif. Tissue Res.* **5**, 270–276.
7. Appleton, J. (1971). Ultrastructural observations on the inorganic/organic relationships in early cartilage calcification. *Calcif. Tissue Res.* **7**, 307–317.
8. Appleton, J., and Blackwood, H. J. J. (1969). Ultrastructural observations on the early mineralization of cartilage. *J. Bone Joint Surg.* **51B**, 385 (AB).

9. Bachra, B. N. (1970). Calcification of connective tissue. *Int. Rev. Cytol.* **5**, 165–208.
10. Bonucci, E. (1969). Further investigation on the organic/inorganic relationships in calcifying cartilage. *Calcif. Tissue Res.* **3**, 38–54.
11. Bonucci, E. (1971). The locus of initial calcification in cartilage and bone. *Clin. Orthop.* **78**, 108–139.
12. Bonucci, E. (1970). Fine structure of epiphyseal cartilage in experimental scurvy. *J. Pathol.* **102**, 219–227.
13. Cameron, D. A. (1963). The fine structure of bone and calcified cartilage. *Clin. Orthop.* **26**, 199–228.
14. Campo, R. D. (1970). Protein-polysaccharides of cartilage and bone in health and disease. *Clin. Orthop.* **68**, 182–209.
15. Campo, R. D., and Dziewiatkowski, D. D. (1963). Turnover of the organic matrix of cartilage and bone as visualized by autoradiography. *J. Cell Biol.* **18**, 19–29.
16. Campo, R. D., Tourtellotte, C. D., and Bielen, R. J. (1969). The protein-polysaccharides of articular, epiphyseal plate and costal cartilages. *Biochim. Biophys. Acta* **177**, 501–511.
17. Chase, W. H. (1959). Cited by I. Gersh. Aging and ground substance of connective tissue. "Research in Aging, VA Prospectus," pp. 5–28. Veterans Administration, Washington, D. C.
18. Dingle, J. T., Fell, H. B., and Coombs, R. R. A. (1967). The breakdown of embryonic cartilage and bone cultivated in the presence of complement-sufficient antiserum. 2. Biochemical changes and the role of the lysosomal system. *Int. Arch. Allergy Appl. Immunol.* **31**, 283–303.
19. Durning, W. C. (1958). Submicroscopic structure of frozen-dried epiphyseal plate and adjacent spongiosa of the rat. *J. Ultrastruct. Res.* **2**, 245–260.
20. Dziewiatkowski, D. D., Tourtellotte, C. D., and Campo, R. D. (1968). Degradation of protein-polysaccharide (chondromucoprotein) by an enzyme extracted from cartilage. *In* "The Chemical Physiology of Mucopolysaccharides" (G. Quintarelli, ed.), pp. 63–79. Little, Brown, Boston, Massachusetts.
21. Eanes, E. D., and Posner, A. S. (1970). Structure and chemistry of bone material. *In* "Biological Calcification: Cellular and Molecular Aspects" (H. Schraer, ed.), pp. 1–26. Appleton-Century-Crofts, New York.
22. Eanes, E. D., Termine, J. D., and Posner, A. S. (1967). Amorphous calcium phosphate in skeletal tissues. *Clin. Orthop.* **53**, 223–235.
23. Eichelberger, L., and Roma, M. (1954). Effects of age on the histochemical characterization of costal cartilage. *Amer. J. Physiol.* **178**, 296–304.
24. Glimcher, M. J., and Krane, S. M. (1968). The organization and structure of bone, and the mechanism of calcification. *In* "Treatise on Collagen" (B. S. Gould, ed.), Vol. 2B, pp. 67–251. Academic Press, New York.
25. Granda, J. L., and Posner, A. S. (1968). Hydrolytic enzymes in different zones of the epiphyseal plate. *J. Bone Joint Surg.* **50A**, 1073Ab.
26. Hirschman, A., and Dziewiatkowski, D. D. (1966). Protein-polysaccharide loss during endochondral ossification: immunochemical evidence. *Science* **154**, 393–395.
27. Hjertquist, S.-O. (1964). Microchemical analysis of aminoglycans (mucopolysaccharides) in normal and rachitic epiphysial cartilage. *Acta Soc. Med. Upsal.* **69**, 23–40.
28. Howell, D. S. (1971). Current concepts of calcification. *J. Bone Joint Surg.* **53A**, 250–258.
29. Howell, D. S., Pita, J. C., Marquez, J. F., and Madruga, J. E. (1968). Partition of calcium, phosphate, and protein in the fluid phase aspirated at calcifying sites in epiphyseal cartilage. *J. Clin. Invest.* **47**, 1121–1132.
30. Jibril, A. O. (1967). Phosphates and phosphatases in preosseous cartilage. *Biochim. Biophys. Acta* **141**, 605–613.
31. Katsura, N., and Davidson, E. A. (1966). Metabolism of connective tissue polysaccharides *in vivo*. IV. The sulphate group. *Biochim. Biophys. Acta* **121**, 135–142.
32. Katz, E. P. (1969). The kinetics of mineralization *in vitro*. I. The nucleation properties of 640-Å collagen at 25°. *Biochim. Biophys. Acta* **194**, 121–129.
33. Kuhlman, R. E. (1965). Phosphatases in epiphyseal cartilage. *J. Bone Joint Surg.* **47A**, 545–550.

34. Kuhlman, R. E., and McNamee, M. J. (1970). The biochemical importance of the hypertrophic cartilage cell area to enchondral bone formation. *J. Bone Joint Surg.* **52A**, 1025–1032.
35. Lucy, J. A., Dingle, J. T., and Fell, H. B. (1961). Studies on the mode of action of excess of vitamin A. 2. A possible role of intercellular proteases in the degradation of cartilage matrix. *Biochem. J.* **79**, 500–508.
36. Marler, E., and Davidson, E. A. (1965). Structure of a polysaccharide protein complex. *Proc. Nat. Acad. Sci. U. S.* **54**, 648–656.
37. Mathews, J. L. (1970). Ultrastructure of calcifying tissues. *Amer. J. Anat.* **129**, 451–458.
38. Mathews, J. L., Martin, J. H., Lynn, J. A., and Collins, E. J. (1968). Calcium incorporation in the developing cartilaginous epiphysis. *Calcif. Tissue Res.* **1**, 330–336.
39. Matukas, V. J., and Krikos, G. A. (1968). Evidence for changes in protein polysaccharide associated with the onset of calcification in cartilage. *J. Cell Biol.* **39**, 43–48.
40. Molnar, Z. (1959). Development of the parietal bone of young mice. I. Crystals of bone mineral in frozen-dried preparations. *J. Ultrastruct. Res.* **3**, 39–45.
41. Molnar, Z. (1960). Additional observations on bone crystal dimensions. *Clin. Orthop.* **17**, 38–42.
42. Morrison, R.-I. G. (1970). The breakdown of proteoglycans by lysosomal enzymes and its specific inhibition by an antiserum to cathepsin D. In "Chemistry and Molecular Biology of the Intercellular Matrix" (E. A. Balazs, ed.), Vol. 3, pp. 1683–1706. Academic Press, New York.
43. Moshier, R. W., and Sievers, R. E. (1965). "Gas Chromatography of Metal Chelates." Pergamon, New York.
44. Quintarelli, G., Sajdera, S., and Dziewiatkowski, D. (1968). Modifications of connective tissue matrices by an enzyme extracted from cartilage. *Histochemie* **15**, 1–20.
45. Scherft, J. P. (1968). The ultrastructure of the organic matrix of calcified cartilage and bone in embryonic mouse radii. *J. Ultrastruct. Res.* **23**, 333–343.
46. Scherft, J. P. (1970). An electron microscopic investigation into the possible presence of periodic acid reactive polysaccharides in the matrix of calcified cartilage and bone. *Proc. Kon. Ned. Akad. Wetesch.* **73**, 414–421.
47. Scherft, J. P. (1972). The lamina limitans of the organic matrix of calcified cartilage and bone. *J. Ultrastruct. Res.* **38**, 318–331.
48. Schubert, M., and Pras, M. (1968). Ground substance proteinpolysaccharides and the precipitation of calcium phosphate. *Clin. Orthop.* **60**, 235–255.
49. Sledge, C. B. (1968). Biochemical events in the epiphyseal plate and their physiologic control. *Clin. Orthop.* **61**, 37–47.
50. Sundström, B., and Takuma, S. (1971). A further contribution on the ultrastructure of calcifying cartilage. *J. Ultrastruct. Res.* **36**, 419–424.
51. Tanaka, S. (1965). Electron histochemical demonstration on the localization of activities of alkaline and acid phosphatases in the cartilage of mice. *Arch. Jap. Chir.* **34**, 587–590.
52. Termine, J. D., Wuthier, R. E., and Posner, A. S. (1967). Amorphous crystalline mineral changes during endochondral and periosteal bone formation. *Proc. Soc. Exp. Biol. Med.* **125**, 4–9.
53. Thyberg, J., and Friberg, U. (1970). Ultrastructure and acid phosphatase activity of matrix vesicles and cytoplasmic dense bodies in the ephphyseal plate. *J. Ultrastruct. Res.* **33**, 554–573.
54. Thyberg, J., Lohmander, S., and Friberg, U. (1971). Ultrastructure of the epiphyseal plate of the guinea pig in experimental scurvy. *Virchow's Arch.* **B9**, 45–57.
55. Tourtellotte, C. D., Campo, R. D., and Dziewiatkowski, D. D. (1963). Degradation of chondromucoprotein by an enzyme extracted from cartilage. *Fed. Proc.* **22**, 413Ab.
56. Woodward, C., and Davidson, E. A. (1968). Structure-function relationships of protein polysaccharide complexes: specific ion-binding properties. *Proc. Nat. Acad. Sci. U. S.* **60**, 201–205.
57. Wuthier, R. E. (1969). A zonal analysis of inorganic and organic constituents of the epiphysis during endochondral calcification. *Calcif. Tissue Res.* **4**, 20–38.

10

Vascularity and Protein–Polysaccharide Complex in Tendons of Young Rats

Isidore Gersh

The vascularity of tendons has been shown to vary from poor to very poor, or there may be none, depending, at least in part, on the site studied, the species, and the technique employed *(1, 2, 5–8)*. It was thought that if tendon was in fact functionally poor in or deprived of circulating blood, then transport of metabolites and nutrients might resemble that in cartilage matrix, which is, in fact, avascular. More specifically, if transport in cartilage is mediated by the protein–polysaccharide complex (P–PC), then one would expect that a similar complex would be prominent also in tendon. This brief report is of a study of vascularity in tendons of young rats, and of the probable distribution of the polysaccharide component of P–PC in tendon.

Methods

The blood vessels of 3-week-old rats were perfused via the aorta with India ink diluted with an equal volume of distilled water. At the end of the perfusion, the major vessels were tied off and the rats were immersed in cold formalin (40% formaldehyde). The skin was incised over the gastrocnemius muscle, the tail, and the abdomen. In the course of 2 days, the specimen was quite hard. The gastrocnemius with its tendon was removed, and the dorsal tendons, muscles, and tips of vertebral spines were removed in one slice. All were embedded in nitrocellulose and sectioned serially 40 μ thick. Every fifth section was stained in dilute hematoxylin, dehydrated, cleared, and mounted in Permount.

The protein–polysaccharide complex was studied in freeze-dried specimens of rat tail tendon frozen ultrarapidly as described in Chapter 4 (this volume). Speci-

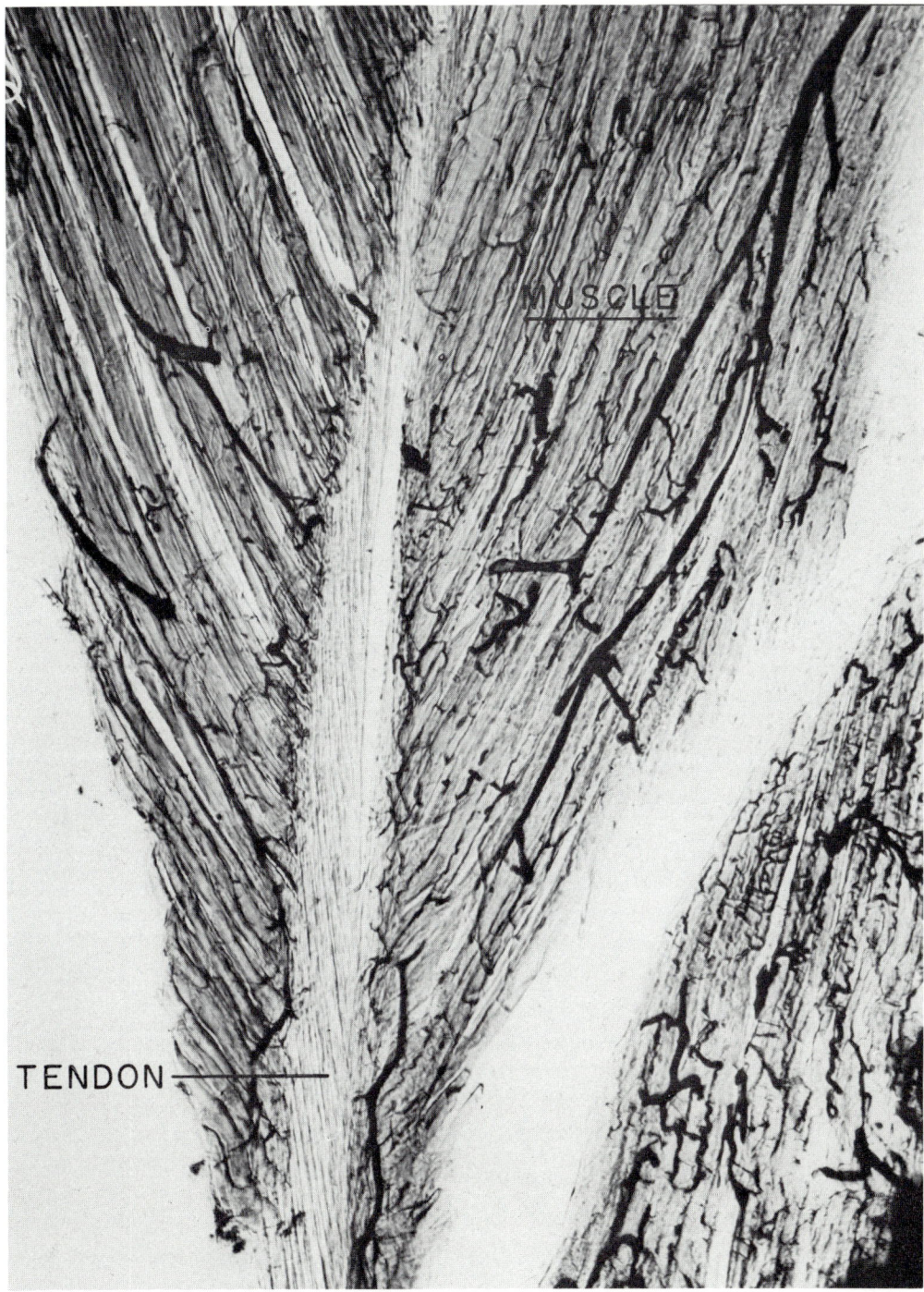

Fig. 10.1. Photomicrograph of section of injected specimen of gastrocnemius muscle and its tendon to show absence of blood vessels from tendon, despite the presence of numerous injected capillaries and larger blood vessels in muscle tissue attached to it. The diameter of the tendon is of the same order as that of the epiphyseal plate in the mouse. × 100.

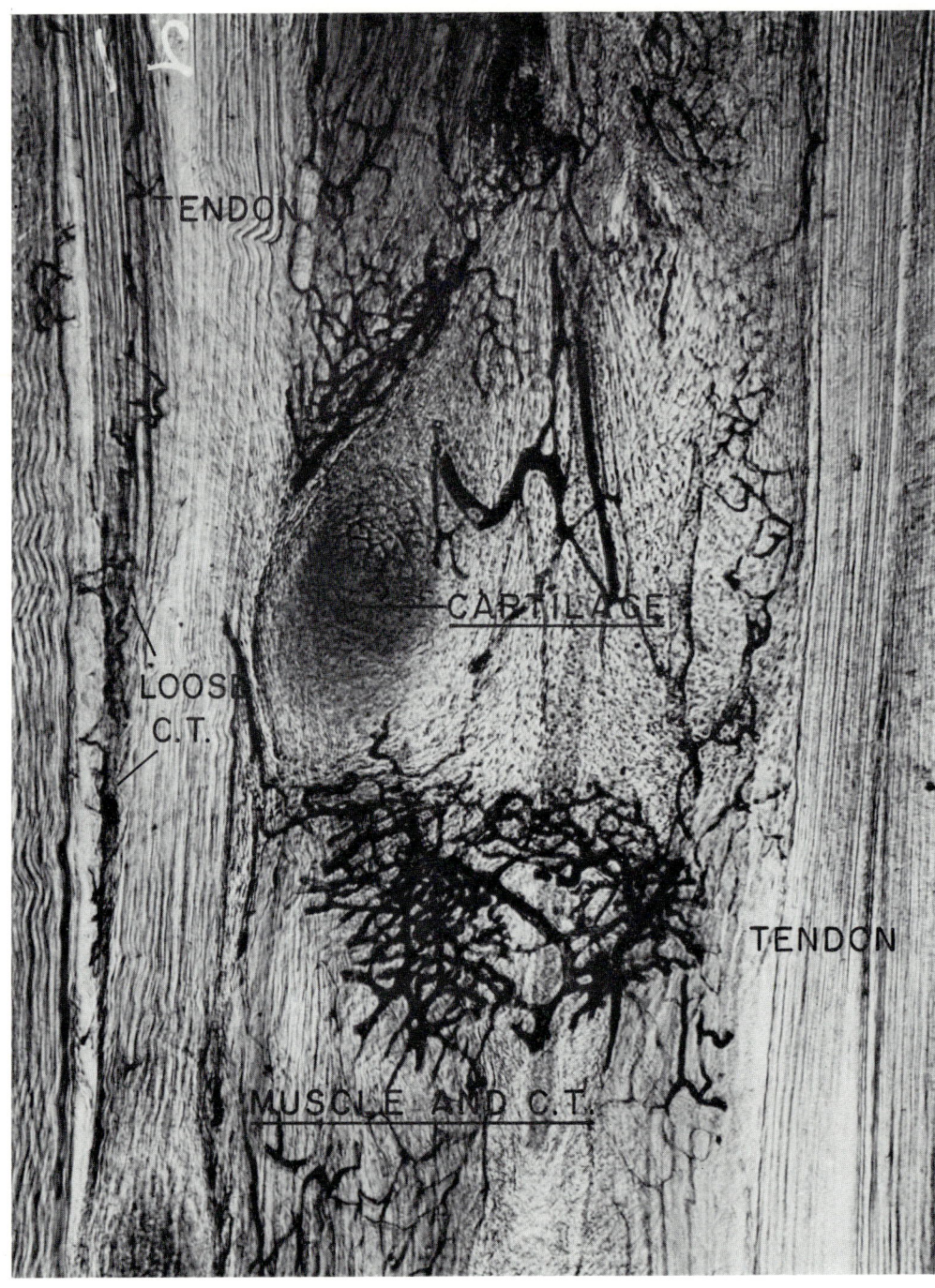

Fig. 10.2. Photomicrograph of section of injected specimen of dorsal tail tissue. Blood vessels of small and large dimensions are present in muscle and loose connective tissue, but are absent from adjacent cartilage and tendon. Some loose connective tissue between tendon slips is well vascularized. × 100.

Methods

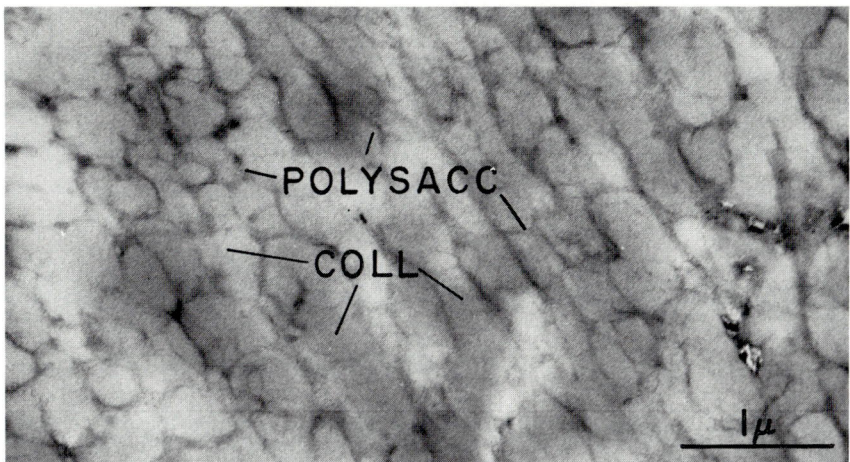

Fig. 10.3. Electron micrograph of tail tendon of 1-day-old rat prepared by freezing and drying with postfixation and staining *in vacuo* with vapors of cobalt(III) trifluoroacetylacetonate. Largely transverse section of dense bundle of collagen, to show the wide-meshed net of dense material thought to represent the polysaccharide component of the protein–polysaccharide complex (POLYSACC) as it extends between collagen fibers (COLL). × 20,000

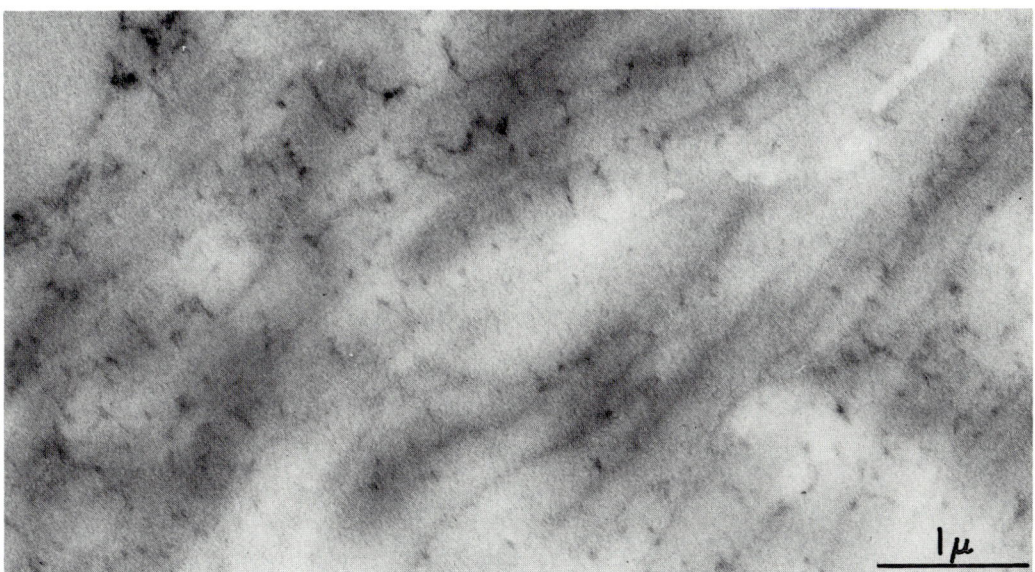

Fig. 10.4. As above, to show the course of the stained material in a largely longitudinal section as it extends between collagen fibers (COLL). × 20,000.

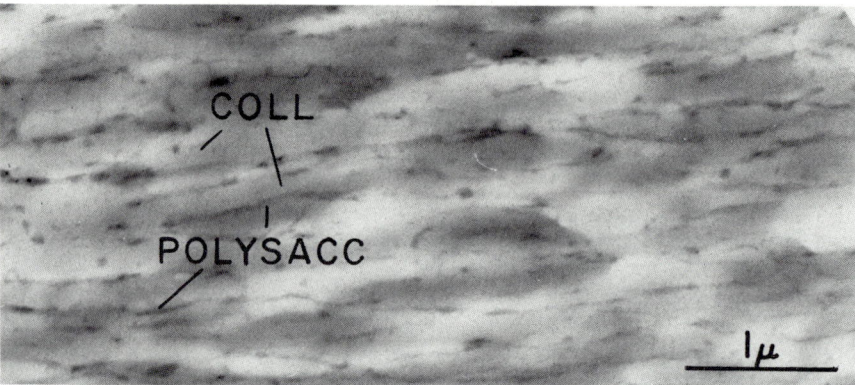

FIG. 10.5. Tail tendon of 8-day-old rat prepared as above. Longitudinal section to show the stained material between the collagen fibers. × 20,000.

mens were dried as usual, and exposed *in vacuo* to vapors of cobalt (III) trifluoroacetylacetonate (Co tfac) as described in Chapter 2 (Vol. I). While cobalt is not specific for the polysaccharide moiety of the P–PC, it serves as a presumptive indicator. Specimens were prepared from rats 1, 8, and 10 days old.

Observations and Discussion

The tendon of the gastrocnemius muscle, as well as the various supraspinous tendons of the tail are indeed avascular, like cartilage, at the same time as adjacent muscle and loose connective tissue are well vascularized (Figs. 10.1 and 10.2). Occasionally, as is in Fig. 10.2, a thin slip of loose connective tissue inter-

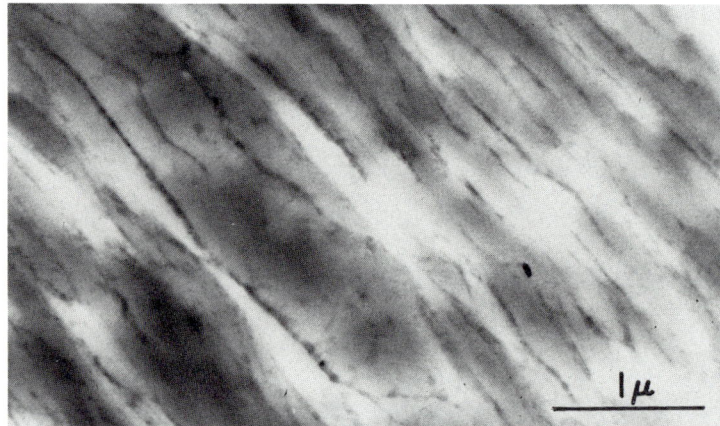

FIG. 10.6. Tail tendon of 10-day-old rat prepared as above. Largely longitudinal section to show the elongated netlike distribution of stained material between collagen fibers. × 20,000.

venes between dense bundles of collagenous tendon, and such tissue is well vascularized.

After postfixation and staining with Co tfac, a widely anastamotic net of submicroscopic dense lines appears in electron micrographs (Figs. 10.3–10.6). The net extends around and along collagen fibers in all directions. The mesh seems to become progressively elongated and oriented as the tendon matures. The dimensions of the stained material vary in thickness from about 200 to 500 Å. The mesh, which encloses the collagen fibers of tendon, seems comparable with the walls of the large matrix compartments of cartilage matrix, where the polysaccharide component appears to be especially prominent. On the other hand, there is a difference. In cartilage, the collagen fibers lie within the polysaccharide component, while in the tendon the polysaccharide moiety is present as strands which loosely embrace the collagen fibers like a wide-meshed net. The presumed polysaccharide component of tendon differs morphologically perhaps from the interfibrillary material described by Fitton Jackson *(3)*; or, if it is the same, the different appearances are caused by differences in fixation. If this is, indeed, the polysaccharide portion of the P–PC in tendon, there is no hint of its exact relation to the protein portion in the material studied. The morphological relations of the presumed polysaccharide component and collagen fibers are quite different from those in a model which are proposed by Jackson and Bently *(4)* for tendon.

I would suggest that the protein–polysaccharide complex in tendon performs the same role in the nonvascularized tendons as it does in nonvascularized cartilage in relation to transport of metabolites and nutrients.

Summary

Blood vessels were absent from rat tail tendon and cartilage after perfusions which successfully demonstrate blood vessels in adjacent muscle and loose connective tissue. The collagen fibers of rat tail tendon are enclosed in a wide-meshed net of a material which is probably polysaccharide, a component of protein–polysaccharide complex. It is suggested that the polysaccharide moiety is involved in diffusion of nutrients and metabolites in tendon, as in cartilage, which is the other major nonvascular tissue.

References

1. Brockis, J. G. (1953). The blood supply of the flexor and extensor tendons of the fingers in man. *J. Bone Joint Surg.* **35B**, 131–138.
2. Edwards, D. A. W. (1946). The blood supply and lymphatic drainage of tendons. *J. Anat.* **80**, 147–152.
3. Fitton Jackson, S. (1956). The morphogenesis of avian tendon. *Proc. Roy. Soc. London* **B144**, 556–572.
4. Jackson, D. S., and Bently, J. B. (1968). Collagen–glucosaminoglycan interactions. *In* "Treatise on Collagen" (B. S. Gould, ed.), Vol. 2A, pp. 189–214. Academic Press, New York.

5. Kölliker, A. (1850). "Mikroskopische Anatomie oder Gewebelehre des Menschen," Vol. 2, 1st half, p. 235. W. Engelmann, Leipzig.
6. Norberg, A. I., Raker, C. W., and Dodd, D. (1967). Equine tendinítis—an angiographic and histologic study. *Proc. 13th Animal Convention Amer. Ass. Equine Pract.* Vol. 13, pp. 243–254.
7. Peacock, E. E. (1959). A study of the circulation in normal tendons and healing grafts. *Ann. Surg.* **149**, 415–428.
8. Smith, J. W. (1965). Blood supply of tendons. *Amer. J. Surg.* **109**, 272–276.

11

Summary, Synthesis, and Speculations

Isidore Gersh

It is axiomatic for nearly all cytochemists that there are general features which can be derived from studies of any single cell type which apply in varying degrees to all cell types. Hence, I feel justified in attempting to extract from the preceding chapters those features which seem to be of general significance. I can justify the choice of each particular cell type as being most suitable for illustrating some general feature; but if I were to do this, it would be without the conviction that the reasoning would stand close introspective scrutiny. The fact is that my colleagues and I chose to work with liver cells, cartilage cells, nerve cells, salivary gland cells of *Drosophila*, etc., because we are more familiar with them than with other cells. We have worked with them for up to 40 years, to which must be added a factor for the experience absorbed from our teachers who had worked with these cells in their time. We came to think of more general problems in terms of the cell types familiar to us.

The same is true of my concentration on one particular method of preparation of cells for cytochemical studies. The logic of freeze-drying with postfixation and staining *in vacuo* with reagents in the vapor phase is transparent and could be the basis for an ideal solution to the problems of fixation and staining. The logic for the ideal situation is this: if only the water phase could be removed instantaneously with no displacements of the solid phase and the solid cell components could react with a reagent in a vapor phase, then the structure would be retained without solution, diffusion, or extraction of specific substances and without other displacements and rearrangements of a grosser sort. This assumes, among other things, that the vapor phase–solid phase reaction is sufficiently sensitive and complete, as well as specific or highly selective, and that the molecular deformations accompanying denaturation are not significant. Of course, this ideal is not attainable, and the method of freezing-drying with postfixation with vapor phase reagents must be regarded as, at most, a practical approximation to the ideal.

There are many biological problems which have been clearly perceived by early biologists of the last century, but have remained insoluble for technical reasons, except by deduction. Most of the problems worked on by the historic giants of embryology, cytology, and genetics are in this category, and many of their concerns recently have been or now are in process of being resolved. An important reason for this happy situation is that the order of magnitude of the units in which the problems are posed is the same as that which is technically feasible at the present time. Now, more so than in the past, there is a constant interplay, frequently in the same laboratory and in the same person, between the development of techniques and the partial solution or attempted solution of certain biological problems. This has been the situation in the research studies reported in this book. What began with a deliberate search for suitable methods for the study of certain problems ended with the redesigning or restatement of the problems to be more in accord with the limits and restrictions of the methods.

State of the Genetic Material

The prime biological problem with which this work deals is the organization and structural basis of protein synthesis in eukaryotic cells. Basic to this problem is the consideration of what constitutes active and inactive genetic material or DNA, and of the changes which take place in the pattern of organization of DNA during activity. Insight on the nature of inactive (heterochromatic) and active (euchromatic) genetic material was obtained by a study of DNA in salivary gland chromosomes of *Drosophila* (Chapter 4, Vol. I). The double helix (first-order DNA helix) of Watson and Crick had been stained in hepatic, exocrine pancreatic and epiphyseal cartilage cells of the mouse, and appeared in electron micrographs as a dense line about 45 Å thick (this volume, Chapter 5 and Vol. I, Chapter 3). Less than one-half of this is the double helix itself, the remainder being the stain which adds on to the phosphate groups of the double helix. The increased mass imparted by this added stain (gallocyanin–chromalum) is responsible for the increased density and thickness of the DNA helix. The dye molecules are spaced closely enough so that with the limited resolution of the electron microscope, they appear as a uniform or solid line. This first-order DNA helix seldom appears as a straight line, but nearly always is wound as a second-order helix with a diameter and period of about 400 Å. This helix appears in ultrathin sections in various planes as circular or elliptical figures or coils. The regularity, uniformity, and near universality of these second-order helices constitute one of the fundamental features of this research. In turn, these second-order helices form patterns of organization which seem to be basic to an understanding of the activity of genetic material in cells.

Changes in the State of the Genetic Material (Figs. 11.1 and 11.2)

The second-order helices may occur singly, or may be aggregated, usually at the nuclear membrane, in crowded groups which may reach light microscopic dimensions. Thus, the second-order DNA helices, when aggregated, are the basis of the microscopically visible component of chromatin when suitably stained. The pattern of distribution of isolated and aggregated second-order helices is relatively constant and characteristic of each cell type, as was found when comparing some forty or more somatic cells of the mouse which had not been stimulated to activity (Vol. I, Chapter 8). This is clearly one of the possibilities which could have been derived from DNA dogma. Every cell type would be expected to make some one or more proteins characteristic of it, in addition, to the many proteins also made by other cells. The new proteins would require prior synthesis of appropriate

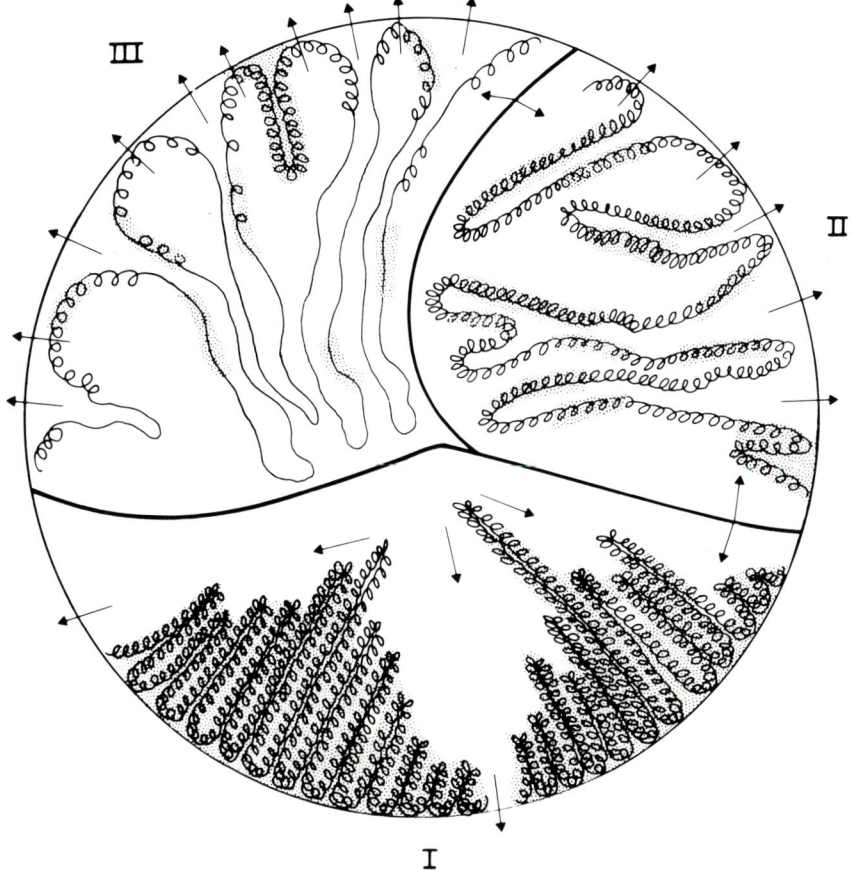

Fig. 11.1. Diagram of changes which take place during stimulated activity (II) and during markedly stimulated activity (III), as compared with the normal nonstimulated state (I), in the manner of distribution of the nuclear DNA, as follows: (1) the degree of aggregation and disaggregation of second-order DNA helices; (2) the uncoiling of second-order helices to form first-order helices; (3) the progressive decrease in repressed DNA sites (stippled), with a corresponding increase in derepressed sites; and (4) increased opportunities for synthesis and transfer of gene products indicated by increased numbers of arrows crossing the nuclear membrane. See text, p. 217.

RNAs, which must involve some change in the availability of some limited number of DNA cistrons. One might expect such adjustments to be made in terms of small molecules, not at all, or not readily, resolvable with the electron microscope if the repressor–derepressor concept of Jacob and Monod *(10)* were alone responsible for the activity of DNA in eukaryotic cells.

The proposal is that the aggregation and disaggregation of second-order DNA helices is a coarse genetic adjustment. In this process, single loops of DNA may be considered euchromatic and potentially active genetically. When they are aggregated the DNA helices may be considered heterochromatic and probably inactive genetically. The aggregation and disaggregation are reversible in most cells. Superimposed on this coarse genetic adjustment is a fine genetic adjustment which consists of small and large molecules acting as repressors and derepressors.

Despite this relatively constant pattern of aggregation of second-order DNA helices in unstimulated cells, the pattern is readily altered in a wide variety of conditions, when studied by the same techniques.

CELL DETERMINATION AND DIFFERENTIATION

It was found in wing disks of third instar larvae of *Drosophila melanogaster* that the second-order DNA helical pattern of nuclei, which are determined as

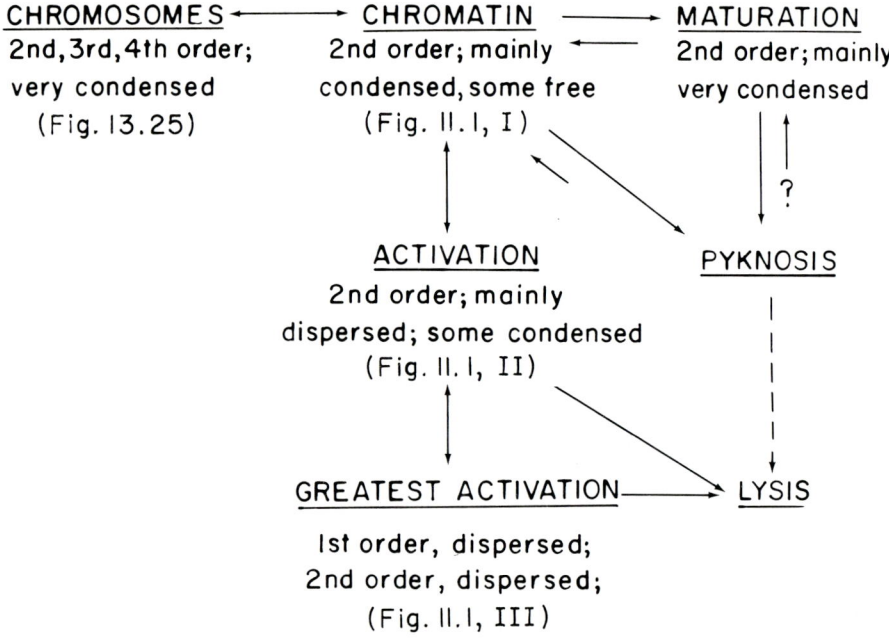

FIG. 11.2. Diagram summarizing the relations of the first-, second-, third-, and fourth-order DNA helices to the state of activity of the nucleus. Some of these changes are illustrated in Fig. 13.25 (Vol. I, Chapter 13) and Fig. 11.1 (I–III). Arrows between the three segments indicate reversibility. See text, p. 217.

wing, differs from that of nuclei of cells which are destined for thorax. When these cells differentiate during early and later stages of pupal development, the patterns of the second-order DNA helix of different cells diverge markedly from the two basic patterns of the larval state, and from each other (Vol. I, Chapter 6). In unpublished work with Dr. J. W. Lash, on developing chick chondrocytes in tissue culture, the DNA molecular patterns of nuclei were clearly distinguishable in four successive states of differentiation: (1) the "undifferentiated" (presecretory) stage: (2) the "differentiated" (secretory stage); (3) the "dedifferentiated" (inactive) stage; and (4) the "redifferentiated" (active) stage. Moreover, the molecular pattern of distribution of second-order DNA helices in stages 2 and 4 resembled closely the characteristic of the differentiated cartilage cells of epiphyseal plate of the mouse.

It has been observed in many different tissues that the second-order DNA helices are commonly more dispersed in embryonic cells than later in differentiation. The assumption is nevertheless that in embryonic cells only a small part of the isolated DNA helices serve as template, the remainder being inhibited by repressor substances. The coarse adjustment through heterochromatization is thus ineffective in these cells. One may assume further that this inhibitory action, which must be quite extensive, is not always completely (100%) effective, or that some "mistakes" take place. One can imagine that under those conditions some genes are active, perhaps at a low level, which are not characteristic of the cell type after differentiation.

This could account for a broad distribution of activities in very early embryos which becomes restricted during development, for example, myosin in chick heart muscle (5), gulonolactone oxidase in the chick kidney which is the last enzyme in a series for the synthesis of ascorbic acid (6), various enzymes necessary for the synthesis of protein–polysaccharide complex of cartilage (15, 16), and the presence of group-specific antigens of the C-type RNA tumor virus (9). It is possible that this phenomenon is very general, awaiting only the application of ultrasensitive and specific methods for the identification of specific proteins. The same phenomenon could be the basis for the finding that some malignant tumor cells produce ectopic hormones not normally associated with the cells of origin of the tumors (8). Finally the same phenomenon could account for the occurrence in a single abnormal cell (as in pancreatic islet cell and adenohypophyseal tumors) of mixed granules of more than one cell type.

STIMULATION OF CELLULAR ACTIVITY

The pattern of distribution of second-order DNA helices of certain cells, is altered in a characteristic way when they are stimulated experimentally (Vol. I, Chapter 10). The cells studied were exocrine cells of the pancreas (stimulated with pilocarpine), epithelial and smooth muscle cells of the uterus (stimulated by the administration of estrogen or during normal pregnancy), chromaffin cells of the adrenal medulla (stimulated with reserpine), epithelial cells of the seminal vesicle (stimulated with testosterone), and spinal ganglion cells (stimulated by crushing of the nerve fibers). In all instances, the peripheral nuclear masses of compact second-order DNA helices (chromatin bodies) were reduced in number and size. There was a corresponding increase in the number of single second-order DNA helices and of aggregates of helices too small to be visible in the light microscope as chromatin bodies. At the same time, the gaps between chromatin aggregates (large and small) became larger and more numerous at the nuclear membrane. The same changes also occurred in certain cells which were in process of being stimulated in the normal course of development. This took place in the cells mentioned above and also in cartilage cells of the epiphyseal plate during secretion of matrix (this volume, Chapter 5). The dispersion of compact chromatin also took place in syncytial cells of the *Drosophila* egg in the brief interphase (Vol. I, Chapter 13), and in puff regions of salivary gland chromosomes of *Drosophila* larvae (Vol. I, Chapter 4). In two (possibly three) instances [cartilage cells (this volume, Chapter 5)], chromatolytic nerve cells (Vol. I, Chapter 12), and somatic nuclei of the *Drosophila* egg (Vol. I, Chapter 13)], some of the second-order DNA helices were obliterated, so that first-order (double) DNA helices were observed as straight lines on electron micrographs.

Cell Maturation

The molecular pattern of distribution of second-order DNA helix is altered in a characteristic way also during the maturation of certain cells (Vol. I, Chapter 9). The cells studied were stratified squamous epithelium, the neutrophilic polymorphonuclear leukocyte, the red blood cell, and the intestinal goblet cell. The consistent pattern in all cells is a progressive reduction in the proportion of single and small aggregates of second-order coils, with a corresponding increase in the proportion of large aggregates, which are visible with the light microscope. The compact chromatin masses aggregate along the nuclear membrane, and there is accordingly a marked reduction in the number and dimensions of the gaps between adjacent DNA aggregates. The progression of events during maturation, as the cells produce less protein, is thus the opposite of that observed when cells are stimulated. Related to this phenomenon are the effects of ovariectomy on the DNA molecular pattern of epithelial and smooth muscle cells of the uterus (Vol. I, Chapter 10), which involve a condensation or compaction of DNA helices like that which takes place during cell maturation.

Cell Metaplasia

This was studied in a poorly understood example—the changeover of the same cell from synthesizing keratin to synthesizing mucin in the mouse vagina (Vol. I, Chapter 11). The marked compaction of the second-order DNA helices, which such stratified cells may undergo (with subsequent lysis), is altered to a state characteristic of mucus-producing goblet cells (Vol. I, Chapter 9), where the DNA helices, though densely packed, are not nearly as compact as in cornifying cells.

Cell Death

The pattern of second-order DNA helices in pyknotic cells was studied in: stratified squamous epithelium of the vagina (Vol. I, Chapter 11), in maturation of the erythrocyte (Vol. I, Chapter 9), in the Barr bodies of cartilage cells (Vol. I, Chapter 7), and in karyolytic cartilage cells of the epiphyseal plate (Vol. I, Chapter 7 and this volume, Chapter 5), all in the mouse. During pyknosis, there is in all cell types studied, a continuation of the DNA changes which take place during maturation, i.e., further compaction of the second-order DNA helices, with the formation of even larger chromatin masses attached to the nuclear membrane, which may be so completely covered as to obliterate nuclear gaps. In cornified vaginal epithelium, the DNA may suddenly disappear through lysis. In the Barr bodies of cartilagenous cells, the DNA helices are broken into coils, fragments of coils, and granules. The latter are probably intermediate stages which are probably difficult to find in cornifying epithelium because of the explosive rapidity with which the changes take place. The changes in the DNA pattern are probably related to the release of lytic enzymes of lysosomes. By contrast with the above, during karyolysis in cartilage cells there is at first a continuation of the DNA changes which take place during excessive activity involving protein synthesis. Then there is a further separation of DNA helices, like that observed during chromatolysis. Then the helices are broken into coils, fragments of coils, and granules, all of which disappear, probably also because of the lytic action of lysosomal enzymes as in pyknotic nuclei. The evidence for lysosomal action in stratified squamous epithelium is mostly morphological, while the evidence in cartilage cells is primarily biochemical.

Cell Duplication (Mitosis) (Vol. I, Chapters 13 and 14)

The second-order DNA helices persist during mitosis, and undergo an extreme degree of compaction as third- and fourth-order coils. During a large part of mitosis (interphase → prophase → metaphase, and telophase → interphase) the chromosomes maintain a close relation to the nuclear membrane, condensing or spreading out on its inner surface. There is evidence which suggests that chromosomes maintain a relation to each other during these stages by way of the nuclear membrane, and that they retain this relationship during interphase, when chromosomal regions are converted to chromatin masses on the surface of the nuclear membrane. The relation of large chromatin masses to other regions presumably is retained during periods of activity, differentiation, etc., since the chromatin masses are reconstituted during recovery periods. The thread, which runs through these varied themes, is that the genetic order of chromosomes persists throughout the life of the cell and cell clone, and that it is somehow related to the nuclear membrane. The high degree of compaction and the steric relations to the nuclear membrane are features which tend to reduce mechanical accidents and thus promote continuity and uniformity of genetic activity during mitosis as well as during interphase. The compacted DNA coils of mitosis are heterochromatic and probably largely inert with respect to synthesis of RNAs. They, thus, resemble other compactions of DNA in chromatin masses occurring in inactive periods, such as maturation, pyknosis, and heterochromatization of salivary gland chromosomes of *Drosophila,* with the difference that the latter group may not be reversible.

To summarize briefly the studies on the patterns of organization of DNA in somatic cells, one is impressed with the constancy of the arrangement of the DNA double helix. The simple double helix exists as such only rarely. Most of the time it is coiled secondarily and this secondary coil is nearly always 400 Å in diameter and periodicity. When inactive or relatively inactive, the second (and higher)-order helices are tightly aggregated to form chromosomes or chromatin masses visible with the light microscope. When active, the second-order DNA helices tend to form small aggregations visible only with the electron microscope or may be separated from each other or even be uncoiled as the first order or basic double helix. The relatively constant and uniform (monotonous) behavior of DNA at this level, which is so essential for duplication and function, seems to be related somehow to its seemingly fixed relations to the nuclear membrane. This is what might be called the coarse adjustment of the genetic apparatus. The fine adjustment is called into play to inhibit the action of inappropriate genes which are exposed at the same time when the appropriate genes respond. That is, when one or more second-order helices are separated from a mass of second-order helices, as during activation, only a small part of the exposed helix may be required for the synthesis of one protein (about three gyres). All the rest of the exposed DNA, or a large part of it, may be inappropriate. The latter must be inhibited or repressed in some way—whether by protein or nucleic acids, acetylation or phosphorylation of the protein, by changes in cross-linking and charge distribution, or by small molecules which may combine with DNA or with protein associated with it. Steric rearrangements may also be significant, as well as changes in the amount or proportion of free and bound water. This kind of activity may be analogous to the repressor–derepressor activities proposed by Jacob and Monod for bacteria (9).

Association of RNA with DNA

The template activity of DNA becomes manifest as the synthesis of various kinds of RNA. In the nucleus, the main visible manifestations are RNA granules associated with chromosomes—RNA granules which might be precursors of heterogeneous RNA, ribosomal RNA, or messenger RNA, and transfer RNA. Chromosomal RNA may be clearly recognizable in salivary gland chromosomes of *Drosophila,* especially in the RNA-rich lamellae of the dark bands (Vol. I, Chapter 4). Granular RNA is especially notable in nucleoli. There is good evidence to show that these granules arise locally from chromosomal regions called nucleolar-associated chromatin, and that the granules are destined to form cytoplasmic ribosomes (2). The morphological representation of this is visible especially in cells stimulated to activity [when nucleoli become enlarged and the associated second-order DNA may become reduced to the first-order (double) helix] and in chromosomal puffs (see above). The obverse may be observed during periods of reduced activity, as during maturation, when the nucleoli became smaller or are not detectable, or during cell death by pyknosis or karyolysis. In many nuclei, there are more RNA granules in the nucleoplasm than in the nucleolus, but there is no way of determining whether they are messenger RNA or ribosomal RNA. It is also not possible to ascertain which are in transit and which (if any) are being stored. Finally, transfer RNA is present as a diffuse, homogeneous background density not resolvable in ordinary preparations and identifiable only because the background density is removed by treating ultrathin stained sections with RNase. It is present in the nucleoplasm, in the nucleolus, and in chromosomes.

Arrangement of Protein in the Nucleus as Pseudovacuoles

The activity of DNA as template requires the presence of a variety of enzymes (depolymerases, etc.). These are included among the basic and acidic proteins extracted from nuclei. None of these can be distinguished by the protein stains used in this study, for they react with acidic or basic groups (among others) indiscriminately, regardless of whether the overall charge is basic or acidic. With most stains the nucleus in freeze-dried cells appears vacuolated, with the less dense contents of the submicroscopic pseudovacuoles enclosed in denser walls. Often the pseudovacuoles appear "empty," depending on the stain. The contents can, however, be stained regularly with some reagents. The term "pseudovacuole" is used for this structural arrangement of protein to replace the term "submicroscopic vacuole," which was used in earlier publications. The possible role of the pseudovacuoles in homeostasis, ion binding, etc. will be discussed on p. 225.

Pseudovacuoles and Transport in the Nucleus

The DNA helices are confined to the denser walls of the nucleoplasm. All the RNA granules and much, if not all, the detectable transfer RNA are also restricted to the nucleoplasmic walls. Many of these RNA particles must be in transit from DNA to the nuclear membrane, and their exclusive presence in the denser walls may point to an important function of the nucleoplasmic walls, i.e., that they may serve as the organic basis for the transport of these important constituents from chromosomes to nuclear membrane. The nucleic acid particles probably pass between the dense chromatin masses attached to the nuclear membrane, and through the nuclear gaps. Whether these correspond with nuclear pores is unknown. Whether the nuclear pores change with increased cellular activity is not known, but it is clear that the nuclear gaps do vary in number and size. Similarly, it seems probable that the denser walls of the pseudovacuoles are also in a state of flux, adapting continuously to, or making possible, changes in the pattern of distribution of DNA with variations in the state of activity of the cells. The location of the nuclear gaps may be significant. In addition to probably being the path by which ribosomal particles enter the cytoplasm, their position may determine the cytoplasmic distribution of specific mRNAs which arise from euchromatic DNA.

As considered later (p. 228), the distribution of active DNA in the nucleus may be regarded as a three-dimensional mosaic. The nuclear pores would seem to extend the range of mosaicism into the cytoplasm, which hence could be regarded as a functional mosaic.

Protein and the Organization of DNA

The proteins of the walls of the pseudovacuoles appear to be homogeneous when viewed with the electron microscope (Vol. I, Chapter 2). Their density decreases in the following order: chromatin, nucleoli, nucleoplasm. This is not to deny that, at the molecular level, the proteins may not be very diverse. An indication of their diversity may be obtained from the specific staining patterns of proteins of chromosome, nucleolus, and nucleoplasm in the polytene nuclei of the salivary glands of *Drosophila* (Vol. I, Chapter 5). If one assumes that each pattern of protein staining is amplified in these nuclei to a corresponding degree to that of the DNA, as compared with diploid nuclei, then it is easy to imagine that there is a similar diversity of reactive protein (at the molecular level) in the diploid nuclei of somatic cells. Among these proteins are three whose existence and function is inferred from the pattern changes of second-order DNA helices: the coiling protein (which retains the second-order DNA helix), the intercoil protein (which binds adjacent second-order DNA helices to form larger aggregates), and the attachment protein (which binds the second-order helix, singly or as aggregates, to the nuclear membrane). These three functions may be mediated by a single protein, or by two or three different proteins, and the chances of distinguishing such molecules in sections with the electron microscope seem quite small. Even if it were possible to selectively stain these particular protein molecules, it is questionable that they would be large enough or have enough contrast to be discriminated from all other nuclear proteins. Their existence is inferred because their presence would make comprehensible a wide range of observations and deductions: the occurrence of peripheral aggregates of chromatin attached to the nuclear membrane, the reversible changes which take place in the pattern of second-order DNA helices during differentiation, activity, maturation, etc., and the perpetuation and constancy during the cell cycle of the relations of the DNA helices and aggregates to each other and to the nuclear membrane. These proteins are also responsible for the maintenance of mosaicism of euchromatic DNA of somatic cells.

Cytoplasm

The Ribosomal Apparatus (Fig. 11.3)

The RNAs produced from template DNA pass into the cytoplasm where they become actively engaged in protein synthesis. Both mRNA and rRNA are granular and indistinguishable while in the nucleus. In the cytoplasm, ribosomes are identified by virtue of their RNA. They are approximately 200 Å in diameter.

Cytoplasm

The internal structure of ribosomes seems to be determined primarily by the distribution of RNA and protein. RNA is present predominantly internally as a central core from which sheets or films extend peripherally. Between them are protein "globules" about 30 Å in diameter, so that the relations of both constituents are very intimate. On careful examination of high-resolution electron micrographs one can see that many ribosomes are enclosed as in a loose cage by filaments about 15 Å in diameter, the approximate diameter of a polypeptide chain. Sometimes the thin, sinuous filaments appear cast off from the ribosomes, and frequently they are identified as highly folded filaments lying in the intercisternal space, surrounded by spherical granules which can be stained and intervene between adjacent cisternae to fill the intercisternal space. The highly folded filaments occupy the central parts of the space. These observations have a bearing on protein synthesis—they indicate that the newly synthesized polypeptide molecules

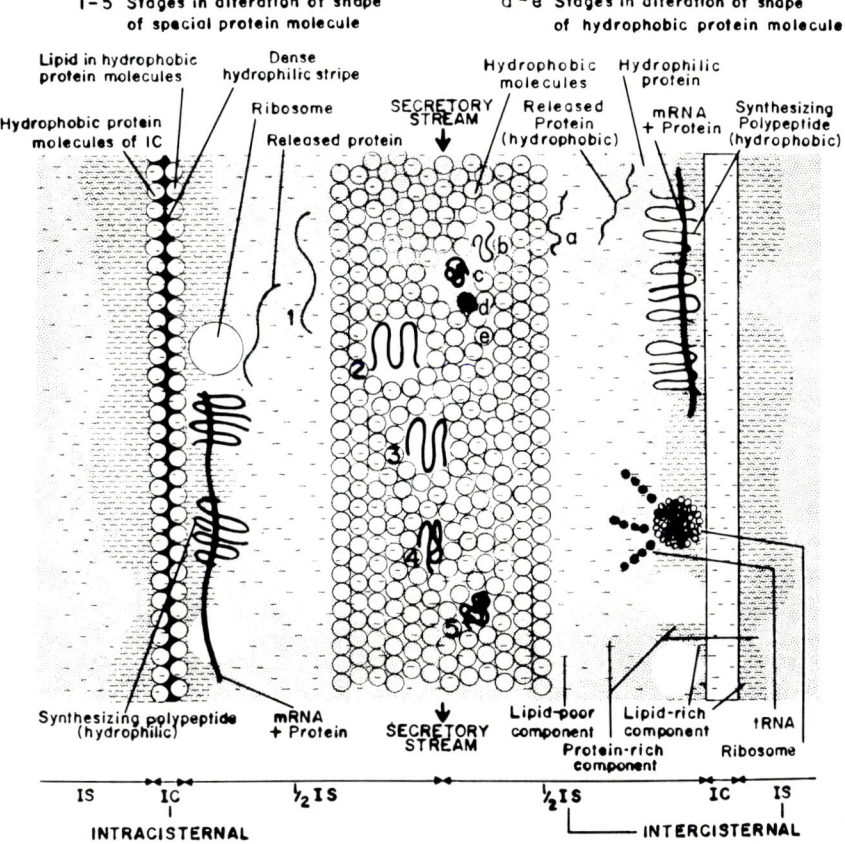

Fig. 11.3. Diagram of a very small portion of the RER to illustrate the distribution of chemical constituents among the components of the RER, and the origin of nascent polypeptide and secretory products. See text p. 222.

are wrapped around ribosome and messenger as they are synthesized and elongated, and that they are shed as a sinuous or folded filament which is enclosed by largely hydrophobic molecules which are perhaps synthesized on adjacent ribosomes. The hydrophobic (insulating) molecules bear the nascent folded molecules as in a stream between adjacent cisternae in the intercisternal space. This interpretation is tentative and requires detailed study by a method which stains one particular protein selectively (or which imparts a particular shape to a specific protein), under various conditions of activity, inhibition, etc.

Some preparations stained for protein have straight, thin filaments which stretch between two to five ribosomes in a single electron micrograph. These are different from the shorter filaments which may be the nascent polypeptides. The longer, straight filaments may be mRNA–protein. This interpretation also is tentative, and requires identification of the RNA component as well as the protein component under various conditions of inhibition of protein synthesis.

Finally, some stained preparations postfixed and stained *en bloc* for nucleic acids have small dense bodies (about 50 Å) which surround ribosomes. Their distribution is not random, as they seem to occur in interrupted columns or rows which radiate from the ribosomes. These may be tRNA-containing particles which are stained densely by virtue of their increased mass attributable to the double staining of nucleic acid and protein. The radiating pattern suggests that tRNA is concentrated in the region of the ribosome through binding with periribosomal proteins. The tRNA may perhaps even be oriented for polarized presentation of the amino acid to the growing polypeptide molecule. This interpretation is very tentative and requires more proof.

THE RER AND THE INTRACISTERNAL REGION (Fig. 11.3)

The whole apparatus described above is part of the RER. It includes ribosomes, probable mRNA, possible tRNA, as well as what might be the nascent polypeptide. The "backbone" of the apparatus is the extended cisterna. Its thickness varies markedly, between about 100 to about 300 Å or more. It is difficult to ascertain whether these differences in thickness are caused by different kinds of denaturation undergone during postfixation and staining or whether they represent different functional states. When about 100 Å thick, the cisterna consists of two plates, each about 50 Å thick, studded with protein globules or molecules. These are about 50 Å in diameter and are separated by a thin line or lines from tightly apposed adjacent molecules. Where the molecules are in contact, the surfaces are more deeply stained and thicker, and, when seen in electron micrographs at lower magnifications, appear to form an irregular line in the middle of the cisterna. In fact, the structural pattern resembles closely that of mitochondrial compartments. There is relatively little lipid in the intracisternal space (this volume, Chapter 2).

THE INTERCISTERNAL SPACE (FIG. 11.3)

By comparison, lipid on the outside of the cisterna is present as a richer, thin, interrupted layer of irregular thickness (this volume, Chapter 1). Infiltrating this layer and extending beyond it is a readily stainable layer of protein which is about three times thicker than the lipid layer. This protein layer also contains lipid, but in lesser amounts. The interval between the protein + lipid walls of adjacent cisternae (the intercisternal space) is filled with hydrophobic protein globules (molecules), which enclose the folded, newly formed protein molecules after they are released from ribosomes. Some lipid is also present in this region.

The functions of the phospholipid are not known; its most important function may be in the formation of a self-assembled, largely lipophilic structure whose orientation and charge effects, in conjunction with proteins, orient the proteins which are on both sides of the double membrane. By analogy with mitochondria, where there is sufficient reason to expect orientation of proteins involved in electron transfer and in certain enzyme systems, it is suggested that proteins adjacent to the double membranes of cisternae are also organized as systems. It is also possible that such systems of the RER are self-assembled from the component parts, as they are perhaps in mitochondria.

Cytoplasmic Pseudovacuoles

After certain vapors are used as postfixatives, cytoplasmic regions characterized by RER appear, in electron micrographs, completely filled and without spaces, which might have been occupied by ice crystals formed during the process of freezing or drying. Similarly, the cytoplasmic regions free of RER are also filled with stainable material, some of which (the wall) is denser than the remainder (the contents of the pseudovacuole). These insights into the structure of protoplasm became possible only after methods were developed for staining the contents of what my associates and I formerly called submicroscopic vacuoles. Their contents were not clearly visible, or appeared pale, in electron micrographs because their density after staining was the same or slightly greater than the embedding plastic. Because the contents are now stainable and fill the spaces with material of sufficient mass to result in increased contrast, the use of the term submicroscopic vacuole seems no longer useful, and should be replaced by the term pseudovacuole.

Pseudovacuolar System and Homeostasis. When submicroscopic vacuoles were described as basic structures of protoplasm, they were proposed as the basis of a two-phase (or multiphase) system in which there is an equilibrium between the dense phase of protoplasm (or protein-rich phase) and the less dense phase (or protein-poor phase). As long as some of each phase persists in effective contact they would be the basis for equilibrium in the ionic concentration and osmotic pressure of different parts of the cytoplasm with each other, of the cytoplasm as a whole with the nucleus, of the various parts of the nucleus with each other, and with the cell as a whole and its environment. The enormous surface of the submicroscopic vacuoles would tend to promote rapid changes for the maintenance of the equilibrium. The same important consequences of this form of organization of protein would apply to pseudovacuolar structure, so long as even trace amounts of the wall macromolecules are in solution and in contact with the denser walls. Finally, if the walls change, as postulated, and enzymes are associated with them, opportunities are presented for maintaining enzymes in an insoluble state, and for changes in availability of active sites of enzymes. Moreover, if enzymes can be kept in an active state while insoluble, it is equally feasible that a series of enzymes could be maintained as a functionally integrated metabolic system like those of mitochondria.

SECRETION

The process of secretion could be considered as beginning with the elaboration from chromosomal DNA of the elements of the ribosomal apparatus and the formation of nascent protein. This is protected from cellular entanglements and contradictions by being surrounded by rather inert (largely hydrophobic) protein molecules with some lipid jointly oriented with the lipophilic core of the protein globules. Together, through the pressure caused by the synthesis of more molecules, these are swept toward the Golgi apparatus, where they may be aggregated as secretory granules and stored. These secretory granules are formed at the periphery of the Golgi apparatus, where the presumably faster moving secretory stream empties into the supposedly slower moving stream, which is assumed to be characteristic of the Golgi apparatus. The secretory granules at the periphery of the Golgi apparatus grow through apposition, and cease growth when more energy is required to add more protein molecules to the outside of the granule than is available locally (i.e., when as many granules fall off as are added). The secretory granule may retain its connection with the Golgi apparatus, even when crowded at the apical pole of the cell, by irregular branching stalks which extend from the membrane which encloses the secretory granule (Vol. I, Chapter 2).

GOLGI APPARATUS

Some of the synthesized protein molecules are not incorporated in secretory granules and remain inert in the branches and plates of the Golgi apparatus. These may leave the cell without ever having been incorporated in secretory granules. The Golgi apparatus varies tremendously in morphology, and is continuous with the SER and the RER. Its globular protein and lipid constituents are distributed throughout the vast interconnected and fenestrated network of lamellae, cords, tubules, granules, and vacuoles, which are all parts of the Golgi apparatus.

From the point of view expressed, the Golgi apparatus could be regarded as a great cellular reservoir of all the proteins synthesized by the ribosomal apparatus and not secreted directly. The catalytic molecules are held together in very close proximity, in a condition of coexistence, by virtue of the lipophilic protein molecules and lipids. These may be considered as self-assembled, as may be those in mitochondrial and cisternal membranes. This view of the Golgi apparatus is consistent with its intimate relations with the RER, and with segregation of various enzymes in limited parts of the Golgi apparatus. The inference is that ribosomal systems, which feed into such segregated parts of the Golgi apparatus, are similarly segregated functionally, depending on the coding of the mRNA, the nature of the protein binding aminoacyl-tRNA, and probably on other factors which specify particular proteins.

This view of the Golgi apparatus is consistent with the formation and existence of lysosomes in cells. It is consistent also with the general finding that the Golgi apparatus is enlarged during increased cellular activity. Finally, this view could be consistent with two general phenomena of secretory activity: (1) that some secretory cells (as epithelial cells in the intestinal tract and cartilage cells) have no secretory granules or lysosomes despite the secretion of enzymes and (2) that direct observation of secretory cells during secretion, as in the pancreas of the living animal, does not reveal any movement of secretory granules out of the cell and into the lumen.

More conventional and contrasting views of the activity of the Golgi apparatus are presented in a recent review *(19)* and in a series of reports *(11–14)*.

Nuclear and Cytoplasmic Mosaicism (Fig. 11.4)

The functional specialization of RER in different parts of the cell, despite morphological similarity, may in turn arise from and reflect a similar functional arrangement of active euchromatic DNA in the nucleus and the distribution of nuclear gaps from which emerge the RNA particles synthesized on the genetically active DNA. The hypothesis is proposed that the active DNA forms a three-dimensional mosaic, and that the RNA derived from it as template is directed into different parts of the cytoplasm by a nucleoplasmic transport system which terminates at nuclear gaps. The RNA products are thus directed into different parts of the cell which synthesize different kinds of proteins. These, in turn, stream out into similar segregated parts of the Golgi apparatus, through which they are secreted, or are secreted directly through certain specific surfaces. The hypothesis that is proposed holds that cells are to a certain extent mosaics, and that ultimately this is attributable to a specific three-dimensional pattern of distribution of second-order DNA helices, both euchromatic and heterochromatic, and to the proteins which retain them in this pattern in relation to the surface of the nucleus.

Only a part of this three-dimensional pattern is operative in determining the corresponding mosaic in the cytoplasm. This part consists of first- and second-order euchromatic helices. This postulated three-dimensional mosaic of first- and second-order DNA helices occupies a restricted portion of the nucleus. First, it is confined exclusively to the pseudovacuolar walls, which represent about 60% of the total volume of the nucleus. Then, it does not extend into some walls, which are occupied by large aggregates of DNA helices (the chromatin masses) and nucleoli. In addition, portions of the walls are occupied by RNA granules in transit or being stored and thus exclude the active DNA helices. Finally, some pseudovacuolar walls do not contain any of these compounds. In the end, the three-dimensional mosaic of euchromatic DNA could occupy a very small portion of the nucleus. A consequence of such confinement is that intercommunication may take place between the various genetically active portions of the DNA of the sort postulated by Britten and Davidson *(1, 4)*, and that opportunities are afforded for the effective action of switch or control genes *(18)*.

The hypothesis of the mosaic nature of the organization of the euchromatic (genetically active) portions of the nucleus and the corresponding mosaic struc-

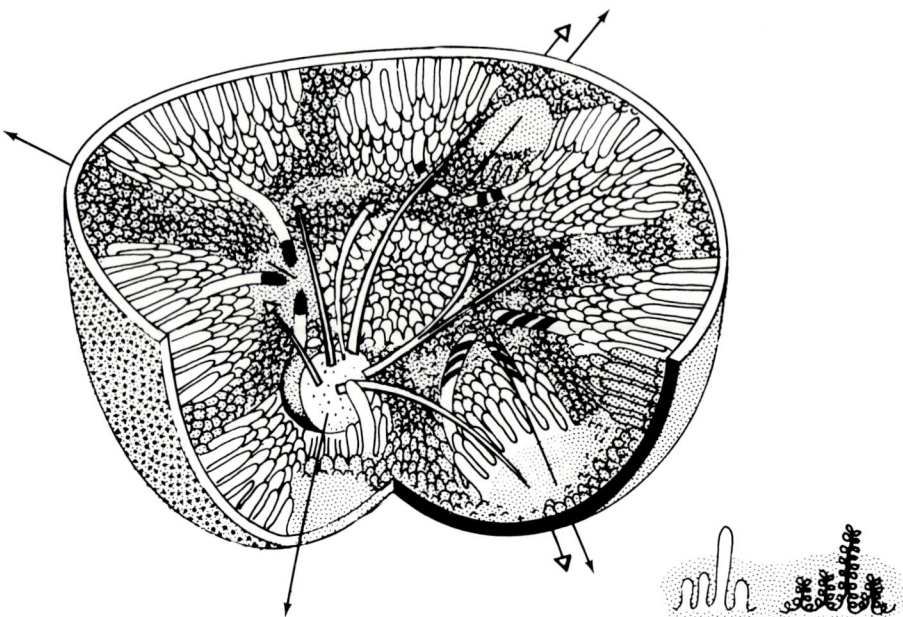

Fig. 11.4. Diagram to illustrate in a three-dimensional representation the relations of euchromatic portions of the second-order DNA helices to illustrate how they may form a mosaic pattern whose products of gene action pass into the surrounding cytoplasm through gaps at the nuclear membrane between compacted heterochromatic DNA aggregates. Systems of genes may be organized in this way through proximity to each other though derived from widely separated chromatin clumps originating from presumably different chromosomes. See text, p. 228.

ture underlying cytoplasmic activities could be applied to a large number of phenomena of experimental embryology associated with morphogenetic determinants in cytoplasm and with regulative and mosaic development. The hypothesis suggests that the various mosaic structures of cytoplasm could arise as a modified extension of the three-dimensional distribution of the euchromatic (active) DNA of the nucleus of the early stages of the developing egg. This topic has been reviewed elegantly by Davidson (3). Modifications imposed on the direct extension of the active nuclear mosaic to cytoplasm could be the morphology of the heterochromatic masses and nucleolus, the number, size, and location of gaps in the nuclear membrane, the occurrence of cytoplasmic streams, etc. This proposed mosaicism of the cytoplasm could in turn be important in determining cell shape and polarity, morphological modifications of the cell surface ranging from cell processes to desmosomes, cellular recognition, and the antigenic mosaic of the cell surface.

MITOCHONDRIAL COMPARTMENTATION

The most striking feature of mitochondria is the extraordinary degree of compartmentation. The basic pattern of organization of mitochondria is the same or nearly the same in hepatic and exocrine pancreatic cells (this volume, Chapter 3). It is related seemingly to the arrangement of protein molecules rather than to that of lipid molecules, though the latter may play an important role in the assemby and function of the structural units of mitochondria. Mitochondria appear as stacks of flattened compartments. Each is delimited by a wall about 50 Å thick, made up of hydrophobic protein molecules or globules, tightly packed to form a plane surface. Lipid components may be oriented in relation to the lipophilic centers of the protein molecules as well as to filling in the interstices between the molecules in an oriented fashion. The walls enclose a narrow compartment which contains proteins (and lipid), some of which may be aggregated, if only temporarily, to form systems of enzymes. Some enzymic activity may also characterize molecules of the membrane wall. When stacked, the double thickness of membrane of adjacent compartments is of about the same dimension as the contents of the compartment or matrix packet. The dimensions are so small that enzymic activity must be regarded as taking place functionally in a plane. The whole stack of compartments is enclosed in a continuous, outer membrane. Whether this pattern of organization is shared by all mitochondria is not known. The full significance of this organization will not be clear until several individual proteins present in mitochondria can be identified *in situ*. The proteins of mitochondria are probably synthesized chiefly by the RER in the cytoplasm, a process probably initiated by nuclear DNA. Mitochondrial DNA is too small to code for the large number of proteins known to be present. They are thought to be self-aggregated and self-assembled. While DNA and RNA have been identified in mitochondria of freeze-dried cells, the methods do not seem promising in elucidating their role in mitochondria.

Connective Tissues

NUCLEIC ACIDS

Cellular organization for protein synthesis is a theme which runs through this work, and various aspects of this topic have been reordered above to emphasize this. A subsidiary theme is the cellular origin of, and the organization of, extracellular components of connective tissues. These last named topics will be treated next.

Collagen and protein–polysaccharide complex (P-PC) are the two major proteins synthesized and secreted by certain cells of the connective tissues. The cells studied were fibroblasts of the developing tail tendon of the rat, chondroblasts of epiphyseal plate of the rat and of joint cartilage of the rabbit and dog, and osteoblasts of the developing parietal bone of the mouse. The pattern of distribution of second-order DNA helices of fibroblast nuclei falls within a certain range of variability, which would probably be narrowed by greater control of the state of activity of the fibroblast (Vol. I, Chapter 8). The equivalent pattern in cartilage cells varies (also) with activity and according to its state of differentiation, but seems different from that of fibroblast nuclei and all other cell nuclei studied in the resting state (this volume, Chapter 5). As secretory activity increases, the second-order DNA helices become separate. Those presumably associated with the formation of nucleolar RNA particles when the cell is at a high level of secretory activity are perhaps uncoiled to form first-order DNA helices. The nucleolar RNA particles presumably pass into the cytoplasm at the peak of synthetic and secretory activity, and the nucleolar and nucleoplasmic RNA particles are thereafter reduced in number. Also presumably secreted (in addition to the two major proteins of extracellular connective tissues) are the lysosomal enzymes which seem to be essential for the continuous remodeling of cartilage matrix during growth (this volume, Chapter 7). One would expect part of the DNA activity to be concerned with the synthesis and secretion of these lytic enzymes. It seems that the peak of secretion of extracellular matrix takes place at a time when the DNA helices of the nucleus have deteriorated by fragmentation of the helices into separate coils and these in turn into granules (this volume, Chapter 5). The continued synthetic activity of the cytoplasmic ribosome system is reminiscent of the continued synthesis of hemoglobin in nonnucleated reticulocytes.

In the cytoplasm, ribosomes are organized in the usual way as RER, with uniformly thin cisternae. These may extend into the cell processes in fibroblasts (this volume, Chapter 4). In these cells, many cell processes contain free or aggregated ribosomes, alone or in addition to those associated with the RER. In cartilage cells of the epiphyseal plate, the amount of RER is richest in the prelytic phase of secretory activity, just preceding the period of maximum activity (this volume, Chapter 5).

Possible Precursor Granules of Tropocollagen

Prominent in the cytoplasm of suitably fixed and stained connective tissue cells is the presence of certain granules. These have been observed only in the cytoplasm of fibroblasts (this volume, Chapter 4), and chondroblasts and osteoblasts (and two related cell types, the osteocyte and the osteoclast) (this volume, Chapter 6), but not in the cytoplasm of any other cell type. (The cells examined include smooth muscle cells, but do not include cells of the renal glomerulus or corneal epithelium.) The granules are large enough to be visible with the light microscope ($\frac{1}{2}$–1 μ). These granules of cartilage cells are concentrated in the peripheral parts of the Golgi apparatus, from where they are presumably eliminated. In the fibroblast, they seem to circulate throughout the cytoplasm and into cell processes, which they may almost fill. The granules visible with the light microscope are third-order granules consisting of a matrix in which are embedded sub(light)-microscopic granules about 1000 Å in diameter, which are the second-order granules. These are, in turn, aggregates of first-order granules which are about 80 Å in diameter. The outer part of each first-order granule stains densely. I suggested that each first-order granule is a precursor of a tropocollagen molecule, and that the tropocollagen molecule is uncoiled or unrolled from the granule when it reaches the extracellular environment. The outstretched or extended molecules are attached at their ends and laterally to form filaments and fibers. The matrix which separates the second-order granules is perhaps composed of largely hydrophobic molecules (this volume, Chapter 6). These are presumably synthesized by nearby ribosomal complexes of the RER and serve to insulate and inactivate the protropocollagen molecules in the cell. The collagen fibers of bone, cartilage, and tendon are closely associated with the polysaccharide component of the protein–polysaccharide complex (P-PC).

PROTEIN–POLYSACCHARIDE COMPLEX

Large and Small Compartments of Cartilage Matrix. The P-PC of connective tissues is also synthesized by connective tissue cells. How these complex molecules are assembled to form the matrix compartments visible with the electron and light microscope has been studied in great detail. The molecular (unit) structure of P-PC consists of a protein core with many side branches to which the polysaccharide polymers are attached. The microscopically visible structures of cartilage matrix formed from the P-PC conform to a single pattern which is best observed in the epiphyseal plate (this volume, Chapter 7). The basic pattern is of large and small compartments. The former are visible with the light microscope, and the latter with the electron microscope. The walls of the large compartments contain primarily the polysaccharide component of the P-PC, as well as the collagen fibers. The contents of the large compartments are further subdivided into smaller compartments, each in turn enclosed in a very thin wall also composed primarily of polysaccharide components. The contents of each small compartment are composed of the protein moiety of the P-PC. The large matrix compartments of cartilarge matrix abut directly on the cartilage cell and its numerous small cell processes. The shape and orientation of the large matrix compartments varies with their position in relation to the cell surface and with their position in longitudinal or transverse plates. The changes in compartmental shape in the matrix during normal turnover of P-PC molecules take place through breakdown into smaller fragments, which can diffuse out of the region, and by replacement through compartment shifts, much as the larger scale adjustments take place in the earth's crust. This kind of movement is different from the movement of small atoms, ions, and molecules, which will be discussed below.

In rat tail tendon, the polysaccharide component is distributed in the earliest stage as filaments or sheets enclosing less dense compartments or vacuoles. In later stages of development, the polysaccharide component is distributed as thin "filaments" which enclose collagen fibers as in a wide-meshed net extended in the direction of the axis of the collagen fibers (this volume, Chapter 10). In the osteoid of bone, the polysaccharide component is similarly distributed.

Calcification and the Walls of Large Matrix Compartments. Calcification in the epiphyseal plate takes place in the walls of the large matrix components (this volume, Chapter 9). It begins with the appearance in the walls of amorphous calcium deposits. As calcification progresses, the amount of amorphous calcium increases and particles are observed in it, which are of the approximate size of crystal nuclei. These dense particles are present in the amorphous calcium which surrounds all later stages of crystal cluster formation. It is presumed that crystals grow from some of the dense particles, and that this is how crystals form in later stages when crystals occur as clusters. The observations were not critical enough to decide whether crystallization is associated with collagen, with the polysaccharide moiety of the P-PC, or with both. The maximum diameter of collagen fibers in cartilage matrix of epiphyseal plate is about 40 Å. This is consistent with hexagonal packing of extended tropocollagen molecules three molecules in diameter. Thinner aggregations have not been identified, but may be presumed to be present. One must conclude that if collagen is necessary for crystal nucleation, then single tropocollagen molecules, or molecules aligned end to end as fibrils two or three tropocollagen molecules thick, are sufficient for nucleation.

Vascularity and Transport in Connective Tissues. Cartilage and bone (osteoid and between Haversian vessels, etc.) are not vascularized, and tendon is poorly or not at all supplied with blood vessels (this volume, Chapter 10). This raises the question as to the mechanism of transfer of metabolites and nutrients to and from the blood vessels. When this problem was probed in cartilage with ferrocyanide, for which a very good cytochemical method is available, it was found that this ion is transported by some nonenzymic process chiefly in the walls of the large matrix compartments (this volume, Chapter 8). If the transfer process in osteoid, bone, and tendon is the same as in cartilage, it would suggest that the organized sites of polysaccharides in these tissues also are involved. Prior findings on osteoid with the light and electron microscope are consistent with the suggestion. It is possible also that the same mechanism is responsible for ion transfer in loose connective tissue of skeletal muscle, which, in contrast with the other connective tissues, is well vascularized. This does not mean that the highly negatively charged chondroitin sulfate is responsible for the movement through cartilage of the strongly negatively charged ferrocyanide ion. It does suggest that the negative charges of the polysaccharide polymer are neutralized temporarily and progressively at a submolecular level of the protein–polysaccharide complex by the protein component. The prime requirement is that this process be progressive, directional, and cyclic or repetitive (as in nerve fibers).

Knowledge of the transport of substances in cartilage, bone, and tendon is of interest for its own sake, as well as for the possibility that it offers a model for transport which could be extended to protoplasm. The subject of transport has arisen several times in the course of this chapter, i.e., in connection with (1) the movement of polynucleotides in the walls of the nuclear pseudovacuoles, (2) the passage of polynucleotides through the nuclear membrane, (3) the movement of RNAs to various parts of the cytoplasm, and (4) the notion of aggregated second-order DNA helices acting as an obstacle to such movement. The topic has also come up in considering the movement of polypeptides along the ribosome and messenger in the RER, the intercisternal stream, and the Golgi apparatus, as well as the passage of a secretion product through the cell. The question is which of these processes could be interpreted with transport in extracellular matrix as a model and to what extent. No immediate answer to this question can be given.

Origin of Protein–Polysaccharide Complex. On morphological grounds, the P-PC seems to fill the intercisternal space (this volume, Chapter 7). Evidence suggests that both protein and polysaccharide components are synthesized in the RER. The structure of the intercisternal space in cartilage cells is indistinguishable from that of the large and small compartments of the extracellular matrix. The material seems to be extruded through small openings in the cell membrane which are about 1000 Å in diameter. These are not numerous. One could speculate that these temporary orifices permit intermittent continuity between the intercisternal spaces of the RER and the cartilage matrix. We are left with an apparent paradox—the precursors of the possible protropocollagen granules seem to be moved centripetally toward the Golgi apparatus, while the precursors of the P-PC seem to move centrifugally toward the cell surface. While it is possible that countercurrent distribution of two different secretory products could take place in a single intercisternal stream, it is also possible that each secretion precursor could be synthesized in different, even adjacent RER units and enter different intercisternal streams. One of them could move centripetally toward the Golgi apparatus (for the formation of third-order tropocollagen granules), while the other stream could move centrifugally toward the cell surface (for the extrusion of P-PC). Such a system would be in accord with the hypothesis of the functionally mosaic nature of cytoplasm and of active DNA regions of the nucleus. More specific methods for the identification of the precursors of secretory products of cartilage cells are necessary before this apparent paradox is resolved. In either case, the process of secretion which is proposed requires that the unit molecules of P-PC are self-assembled in the cytoplasm (probably in the intercisternal space) where they form small and large matrix compartments. These are then secreted as preassembled units whose precise shape and relations in the extracellular matrix are attained after spatial adjustments with preexisting, intact large matrix compartments.

Coda

In closing, I will reiterate one theme with which this work opened, i.e., the necessity of technical innovation and developments if cytochemistry is to be advanced at the submicroscopic level. Such developments in cytochemistry are necessary because of the narrow limits of error imposed on this field by the high resolution of the electron microscope. I have made the point earlier in this chapter that many fundamental biological problems which have been subjects of concern to biologists for the last century eventually come down to analysis at molecular or supramolecular dimensions. This has only recently become attainable with electron microscopes and other instruments. For this reason, it is of the greatest importance to ascertain as far as possible what we do to cellular components when we preserve or fix them, when we stain them for any specific cell constituents, and when we prepare them during embedding, sectioning, subsequent staining, drying, and finally while photographing selected fields in the electron microscope. In this work, I have concentrated on fixation and staining of some cell constituents; by keeping all other factors constant as far as possible, I hope that I have thereby minimized the possibilities that significant artifacts take place at subsequent stages of preparation. Even with this limited view, some fundamental problems have become accessible to study, and certain aspects of these problems have been summarized in this chapter.

It is pleasing that the methods used to elucidate a large proportion of the findings are confirmed by other, independent criteria. These include DNA, RNA, lipid, ferrocyanide, and crystalline and noncrystalline (amorphous) calcium deposits. Other methods and topics lacking this kind of independent check, such as the method for acid mucopolysaccharide, mitochondrial compartmentation, the intracisternal cavity of the RER, and the pseudovacuolar structure of protoplasm, must be regarded as less certain than the list in the preceding sentence, and come to be evaluated as more or less plausible, possible, or probable. Certainly, in the less certain fields, consistency, capability of repetition, and a good fit with pertinent literature are all necessary. The conclusion seems unavoidable that not only the work reported here, which has not been confirmable by other independent criteria, but all other work of a similar nature, has about it an element of uncertainty, and that the best to be hoped for is that there be a high degree of probability that the observations and deductions are meaningful and useful in understanding cellular behavior.

The question has been raised as to how certain one can be of the identification of cell structure in freeze-dried specimens. In the first place, it should be stated that in properly prepared freeze-dried specimens, ice crystal formation most probably does not take place, at least to a resolution of molecular dimensions. In the second place, there is no difficulty whatever in identifying most, if not all, cell structures in freeze-dried specimens with the light microscope, or are there any great discrepancies at the level of the electron microscope in most structures, such as the various membranes in the RER, in the nucleus, and in the cell surface, the mitochondrial compartmentation, chromatin particles, nucleoli, and Golgi structures. To be sure, there may be some morphological differences in specimens prepared by freezing-drying as compared with those prepared by the use of fluid fixatives. Such differences are indeed to be expected: similar differences occur when different fluid fixatives are used, without throwing doubt on the preexistence in cells of RER, mitochondrial compartmentation, etc. Such deviations from accepted norms of structure do indeed occur after freeze-drying. Whether these differences are accepted as very significant, or of little significance, is not simply a matter of taste, but has to be weighed carefully in the light of whatever other information is available from other fields. In other words, the available evidence bearing on the total picture must be considered in evaluating the probable reality of structural organization at the level of the electron microscope. This includes not only the pertinent biochemical, genetical, and physiological evidence, say, but also the possibilities of technical errors, such as extraction, diffusion, displacement and rearrangement, deformation, recombination, swelling, shrinkage, and denaturation.

This leads to a consideration of possibilities of error at the molecular level, where there are few guide lines. At this level it is difficult or impossible to ascertain exactly how much change takes place in protein molecules in cells during denaturation by fluid fixatives or by freezing-drying with subsequent treatment with vapor reagents. Certainly, in molecules extracted from fresh tissues, there may be significant, though often reversible, molecular deformations of proteins, as there may be in nucleic acids. For this reason I have tried to avoid taking a strong stand on the significance of measurements of molecular dimensions, and have even cautioned against accepting them as significant. For the same reason, similarities in size and shape at the molecular level are not considered equivalent to chemical identity. This would be a repetition at the molecular level of errors in evaluation which were made at the microscopic level by Fischer (7) and Mann (17).

With these points in mind, it is surprising how close the agreement is between cell structures preserved by freeze-drying and by fluid fixatives. The major difference is the occurrence of pseudovacuoles in the nucleus and cytoplasm of freeze-dried cells, and their absence in cells fixed by immersion in fluid fixatives. If ice crystals are not formed during the preparation of freeze-dried specimens, as I believe is most probable from the evidence cited earlier, then there is a strong probability that the pseudovacuoles are real and that they are destroyed during fixation by immersion in fluid fixatives. While this does not seem to be a major difference, it may be a very significant one, for the pseudovacuolar structure of cells may be important not only in maintaining homeostasis in relation to their environment, but also in the fixation of certain enzymes in an active state. The pseudovacuolar structure of protoplasm may be the physical basis which underlies the organization of certain enzymic systems in cell metabolism. The kind of organization would be the equivalent for ground substance of protoplasm, of organized systems of enzymic activities in mitochondria in relation to mitochondrial compartmentation.

In cytochemistry, the relations between techniques and their applicability to biological problems is very intimate. Likewise, an interest in certain biological problems should impel toward technical innovation. An appreciation of this mutual dependence is part of what this work is about.

References

1. Britten, R. J., and Davidson, E. H. (1969). Gene regulation for higher cells: a theory. *Science* **165**, 349–357.
2. Busch, H., and Smetana, K. (1970). "The Nucleolus." Academic Press, New York.
3. Davidson, E. H. (1968). "Gene Activity in Early Development." Academic Press, New York.
4. Davidson, E. H., and Britten, R. J. (1971). Note on the control of gene expression during development. *J. Theoret. Biol.* **32**, 123–130.

5. Ebert, J. D. (1953). An analysis of the synthesis and distribution of the contractile protein, myosin, in the development of the heart. *Proc. Nat. Acad. Sci. U.S.* **39**, 333–344.
6. Fabro, S. P., and Rinaldini, L. M. (1965). Loss of ascorbic acid synthesis in embryonic development. *Develop. Biol.* **11**, 468–488.
7. Fischer, A. (1899). "Fixirung, Färbung und Bau des Protoplasmas." G. Fischer, Jena.
8. Gellhorn, A. (1969). Ectopic hormone production in cancer and its implication for basic research on abnormal growth. *Advan. Intern. Med.* **15**, 299–316.
9. Huebner, R. J., Kelloff, G. J., Sarma, P. S., Lane, W. T., Turner, H. C., Gilden, R. V., Oroszlan, S., Meier, H., Myers, D. D., and Peters, R. L. (1970). Group-specific antigen expression during embryogenesis of the genome of the C-type RNA tumor virus: implications for ontogenesis and oncogenesis. *Proc. Nat. Acad. Sci. U.S.* **67**, 366–376.
10. Jacob, F., and Monod, J. (1961). On the regulation of gene activity. *Cold Spring Harbor Symp. Quant. Biol.* **26**, 193–211.
11. Jamieson, J. D., and Palade, G. E. (1967). Intracellular transport of secretory proteins in the pancreatic exocrine cell. I. Role of the peripheral elements of the Golgi complex. *J. Cell Biol.* **34**, 577–596.
12. Jamieson, J. D., and Palade, G. E. (1967). II. Transport to condensing vacuoles and zymogen granules. *J. Cell Biol.* **34**, 597–615.
13. Jamieson, J. D., and Palade, G. E. (1968). III. Dissociation of intracellular transport from protein synthesis. *J. Cell Biol.* **39**, 580–588.
14. Jamieson, J. D., and Palade, G. E. (1968). IV. Metabolic requirements. *J. Cell Biol.* **39**, 589–603.
15. Lash, J. W. (1968). Chondrogenesis: genotypic and phenotypic expression. *J. Cell Physiol.* **72**, (Suppl. 1), 35–46.
16. Lash, J. W. (1968). Somitic mesenchyme and its response to cartilage induction. *In* "Epithelial-Mesenchymal Interactions" (R. Fleischmajer and R. E. Billingham, eds.), pp. 165–172. Williams & Wilkins, Baltimore, Maryland.
17. Mann, G. (1902). "Physiological Histology." H. Hart, Oxford.
18. McClintock, B. (1967). Genetic systems regulating gene expression during development. *Develop. Biol. Suppl.* **1**, 84–112.
19. Whaley, W. G., Dauwalder, M., and Kephart, J. E. (1972). Golgi apparatus: influence on cell surfaces. *Science* **175**, 596–599.

Author Index

Numbers following the names of authors indicate the pages where complete references are listed. Numbers in italics refer to reference listings for senior or sole authors. Roman numbers refer to listings for co-authors.

A

Akers, C. K., 29
Ali, S. Y., *124, 203*
Ames, A., 3rd., 27
Anderson, C. E., *84, 147*
Anderson, H. C., 125, *203*
Appleton, J., *203*
Attardi, B., 61
Attardi, G., *61*

B

Bachra, B. N., *204*
Bailey, P. S., 27
Baker, R. F., 30, 62
Barajas, L., 62, 230
Barland, P., *174*
Barnabei, O., 30
Barnett, C. H., 174, *174*
Barrett, A. J., 125
Benditt, E. P., 85, 148
Benedetti, E. L., 27
Benson, A. A., 27
Bently, J. P., 211, *211*
Bergman, B., *185*
Bhatmagar, R. S., *84*
Bielen, R. J., *204*
Blackwood, H. J. J., 203
Bloom, W., *147*
Bondareff, W., 61, 74, *84,* 84, *147, 185*
Bonucci, E., *204*
Borst, P., *61*
Bramley, T. A., 27
Brandt, K. D., *174*
Britten, R. J., 227, 235, *235*
Brockis, J. G., *211*
Brower, T. D., *185*

Brown, C. D., 176
Brown, H. D., 27
Buckley, I. K., *125*
Bütschli, O., 124, *125, 174*
Bullivant, S., *27,* 61
Bullough, P., 186
Busch, H., *235*

C

Cameron, D. A., *174, 204*
Campo, R. D., 125, 172, *174,* 199, *204,* 205
Catchpole, H. R., 84, 125, *185*
Chance, B., 29
Chapman, D., 27
Chase, W. H., *61,* 184, *185*
Chattopadhyay, S. K., 27
Cifonelli, J. A., 175
Clark, A. E., *174*
Cochrane, W., 174
Coleman, R., 513
Collins, E. J., 205
Coombs, R. R. A., *27,* 125, 204
Cooper, G. W., *84*
Crane, F. L., 27
Criddle, R. S., 28
Curran, R. C., 174

D

Daniel, E. E., 28
Danielli, J. F., 27
Dauwalder, M., 236
Davidson, E. A., 176, 204, 205
Davidson, E. H., 227, *228, 235, 235*
Davies, D. V., *174*
Davis, H. F., 175
DeBruijn, W. C., 29

DeLuca, S., 176
Dennis, J. B., *185*
De-Thé, G., *61*
Deuel, H. J., 27
Dingle, J. T., 125, *125, 204*, 205
Dodd, D., 212
Dorfman, A., 172, *174*, 175, 176
Durning, W. C., *125*, 146, *147*, 169, *174*, 184, *185, 204*
Dziewiatkowski, D. D., *125*, 126, 172, 174, 176, 204, *204*, 205

E

Eanes, E. D., *204*
Ebert, J. D., *235*
Edwards, D. A. W., *211*
Eichelberger, L., *204*
Eisenstein, R., *174*
Elfvin, L.-G., 30
Emmelot, P., 27
Engel, M. B., 185
Engelman, D. M., 30
Engfeldt, B., *174*
Ernster, L., *61*
Evans, H., 186
Evans, L., 124

F

Fabro, S. P., *236*
Falkmer, S., 175
Farin, I., 186
Fawcett, D. W., 27, 147
Fell, H. B., 125, *125*, 204, 205
Fernando, N. V. P., 85
Fessenden-Raden, J. M., *61*
Finean, J. B., *513*
Fischer, A., *174, 236*
Fitton Jackson, S., 74, *84, 147, 175*, 211, *211*
Flickinger, C. J., 27
Frame, J., 176
Freeman, M. A. R., 186
Friberg, U., 205

G

Gellhorn, A., *235*
Gersh, I., 27, *61, 84, 125, 147*, 184, 185, 204
Ghadially, F. N., *175*
Gilden, R. V., *236*
Glauert, A. M., 27, 28
Glimcher, M. G., 187, *204*
Godman, G. C., *85, 147, 175*

Goldberg, B., *85*
Goldberger, R. F., 28
Gordon, A. S., 30
Granda, J. L., *125, 204*
Grant, M. E., *84*
Green, D. E., *28*, 30
Green, H., 85
Greenwald, A. S., *185*
Greep, R. O., *147*
Greer, R. B., *125*
Gregory, J. D., 176
Gulek-Krzywicki, T., 28
Gurd, F. R. N., *28*
Gustafson, G. T., 175

H

Hammerman, D., 185
Hardy, W. B., *175*
Hascall, V. C., *175*, 176
Havivi, E., *125*
Hay, E. D., 85, 148
Haynes, D. W., 185
Helfet, A. J., 176
Heller-Steinberg, M., *85, 148*
Hellman, W., 176
Hernandez, W., 175
Highton, T. C., 175
Hilz, H., 175
Hinckley, A., 29
Hirschman, A., *85, 148, 204*
Hjertquist, S.-O., 174, *204*
Hodge, J. A., *185*
Holland, J. J., 28
Honner, R., *186*
Horwitz, A. L., *175*
Howell, D. S., *204*
Hranisaoljevic, J., 176
Huebner, R. J., *236*

I

Isenberg, I., 61, 84, 147
Ito, S., *125*
Iwano, K., *125*

J

Jackson, D. S., 211, *211*
Jacob, F., *236*
Jamieson, J. D., 29, *236*
Janicke, G. H., 125
Janis, R., 174
Jensen, W. A., 43, *61*

Author Index

Jibril, A. O., *125*
Joseph, N. R., 185
Józsa, L., *175*

K

Kaplan, D. M., *28*
Katsura, N., *204*
Katz, E. P., *204*
Kelloff, G. J., *236*
Kember, N. F., 125
Kephart, J. E., *236*
Khan, T., *175*
Kharchuk, L. N., 125
Kidwai, A. M., *28*
Kiehn, E. D., *28*
Kirsig, H. J., 175
Kleine, T. O., *175*
Kleinschmidt, A. K., 176
Knese, K.-H., *125*
Knoop, A. M., 125
Koehler, J. K., *28*
Kölliker, A., *212*
Korn, E. D., *28*
Krane, S. M., 187, 204
Krikos, G. A., 205
Kroon, A. M., 61
Kuettner, K. E., 174
Kuhlman, R. E., *125, 204*
Kuylenstierna, B., 61

L

Lachmann, P. J., 27
Lane, N., 175
Lane, W. T., 236
Lash, J. W., 236
Leblond, C. P., 175
Lehninger, A. L., *61*
Lenard, J., *28*
Leslie, R. B., 27
Lippiello, L., 125, 186
Lison, L., *28*
Lohmander, S., 205
Lovelock, J. E., *28*
Lovern, J. A., *28*
Lucy, J. A., 28, *28, 125*
Luft, J. H., *175*, 176
Luzzati, V., *28*
Lynn, J. A., 205

M

McCabe, D. M., 85, 148
McClintoch, B., *236*

McCutchen, C. W., 177, *186*
McGee-Russell, S. M., *29*
McKibbin, B., 185
McNamee, M. J., 125
Maddy, A. H., *27*
Madruga, J. E., 204
Malhotra, S. K., *28, 61*
Mankin, H. J., *125, 186*
Mann, J. G., *236*
Marler, E., 205
Maroudas, A., 177, *186*
Marquez, J. F., 204
Martin, J. H., 204
Mason, R. J., 30
Mathews, M. B., *175, 205*
Matsuzawa, T., *126*
Matukas, V. J., *175, 205*
Mazhuga, P. M., 125
Meachim, G., *126*
Meier, H., 236
Meldolesi, J., *29*
Mollenhauer, H. H., *29*
Molnar, Z., *148, 184, 186, 205*
Monod, J., 236
Moor, H., *29*
Moretz, P. C., *29*
Morré, D. J., 29
Morrison, R. I. G., *126, 205*
Moshier, R. W., 205
Movat, H. Z., *85*
Muir, H., 174, *186*
Myers, D. B., *175*
Myers, D. D., 236

N

Nass, M. M. K., *62*
Neufeld, A., 29
Nicolson, G. L., *175*
Noller, C. R., *29*
Norberg, A. I., *212*
Nowikoff, M., 124, *126, 175*

O

Orbison, J. L., 175
Oroszlan, S., 236
Overton, J., 175

P

Palade, G. E., 29, 236
Palfrey, A. J., 174
Panner, B. J., 175

Parker, J., 84, 147
Parsons, D. F., 29, *29*
Partridge, S. M., *175*
Peacock, E. E., Jr., *212*
Peters, R. L., 236
Peters, T. J., 176
Pihl, E., *175*
Pita, J. C., 204
Pollard, T. D., 30
Pomerat, C. M., 126
Poole, A. R., 126
Porter, K. R., *84*, 126, 147
Posner, A. S., 125, 204, 205
Pras, M., 205
Prezbindowski, K. S., 27
Priest, R. E., 172, *175*
Prockop, D. J., 84, *85*

Q

Quintarelli, G., *126*, *175*, *205*

R

Rabinowitz, M., *62*
Racker, E., *29*, 61
Radcliffe, M. A., 28
Raker, C. W., 212
Ralston, A. W., 29
Rambourg, A., *175*
Rayns, D. G., 175
Reiss-Husson, F., 28
Revak, C., 125
Revel, J-P., *85*, *148*, 176
Richmond, M. E., *176*
Rinaldini, L. M., 236
Rivas, E., 28
Robertson, J. D., 29
Robinson, H. C., 176
Roma, M., 204
Romeo, D., 29
Rose, G. G., *126*
Rosenberg, L., *176*
Ross, R., *85*, *148*
Rothfield, L., *29*
Roy, S., 175
Ruppricht, W., 124, *126*, 176
Ruzicka, F. J., 27

S

Sajdera, S., 126, *176*, 203
Salpeter, M. M., *176*
Saludjian, P., 28

Sandson, J., 174
Sarma, P. S., 236
Sasai, Y., *176*
Saunders, A. M., 176
Schaffer, J., 121, *126*, 146, *148*
Schatz, G., 62
Scherft, J. P., *205*
Schubert, M., *176*, 205
Seno, S., *29*
Serafini-Fracassini, A., 176, *176*
Shea, S. M., *176*
Sievers, R. E., 205
Silbert, J. E., 176
Silverman, L., 176
Singer, S. J., *29*, 34
Sjöstrand, F. S., xiv, *29*, *30*, 62, 216
Sledge, C. B., *126*, 205
Smith, J. W., 176, *176*, *212*
Smith, R. H., 147
Sorgente, N., 174
Spira, E., *186*
Stephenson, J. L., 61, 84, 147
Stiles, J. W., 27
Stockwell, R., *126*
Stoeckenius, W., *30*, 62
Stossel, T. P., *30*
Sun, F. F., 27
Sundström, B., *205*
Swanson, S. A. V., 186
Swift, H., 62
Szederkényi, G., 175
Szubinska, B., *176*

T

Takuma, S., 205
Tanaka, S., *126*, *205*
Tardieu, A., 28
Telser, A., *176*
Termine, J. D., 204, *205*
Thomas, L., *126*
Thompson, R. C., 186
Thompson, W., 29
Thyberg, J., *126*, *205*
Tourtellotte, C. D., 125, 204, *205*
Tria, E., *30*
Turner, H. C., 236

V

van Deenen, L. L. M., *62*
Vanderkooi, G., 28, *30*
Van Sickle, D. C., 186
Vaughan, M., 30

Author Index

W

Wallach, D. F. H., *30*
Wassermann, F., *85, 148*
Weiss, C., *176*
Weiss, L., *30*
Wells, P. G., 176
Weston, P. D., 125
Whaley, W. G., *236*
Whiting, A. H., 175
Williams, G. R., 29
Wilsman, N. J., *186*
Wilson, D., 29
Wittcoff, H., *30*

Woessner, J. F., Jr., 126
Woodward, C., *205*
Wuthier, R. E., 205, *205*

Y

Yardley, J. H., *176*
Yoshizawa, K., 29

Z

Zelander, T., *176*
Zhitnikov, A. Y., 125
Zito, R., 175

Subject Index

A

Amorphous calcium in cartilage matrix, 181–205

B

Barr body, in chondrocytes, 195–220 (Vol. 1)

C

Calcium deposition in cartilage matrix, 187–205
Cartilage matrix, 119–124, 149–176
   cellular origin of compartments, 159–173
   compartmentalization, 121, 124
   distribution of components in compartments, 149–176
   mode of extrusion from cytoplasm, 162–166
   movement of ferrocyanide in, 177–186
   relation of calcium crystal formation to walls of compartments, 187–205
Cell lysis, cartilage, 86–126
Cell membranes, 1–30
Cell processes
   of chondrocytes, 133
   of fibroblasts, 64, 68
Chondrocytes, 86–126
   cell matrix compartments, 149–176
      inadequacy of conventional methods of fixation, 169–172
      mode of extrusion from cytoplasm, 162
      origin of, in intercisternal space, 162–163, 172
      relation to RER, 173
      self-assembly in cytoplasm, 162–163
   fixation of, for nucleic acids, 87
   nucleus
      chromatin of, in light microscope, 113
      DNA molecular pattern in electron microscope, 88–98
         cells of post-proliferative stage, 90–91
            preparatory stage, 92
            prelytic stage, 95
            lytic stage, 97
      DNA helices, 90–97
      DNA segments and granules, 96–97
      measurements of nuclear volume, 117
      nucleolus of, 92
      nucleoplasm in cells in various stages, 92–95
   possible tropocollagen precursor granules, 127–148
      first- and second-order granules, 132–137
      methods of preparation, 127–129, 133–136
      third-order granules, 132–133, 137
         distribution in cell, 137
         matrix of, 133, 144
         relation to DNA and RNA, 145
            Golgi apparatus, 137
            growth of epiphyseal plate, 145
            RER, 144
            secretion, 145–146
   post-fixation, with cross-linking vapors, 87
   RER in cells of various stages
      appearance in electron microscope, 92–97
      appearance in light microscope, 115
      intercisternal space, 92–97
      intracisternal space, 92–97, 124
      measurement of, 117
      ribosomes in, 90–97
   staining of nucleic acids, 87
      identification, 97
      use of DNase and RNase, 97–102
Chromatin in various stages
   changes during cell death, 86–126
   changes during karyolysis, 86–126
   in chondrocytes
      appearance in electron microscope, 90–97
      in light microscope, 113
      pattern, persistence during interphase in all reversible changes, 219
      relation to nuclear membrane, 220
Chromosomal pattern
   importance of nuclear membrane, 220
   persistence during interphase in all reversible changes, 219
Compartmentation
   cartilage cell matrix compartments, 154–172
   Golgi apparatus, 226–227

Subject Index

mitochondria, 48–59, 229
RER, 222–224
Connective tissue cells, 229–231
   DNA and cellular activity, 67, 69, 86–126, 230
   possible precursor granules of protropocollagen, 63–85, 127–148, 229–230
   protein–polysaccharide complex, 231–233
      extracellular distribution of, 119–124, 149–176, 231–232
      origin, organization and secretion, 154–172, 233
      relation to calcification, 187–205, 231–232
      to transport, 177–186, 232–233
   RER and cellular activity, 67, 69, 86–126, 230
Cytochemical staining
   cytochemical tests
      calcium, amorphous and crystalline, 189–190
      di- and polyvalent cation, 189–190
      ferrocyanide, 178–181
      phospholipid, 4–16, 34
      protein–polysaccharide complex, 150–153
Cytoplasmic mosaicism, 227–228

D

DNA molecular patterns during
   cell death, 86–126
   karyolysis, 86–126
DNA molecular patterns of chondrocytes, 86–126
DNA helices
   breakdown as granules, 96–97
   coils and segments, 96–97
   first order, 90–97, 214, 219
   fourth order, 219
   second order, 214
   third order, 219–220

E

Epiphyseal plate
   growth stages, 88–98
   lytic stage, 97
   postproliferative stage, 92
   prelytic stage, 95
   preparatory stage, 92
   proliferative stage, 90
   relation to nucleic acids, 118–120, 122
Euchromatic mosaicism, 227–228

F

Ferrocyanide, movement in various kinds of connective tissues, 177–186
Fibroblasts
   intracisternal space of, 69
   possible tropocollagen precursor granules, 63–85
      first-, second- and third-order granules, 64–68
      matrix of third-order granules, 74
      method of preparation, 64
      possible mode of extrusion, 74
   pseudovacuoles, 84
   rat tail tendon, 63–85
   RER, 67–69
Fixation, 3–15 (Vol. 1)
Freeze-drying, 4–5, 16–29 (Vol. 1)
Functional mosaicism of cytoplasm and nucleus, 227–228

G

G2 print, xiv
Genetic material
   association of DNA with various RNAs, 220–221
   changes in pattern during
      cell determination and differentiation, 215–216
      cell duplication, 219–220
      cell maturation, 218
      cell metaplasia, 218
      stimulation of cellular activity, 217–218
   chromatin, or aggregates of, 214–216
   constancy of chromatin pattern in unstimulated cells, 217
   relation to nuclear proteins, 222
   significance of nuclear gaps for transport of DNA products to cytoplasm, 221–222
   state in eukaryotic cells, 214–220
Golgi apparatus, 226–227, *see also* Compartmentation
   lipid components, 20
   relation to protein synthesis and storage, 226–227
      to secretory granules, 227

H

Hepatic cell mitochondria, 42–62

L

Large matrix compartments of cartilage matrix
   description, 121, 124
   distribution of collagen, protein and polysaccharide components, 149–176

movement of ferrocyanide in walls of, 177–186
  relation to calcium deposition, 187–205
Lipid components of hepatic cells, 20
  quantitative aspects, 31–41
Lipid components of membranes
  cell membrane, 16–17
  Golgi apparatus, 18
  methods of study, 4–9
  quantitative study of lipid components, 31–41
    in Golgi apparatus, 38–41
    in intracisternal space, 34–38
    in nuclear membrane, 38
    in wall of RER, 34–38
  walls of nuclear membrane, 16, 20
  of RER, 16
Lipid components
  of mitochondria, 42–62
  of pancreatic exocrine cells, 17–18, 34–41

M

Matrix of cartilage, 149–176
  compartmentalization, 121, 124
  lysis, 119–121
  movement of ferrocyanide, 177–186
  pericellular space, absence of, 123
  remodeling, 184
Matrix of epiphyseal plate cartilage, 149–186
  cellular origin, 149–176
  description of small and large matrix compartments, 121, 124
  disposition of collagen and protein-polysaccharide complex in large and small matrix compartments, 154
  identification of protein–polysaccharide complex, 150–153
  inadequacy of conventional methods of fixation, 169–172
  relation of walls of large matrix compartments to calcium deposition, 195–202, 231–232
    absence of globules from matrix, 202–203
    identification of calcium, 193–195
    method, 188–191
    occurrence of amorphous calcium in walls
      first occurrence, 191
      relation to protein–polysaccharide complex, 198–200
    occurrence of crystals in walls, 190–193
Matrix of joint cartilage, 173–174

  distribution of collagen and protein–polysaccharide complex in compartments, 173–174
  movement of ferrocyanide in matrix, 177–186
    method, 178–181
    relation to movement in other kinds of connective tissue (osteoid, rat tail tendon, loose connective tissue, reticular connective tissue, 182–183
    relation to walls of large matrix compartments, 184–185
  pericellular zone, 173–174
Membranes, 1–30, 31–41
  globular (protein) molecules, 21–23
  lipid components, 16–20
    cell membrane, 17
    Golgi apparatus, 17–18
    methods for study, 4–16
      controls, 8–10
      list of reagents tested, 8
      mechanism of staining, 5–6
    nuclear membrane, 16
    quantitative study, 31–41
      Golgi apparatus, 38–41
      intracisternal space, 34–38
      method of study, 33
      nuclear membrane, 38
      walls of RER, 34–38
    walls of RER, 16–17
      relation to proteins and ribosomes, 23
  protein components
    cell membrane, 23
    methods of study, 16
    nuclear membrane, 23
    RER, 21–23
Mitochondria, 42–62
  cristae (?), 59
  compartmentation, 48–59
    contents of compartments, 48–54, 57–59
    membranes, 48–54, 57–59
  lipid components, 48
    distribution, 47
    membranes, 57
    methods of study, 44
    nucleic acids, 54–55

N

Nuclear membrane
  importance for chromosomal persistence during interphase, 220
  relation to Barr body, 195–220 (Vol. 1)
    to chromatin, 365–375 (Vol. 1)
  significance for mitosis, 365–375 (Vol. 1)

Nuclear mosaicism, 227–228
Nuclear proteins, 221
  relation to chromatin, chromosomes, mitosis, and differentiation, 222
Nucleic acids in
  chondrocytes, 86–126
  fibroblasts, 63–85, 231 (Vol. 1)
Nucleus, 214–222

O

Osteoblasts of young parietal bone, 127–148
  possible tropocollagen precursor granules, 127–148
    first-, second- and third-order granules, 139–141
    method of preparation, 137–139
    relation to PAS positive granules, 141
    to RER, 144
Osteoclasts and osteocytes—possible tropocollagen precursor granules, 139–141

P

Pancreatic exocrine cell, 42–62
  lipid component, 48
  mitochondria
    nucleic acids, 54–55
    protein component, 48–53
Perinucleolar chromatin
  changes during cell metaplasia, 218
    during stimulation of cell activity, 217–218
Proteins of
  chondrocytes, 153–154, 159–163
  membranes
    cell membrane, 23
    methods of study, 16
    nuclear membrane, 23
    RER, 21–23
      intracisternal globules, 21–23
      walls of, 223–224
  mitochondria
    distribution of proteins, 222
    vapor stains for proteins, 16–52 (Vol. 1)
Protein–polysaccharide complex
  cartilage matrix, 119–124, 149–176
  tendon, 206–212
Pseudovacuolar structure of cytoplasm, 225–226
  significance for homeostasis, 225–226

R

RER
  cisternal wall, lipid and protein components, 224–225
  globular structure of intracisternal space, 224
  intercisternal space, 224–225
  possible functional compartmentation, 225
  relation to cell matrix compartments in chondrocytes, 172–173
Ribosomal apparatus
  internal structure of ribosomes, 222–224
    distribution of ribosomal proteins and rRNA, 222–223
  relations of messenger, 224
  relation to tRNA, 224
  relations to nascent protein filaments, 222–224

S

Secretion, 226
  origin of secretory granules, 226
  possible precursor tropocollagen granules, 145–146
  protein–polysaccharide complex, 154–163, 166–172
  relation to Golgi apparatus, 137
Submicroscopic vacuoles, 221–222

T

Tendon, 206–212
  noncollagenous components, 206–212
  possible importance of protein–polysaccharide complex in transport of ferrocyanide, 211
  vascularity of, 206–212
Transport of ferrocyanide
  in cartilage matrix, 177–186
  in tendon, 206–212
Tropocollagen, possible precursor granules
  in chondrocytes, 127–148
  in fibroblasts, 63–85
  in osteoblasts, osteoclasts, and osteocytes, 127–148

W

Wall of large and small compartments of cartilage matrix, 231